Tellurium in Organic Synthesis

BEST SYNTHETIC METHODS

Series Editors

A. R. Katritzky
University of Florida
Gainesville, Florida
USA

O. Meth-Cohn
Sunderland Polytechnic
Sunderland
UK

C. W. Rees
Imperial College of Science and Technology
London, UK

Richard F. Heck, *Palladium Reagents in Organic Syntheses*, 1985
Alan H. Haines, *Methods for the Oxidation of Organic Compounds: Alkanes, Alkenes, Alkynes, and Arenes*, 1985
Paul N. Rylander, *Hydrogenation Methods*, 1985
Ernest W. Colvin, *Silicon Reagents in Organic Synthesis*, 1988
Andrew Pelter, Keith Smith and Herbert C. Brown, *Borane Reagents*, 1988
Basil Wakefield, *Organolithium Methods*, 1988
Alan H. Haines, *Methods for the Oxidation of Organic Compounds, Alcohols, Alcohol Derivatives, Alkyl Halides, Nitroalkanes, Alkyl Azides, Carbonyl Compounds, Hydroxyarenes and Aminoarenes*, 1988
H. G. Davies, R. H. Green, D. R. Kelly and S. M. Roberts, *Biotransformations in Preparative Organic Chemistry: The Use of Isolated Enzymes and Whole Cell Systems*, 1989
I. Ninomiya and T. Naito, *Photochemical Synthesis*, 1989
T. Shono, *Electroorganic Synthesis*, 1991
William B. Motherwell and David Crich, *Free Radical Chain Reactions in Organic Synthesis*, 1991
N. Petragnani, *Tellurium in Organic Synthesis*, 1994
T. Imamoto, *Lanthanides in Organic Synthesis*, 1994
A. J. Pearson, *Iron Compounds in Organic Synthesis*, 1994

Tellurium in Organic Synthesis

N. Petragnani

Institute of Chemistry,
University of São Paulo, Brazil

ACADEMIC PRESS

Harcourt Brace & Company, Publishers
London San Diego New York
Boston Sydney Tokyo Toronto

ACADEMIC PRESS LIMITED
24–28 Oval Road
London NW1 7DX

US Edition published by
ACADEMIC PRESS INC.
San Diego, CA 92101

This book is a guide providing general information concerning its subject matter; it is not a procedural manual. Synthesis of chemicals is a rapidly changing field. The readers should consult current procedural manuals for state-of-the-art instructions and applicable government safety regulations. The publisher and the authors do not accept responsibility for any misuse of this book, including its use as a procedural manual or as a source of specific instructions.

British Library Cataloguing in Publication Data

A catalogue record for this book is available from the British Library

ISBN 0-12-552810-8

Typeset by Photo·graphics, Honiton, Devon.
Printed in Great Britain by Hartnolls Limited, Bodmin, Cornwall

Contents

Foreword

There is a vast and often bewildering array of synthetic methods and reagents available to organic chemists today. Many chemists have their own favoured methods, old and new, for standard transformations, and these can vary considerably from one laboratory to another. New and unfamiliar methods may well allow a particular synthetic step to be done more readily and in higher yield, but there is always some energy barrier associated with their use for the first time. Furthermore, the very wealth of possibilities creates an information-retrieval problem. How can we choose between all the alternatives, and what are their real advantages and limitations? Where can we find the precise experimental details, so often taken for granted by the experts? There is therefore a constant demand for books on synthetic methods, especially the more practical ones like *Organic Syntheses*, *Organic Reactions*, and *Reagents for Organic Synthesis*, which are found in most chemistry laboratories. We are convinced that there is a further need, still largely unfulfilled, for a uniform series of books, each dealing concisely with a particular topic from a *practical* point of view—a need, that is, for books full of preparations, practical hints and detailed examples, all critically assessed, and giving just the information needed to smooth our way painlessly into the unfamiliar territory. Such books would obviously be a great help to research students as well as to established organic chemists.

We have been very fortunate with the highly experienced and expert organic chemists, who, agreeing with our objective, have written the first group of volumes in this series, *Best Synthetic Methods*. We shall always be pleased to receive comments from readers and suggestions for future volumes.

A. R. K., O. M.-C., C. W. R.

Preface

Tellurium is the fourth element of the VI^b family of the periodic table, which starts with oxygen. Since tellurium exhibits an electronic configuration similar to that of selenium and sulphur, the chemical behaviour of these elements is obviously closely related. This similarity was a hindrance to the greater development of tellurium chemistry. During several decades, research was restricted to an extrapolation of well-established reactions for the preparation and use of organic sulphur compounds to selenium, and mainly from selenium to tellurium.

Although over a quarter of a century ago it would not have been predicted that the importance of the VI^b family of the periodic table would exceed that of its second element, sulphur, the development of organoselenium chemistry has been so explosive that little comment is necessary, as illustrated by the impressive number of papers as well as by several books [1–3] on the field. Tellurium compounds have taken an even longer time to rise from being considered an exotic and perverse element to a useful tool in organic chemistry.

In the mid-1950s the German chemist Heinrich Rheinboldt reviewed the preparation and reactivity of the organic tellurium compounds by comparison with selenium analogues in *Houben–Weyl—Methoden der Organischen Chemie*, Vol. IX. During the following years, two books and several review articles were published on the organic chemistry of tellurium, but these were still within the areas of preparation and reactivity. It was in 1971 that the first "Symposium on the Selenium and Tellurium Chemistry" took place, and since then symposia have been held every 4 years. In the last few years tellurium compounds have begun to be employed in organic synthesis, giving rise to a significant number of publications and to several review articles. Finally, in 1990, a *Houben–Weyl* volume of more than 1000 pages was published covering all aspects of tellurium chemistry (author, K. Y. Irgolic).

At present the author feels that the remarkable development attained by the organotellurium chemistry is a clear reason for a monograph from the point of view of the organic chemist.

The aim of this monograph is to provide a comprehensive overview of the preparation and synthetic applications of tellurium compounds. It will focus on the preparation of selected inorganic tellurium compounds and on the main classes of organotellurium compounds. The major interest of the volume probably resides in the use of these inorganic and organic compounds as reagents in organic syntheses as well as the conversion of organotellurium into free organic compounds. In this regard it is sufficient to emphasize that extremely useful reactions achieved with tellurium reagents, such as selective reductions and oxidations of a large variety of organic functionalities, have not been mentioned until now in textbooks specifically devoted to organic synthetic methodology.

This monograph is appropriate for chemists who, although not experts in tellurium chemistry, have had a basic grounding to graduate level in organic chemistry and with sufficient experience of typical experimental operations.

A considerable number of "experimental procedures" have been included. These have been described in accordance with the original papers, and therefore the reader will find significant differences in the description of these procedures, some rich and others very poor in detail. In the last case, the practitioner who is used to overcoming the usual laboratory difficulties will need to use a certain degree of initiative.

The author will feel very gratified if this volume helps some chemists to become familiar with tellurium which up till now they may have considered only as a useless and unpleasant element.

N. PETRAGNANI

REFERENCES

1. D. L. Klayman and W. H. H. Gunther (ed.), *Organic Selenium Compounds: Their Chemistry and Biology*. Wiley, New York, 1973.
2. C. Paulmier, *Selenium Reagents and Intermediates in Organic Synthesis*. Pergamon Press, Oxford, 1986.
3. D. Liotta (ed.), *Organoselenium Chemistry*. Wiley, New York, 1987.

Detailed Contents

Acknowledgement

The author and Academic Press gratefully acknowledge the financial help provided by FAPESP (Fundação de Amparo à Pesquisa do Estado de São Paulo, Brazil).

Abbreviations

AIBN	2,2′-azobisisobutyronitrile
18-crown-6	1,4,7,10,13,16-hexaoxacyclooctadecane
DMAP	4-dimethylaminopyridine
DME	1,2-dimethoxyethane
DMF	*N*,*N*-dimethylformamide
DMSO	dimethyl sulphoxide
E	electrophile
HMPA	hexamethylphosphoramide (CAUTION—CANCER-CAUSING SUSPECT AGENT)
LDA	lithium diisopropylamide
LTMP	lithium 2,2,6,6-tetramethylpiperidide
MCPBA	*m*-chloroperbenzoic acid
NBS	*N*-bromosuccinimide
NBSP	*p*-nitrobenzenesulphonyl peroxide
NCS	*N*-chlorosuccinimide
NMP	*N*-methyl-2-pyrrolidone
Nu	nucleophile
ox.	oxidation
red.	reduction
r.t.	room temperature
SKR	Sharpless kinetic resolution
TBTH	tributyltin hydride
TFA	trifluoroacetic acid
THF	tetrahydrofuran
THP	tetrahydropyranyl
TMEDA	*N*,*N*,*N*′,*N*′-tetramethylenediamine
TUDO	thiourea dioxide

1

Introduction

1.1 SOME PHYSICAL PROPERTIES OF TELLURIUM

The Pauling electronegativities of carbon and tellurium are, respectively, 2.5 and 2.1. This, in addition to the large volume of the tellurium atom (atomic radius 1.37, ionic radius 2.21), promotes easy polarization of Te—C bonds. The ionic character of the bonds increases in the order $C(sp^3)—Te^-$ > $C(sp^2)$—Te > C(sp)—Te in accordance with the electronegativity of carbon accompanying the s character.

The Te—C bond therefore exhibits a high reactivity, as demonstrated typically by the easy heterolytic cleavage towards nucleophilic reagents.

TABLE 1.1

Some physical properties of the VI^b family of elements

Element	Atomic number	Atomic weight	Electronic configuration	Pauling electronegativity	Ionization potential	Ionic radius	Atomic radius
O	8	15.99	$1s^22s^22p^4$	3.5	13.61	1.40	0.66
S	16	32.06	$2s^22p^63s^23p^4$	2.5	10.36	1.84	1.04
Se	34	78.96	$3s^23p^63d^{10}4s^24p^4$	2.4	9.75	1.98	1.17
Te	52	127.6	$4s^24p^64d^{10}5s^25p^4$	2.1	9.01	2.21	1.37

1.2 MONOGRAPHS AND REVIEW ARTICLES

1. W. C. Cooper (ed.), *Tellurium*. Van Nostrand Rheinhold, New York, 1971.
2. K. Y. Irgolic, *The Organic Chemistry of Tellurium*. Gordon and Breach, New York, 1974.
3. N. Petragnani and M. Moura Campos, *Organomet. Chem. Rev.* **2**, 61 (1967).
4. K. J. Irgolic and R. Zingaro, in *Organometallic Reactions*, (ed. E. Becker and M. Tsutsui). Wiley, New York, 1971.

5. K. J. Irgolic, *J. Organomet. Chem.* **103**, 91 (1975); K. J. Irgolic, *J. Organomet. Chem.* **130**, 411 (1977); K. J. Irgolic, *J. Organomet. Chem.* **158**, 235 (1978); K. J. Irgolic, *J. Organomet. Chem.* **189**, 65 (1980); K. J. Irgolic, *J. Organomet. Chem.* **203**, 368 (1980).
6. S. Uemura, *Nippon Kagaku Kaishi* **36**, 381 (1981).
7. N. Petragnani and J. V. Comasseto, *Proceedings of the 4th International Conference of the Organic Chemistry of Selenium and Tellurium* (ed. F. Y. Berry and W. R. McWhinnie), pp. 98–241. The University of Aston in Birmingham, Birmingham, 1983.
8. S. Uemura, *J. Synth. Org. Chem. Jpn* **41**, 804 (1983).
9. L. Engman, *Acc. Chem. Res.* **18**, 274 (1985).
10. N. Petragnani and J. V. Comasseto, *Synthesis* **1** (1986).
11. H. Suzuki, *J. Synth. Org. Chem. Jpn* **45**, 603 (1987).
12. J. D. Sadekov, B. B. Rivkin and V. Y. Minkin, *Russ. Chem. Rev.* **56**, 343 (1987).
13. S. Patai and Z. Rappoport (ed.), *The Chemistry of Organic Selenium and Tellurium Compounds,* Vols I and II. Wiley, New York, 1986, 1987.
14. L. Engman, *Phosphorus Sulfur* **38**, 105 (1988).
15. N. Petragnani and J. V. Comasseto, *Synthesis* 793 (1991).
16. N. Petragnani and J. V. Comasseto, *Synthesis* 897 (1991).
17. H. Rheinboldt, in *Houben–Weyl—Methoden der Organischen Chemie* (ed. E. Muller), 4th edn, Vol. IX. Georg Thieme, Stuttgart, 1955.
18. K. Y. Irgolic, *Houben–Weyl—Methoden der Organischen Chemie* (ed. D. Klamann), 4th edn, Vol. E12b. Georg Thieme, Stuttgart, 1990.

2

Preparation of the More Important Inorganic Tellurium Reagents

2.1 TELLURIUM TETRACHLORIDE

Tellurium tetrachloride is prepared directly from the elements [1].

$$Te + 2Cl_2 \longrightarrow TeCl_4$$

Experimental procedure [1]

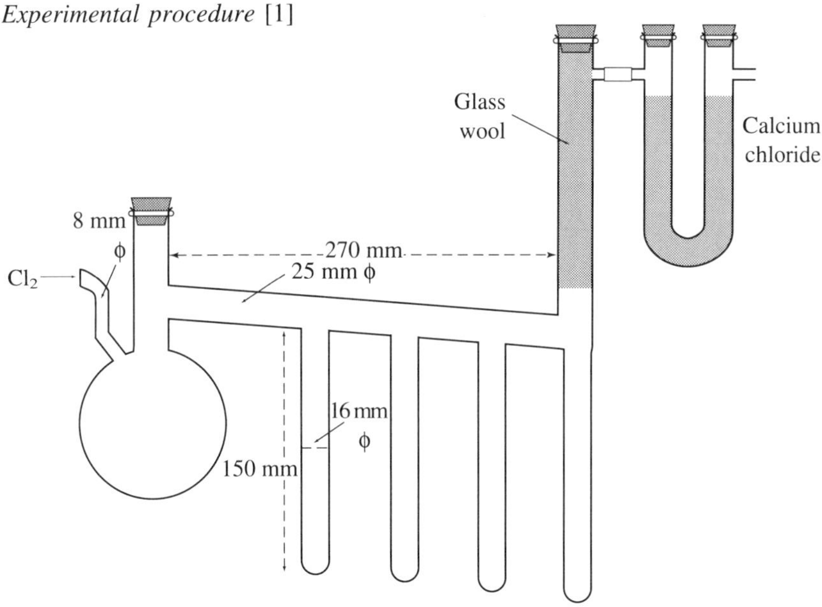

The apparatus in the figure is charged with 100–150 g of fine powdered Te (the apparatus and the Te have been previously heated overnight at 110°C to ensure dryness). A stream of Cl_2 (dried by bubbling in concentrated H_2SO_4) is slowly introduced by means of a Tygon tube; meanwhile, the apparatus is heated with a small burner flame. After few minutes the solid Te begins to be converted into a black liquid. The reaction is exothermic (and proceeds spontaneously)

but the absorption of Cl_2 is accelerated by heating the mixture from time to time. After a while the mixture becomes clearer, giving finally an amber-coloured liquid, which by increased heating forms a vapour of the same colour. At that time the absorption is complete and all the Te has been converted into $TeCl_4$. By vigorous burner heating the product is transferred by distillation, under a continuous Cl_2 stream, into the tubes, where it solidifies. The tubes are detached from the apparatus by melting. The yield is 90%.

Tellurium tetrachloride is a pale yellow crystalline solid, m.p. 225°C, highly hygroscopic, soluble in dioxane, acetone, ether, methanol and ethanol but less soluble in benzene and chloroform.

Tellurium tetrachloride is instantaneously hydrolyzed by water, giving tellurium oxochloride, which is soluble in concentrated hydrochloric acid, forming $HTeCl_5$ and H_2TeCl_6.

$$TeCl_4 + H_2O \longrightarrow TeOCl_2 + 2HCl$$

The Raman data suggest the ionic structure $TeCl_3^+Cl^-$ in both the solid and liquid states [2].

2.2 TELLURIUM DIOXIDE

Tellurium dioxide is prepared by oxidation of elemental tellurium with concentrated nitric acid [3].

$$Te \xrightarrow{\text{conc}\cdot\ HNO_3} 2TeO_2 \cdot HNO_3 \xrightarrow[400°C]{\Delta} TeO_2$$

Experimental procedure [3] Commercial Te (20 g, finer than 60 mesh) is weighed into a 1000 ml beaker, covered with H_2O (200 ml) and treated by the slow addition of 95 ml of concentrated HNO_3 (95 ml, specific gravity 1.42). The reaction is allowed to continue for 5–10 min with occasional agitation. Insoluble impurities are removed immediately by filtration on a Buchner funnel. The filtrate is transferred to a 600 ml beaker. Concentrated HNO_3 (65 ml) is then added and the solution boiled until oxides of nitrogen have been expelled. Basic nitrates of antimony and bismuth precipitate at this point if these substances are present as impurities. These are removed by filtering through asbestos, after which the clear liquor is evaporated gently on a water bath under a hood in an open 600 ml beaker. Basic tellurium nitrate is deposited. The evaporation is continued until the solution volume has been reduced to 100 ml. The solution is then cooled. The crystalline deposit is filtered, washed with H_2O on a suction filter, and air dried. The dry crystals are placed in a 400 ml beaker, covered by inverting a 1000 ml beaker over the smaller beaker, and heated at a hot-plate temperature of 400–430°C for 2 h. The TeO_2 (21 g, 84%) is cooled and bottled immediately to avoid slow darkening due to reduction by organic matter from the atmosphere.

Tellurium dioxide is a white solid existing in two crystalline forms. It is soluble in water and in both aqueous sodium hydroxide and hydrochloric acid.

2.3 ALKALI METAL TELLURIDES ($Te^{2-}Cat_2^+$)

These reagents are usually prepared *in situ*.

The preparative methods are outlined in sequence. The experimental procedures are described in a later chapter.

2.3.1 From the elements ($2Na + Te \longrightarrow Na_2Te$)

2.3.1.1 In liquid ammonia [4]

2.3.1.2 In DMF [5]

2.3.1.3 In the presence of naphthalene [6]

2.3.2 From tellurium and reducing agents ($Te \xrightarrow{red} Te^{2-}$)

2.3.2.1 Rongalite ($HOCH_2SO_2Na/NaOH^-$) [7]

2.3.2.2 Thiourea dioxide (TUDO, $HN=C(NH_2)S(O)OH$)/NaOH [8]

2.3.2.3 Hydride transfer reagents

KBH_4/NaOH [9]
Me_4NBH_4 (giving tetramethylammonium telluride) [10]
$NaBH_4$/DMF [11]
$NaBH_4/H_2O$ [12]
$LiHBH_3$ [13]

2.3.2.4 Tin(II) chloride/KOH/DMSO [14]

2.3.2.5 Hydrazine hydrate/NaOH [15]

2.3.3 From tellurium and non-reducing bases

2.3.3.1 NaH/DMF [16]

2.3.3.2 NaH/DME [17]

2.3.3.3 NaH/*N*-methyl-2-pyrrolidone [18]

2.4 ALKALI METAL DITELLURIDES (Na_2Te_2)

2.4.1 From the elements ($2Na + 2Te \longrightarrow Na_2Te_2$)

2.4.1.1 In liquid ammonia [19]

2.4.1.2 In HMPA [20]

2.4.1.3 In the presence of naphthalene [21]

2.4.2 From tellurium and reducing agents ($2Te \xrightarrow{red} Te_2^{2-}$)

2.4.2.1 Rongalite [22]

2.4.2.2 TUDO/NaOH [23]

2.4.2.3 Hydride transfer reagents

$NaBH_4$/EtONa/EtOH [24]
$NaBH_4$/DMF [25]

2.4.2.4 Hydrazine hydrate/NaOH [26]

2.5 HYDROGEN TELLURIDE (H_2Te)

Hydrogen telluride is prepared *in situ* by hydrolysis of aluminium telluride [27].

$$Al_2Te_3 + 3H_2O \longrightarrow 3TeH_2 + Al_2O_3 \qquad [27]$$

2.6 SODIUM HYDROGEN TELLURIDE (NaHTe)

Sodium hydrogen telluride is prepared by reduction of tellurium with $NaBH_4$ under several conditions. The original procedure uses ethanol as the solvent, adding, after complete reduction of the tellurium, an appropriate amount of acetic acid (see Section 4.1.2, ref. [9]) [28].

REFERENCES

1. Y. F. Suttle and C. R. F. Smith, *Inorg. Synth.* **III**, 140 (1956).
2. H. Gerding and H. Houtgraff, *Rec. Trav. Chim.* **73**, 737 (1954).
3. H. Marshall, *Inorg. Synth.* **III**, 143 (1950).
4. Section 3.1.1.1 (a), ref. [1].
5. Section 3.1.1.1 (a), ref. [8].
6. Section 3.1.1.1 (a), refs [10, 11].
7. Section 3.1.1.1 (b), refs [14, 15]; Section 4.1.6, ref. [19]; Section 4.1.9, ref. [33]; Section 4.1.11, ref. [41]; Section 4.2.3, ref. [6].
8. Section 3.1.1.1 (b), ref. [23].
9. Section 3.1.1.1 (b), ref. [24].
10. Section 3.1.1.1 (b), ref. [15].
11. Section 3.1.2.2, ref. [8]; Section 4.1.10, ref. [35]; Section 4.4.1.1, ref. [2].
12. Section 3.1.3.1, ref. [1]; Section 4.1.12.6, ref. [58].
13. Section 4.3, ref. [1].
14. Section 3.1.1.1 (b), ref. [26].
15. Section 3.1.1.1 (b), ref. [27].
16. Section 3.1.2.1, ref. [5]; Section 3.11.2.1, ref. [4]; Section 4.1.5, ref. [17]; Section 4.2.1, ref. [1]; Section 4.4.2.2, ref. [5].
17. Section 4.2.3, ref. [5].
18. Section 3.1.2.1, ref. [7].
19. Section 3.2.1.1, ref. [1].
20. Section 3.2.1.2, ref. [10].
21. Section 3.2.1.1, ref. [3].
22. Section 3.2.1.1, ref. [6].
23. Section 3.2.1.1, ref. [8].
24. Section 3.2.1.1, ref. [5].
25. Section 4.4.1.1, ref. [2].
26. Section 3.2.1.1, ref. [9].

27. Section 4.1.1.1, ref. [1]; Section 4.1.4, ref. [14]; Section 4.1.6, ref. [18].
28. Section 4.1.2, refs [9, 11, 13]; Section 4.1.3, ref. [3]; Section 4.1.4, ref. [16]; Section 4.1.9, ref. [32]; Section 4.1.11, ref. [40]; Section 4.1.12.1, ref. [49]; Section 4.1.12.3, ref. [56]; Section 4.1.12.5, refs [12, 57]. Section 4.1.12.6, ref. [58]; Section 4.4.1.1, ref. [2]; Section 4.4.1.2, ref. [3]; Section 4.4.1.3, ref. [4]; Section 4.4.1.4, refs [6, 7]; Section 4.4.2.1, refs [8, 9].

3

Preparation of the Principal Classes of Organic Tellurium Compounds

This chapter is devoted mainly to the preparation of those classes of organotellurium compounds which have been more systematically investigated in the past years, such as diorganyl tellurides, diorganyl ditellurides, organyltellurium trihalides and diorganyltellurium dihalides.

These classes include compounds exhibiting structural, biological or theoretical interest and compounds particularly valuable as reagents or intermediates in organic synthesis.

The preparation of other classes of compounds such as telluroketals, tellurenyl halides, aryl tellurocyanates and the exceptionally important synthetic reagents diaryl telluroxides and aryltellurinic anhydrides are also described.

The principal preparative methods of these compounds are outlined in the Schemes 3.1–3.3. The physical properties and stability of the mentioned compounds are also described briefly.

Diorganyl tellurides (Section 3.1)

The main routes to symmetrical diorganyl tellurides involve the direct reaction of nucleophilic telluride dianions (usually as Na_2Te) with alkylating or arylating reagents. Otherwise the electrophilic reagents tellurium tetrahalides react with arylmagnesium reagents, giving diaryl tellurides.

Unsymmetrical tellurides are obtained as follows:

- starting from nucleophilic tellurolate anions (easily generated by telluration of Grignard or lithium reagents or by reduction of diorganyl ditellurides) by reaction with alkylating or arylating reagents or by addition to acetylenes;
- by cleavage of diaryl ditellurides with arylmagnesium reagents, diazonium salts or arenesulphonylazo compounds;
- by the reaction of electrophilic tellurenyl halides with Grignard reagents.

Low-molecular-weight diorganyl tellurides are colourless or yellowish liquids with an unpleasant and penetrating odour. Dimethyl telluride is a metabolite of

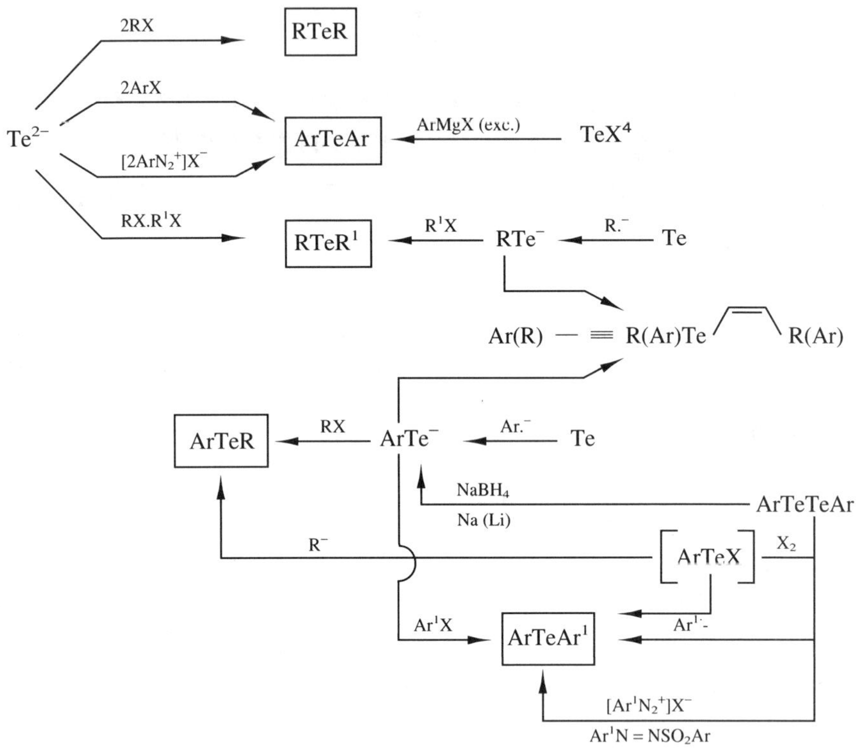

SCHEME 3.1

tellurium and tellurium compounds in a variety of living organism, including humans. Higher dialkyl tellurides and most diaryl tellurides are low-melting-point solids (diphenyl telluride is a liquid).

Diorganyl tellurides are soluble in the common organic solvents.

Due to the above-mentioned organoleptic properties it is recommended that contact between dialkyl tellurides and the skin is avoided, and that, in general, all work involving diorganyl tellurides is carried out under a well-ventilated hood.

The thermal stability and sensitivity to the atmosphere of diorganyl tellurides is dependent on the organic moiety. Many aliphatic, cycloaliphatic, as well as vinylic and acetylenic, tellurides are reported to be decomposed by light, and therefore some authors recommend preparation of these compounds in the dark or under a red light [1]. These tellurides are also sensitive to exposure to the open atmosphere, easily undergoing oxidation to mixtures containing the corresponding telluroxides in addition to the other products. These oxidations are especially effective in solution, where the oxidation products separate as amorphous insoluble solids. Benzyl groups in tellurides are characterized by a peculiar lability, as demonstrated by the oxidative cleavage of aryl benzyl tellurides leading to aromatic carbonyl compounds [2]. Diaryl tellurides are generally more stable and can be handled in the open atmosphere.

Diorganyl ditellurides (Section 3.2)

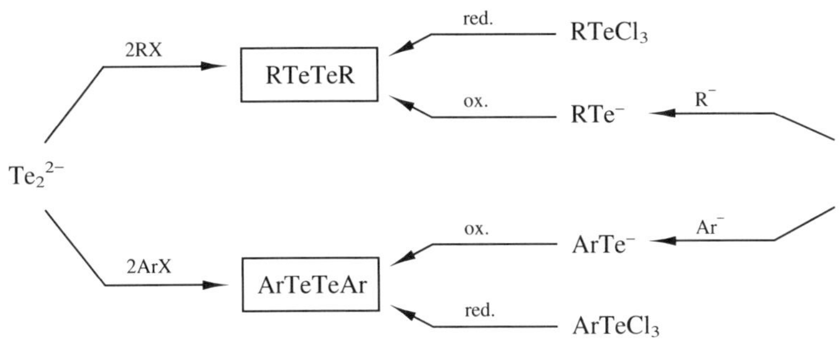

SCHEME 3.2

Diorganyl ditellurides are prepared by three routes:

- alkylation or arylation of the ditelluride dianion (usually as Na_2Te_2);
- oxidation of tellurolate anions;
- reduction of the corresponding organyltellurium trichlorides.

The Te_2 group is a chromophor, and aliphatic and aromatic ditellurides exhibit a characteristic orange to pale or dark red colour (absorption maximum at about 400 nm).

The short-chain aliphatic ditellurides are liquids with a pungent odour, whereas the higher members (more than ten C chains) are low-melting-point solids.

Diaryl ditellurides, which are more important as synthetic intermediates, are solids, highly soluble in solvents such as petroleum ether, benzene, chloroform, ether and tetrahydrofuran, but are less soluble in methanol and ethanol.

Dibenzyl ditelluride exhibits the peculiar lability of the tellurium–benzyl bond already mentioned for the corresponding tellurides. By heating the solid at 120°C, or by exposure in solution to ordinary incandescent lighting or to a Hanovia lamp, a rapid decomposition into elementary tellurium and dibenzyl telluride occurs [3].

Organyltellurium trichlorides and diorganyltellurium dihalides (Sections 3.5 and 3.9)

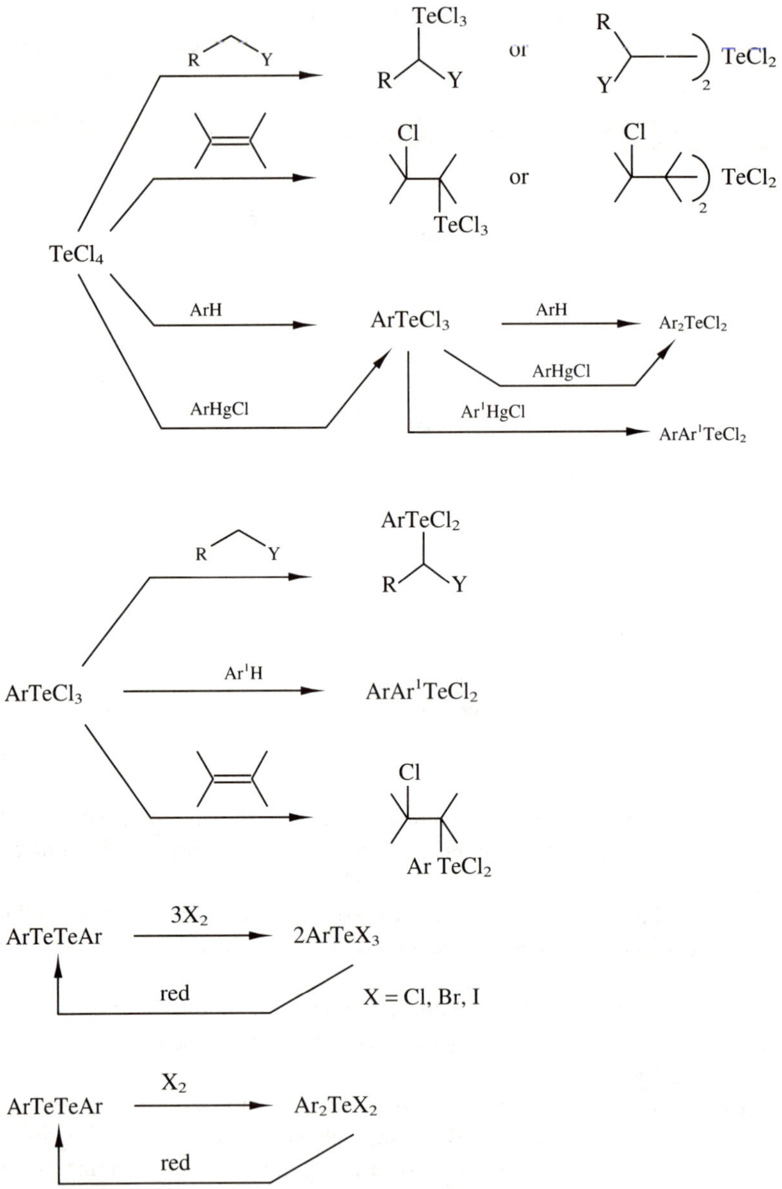

SCHEME 3.3

Organyltellurium trichlorides and diorganyltellurium dichlorides are prepared starting from the electrophilic tellurium tetrachloride (or aryltellurium trichlorides) by:

- condensation reactions with active methylene compounds;
- addition to a C=C bond (or a C≡C bond, not shown in the scheme);
- electrophilic substitution in aromatic hydrocarbons;
- reaction with organomercury chlorides.

Organyltellurium trichlorides and the not directly accessible tribromides and triiodides are obtained by the halogenolysis of the Te—Te bond of the corresponding ditellurides.

Diorganyltellurium dihalides can also be prepared by the addition of halogens to the parent tellurides.

The trihalides and dihalides are crystalline compounds. The chlorides are colourless (or yellow such as some aryltellurium trichlorides), the colour changing to orange and red (or deep red) for the bromides and iodides.

The aryltellurium trihalides are generally more stable than the alkyltellurium trihalides (alkyltellurium trichlorides, produced by the addition of $TeCl_4$ to olefins, easily liberate elemental tellurium).

The reactivity of aryltellurium trihalides decreases on going from the chlorides to the iodides, the same trend occurring for hydrolysis. Aryltellurium trichlorides are very sensitive to water and moisture and are easily hydrolysed, the tribromides being more stable, while the triiodides are unaffected by cold water and can be prepared even by aqueous procedures. Diaryltellurium dihalides are stable in water, and ionic exchange reactions allow the conversion of dichlorides into dibromides and diiodides.

Aryltellurium trichlorides are highly soluble in methanol and ethanol but less soluble in benzene. Diaryltellurium dichlorides exhibit inverse solubilities, being more soluble in benzene than in methanol or ethanol. These properties allow an easy separation of diaryl tellurides from diaryl ditellurides (frequently formed as by-products in the preparation of tellurides): the mixture is treated with SO_2Cl_2 and the obtained mixed dichlorides and trichlorides are separated by the appropriate solvents and reduced back into the pure tellurides and ditellurides.

REFERENCES

1. D. L. J. Clive, G. J. Chittattu, V. Farina, W. A. Kiel, S. M. Menchen, C. G. Russell, A. Singh, C. K. Wong and N. J. Curtis, *J. Am. Chem. Soc.* **102**, 4438 (1980).
2. J. T. B. Ferreira, A. R. M. Oliveira and J. V. Comasseto, *Tetrahedron Lett.* **33**, 915 (1992).
3. H. K. Spencer and M. P. Cava, *J. Org. Chem.* **42**, 2937 (1977).

3.1 DIORGANYL TELLURIDES

Diorganyl tellurides, compounds with two organic groups linked to a tellurium atom, constitute the most abundant and familiar class of organic tellurium compounds. The organic groups, of the most differentiated types, can be identical or different, giving rise to symmetrical or unsymmetrical tellurides.

Since symmetrical dialkyl and diaryl tellurides are the most employed in organic synthesis, and often exhibit structural or biological interest, their preparation will be examined in detail, focusing on the methods and procedures that are considered today as the most popular.

3.1.1 Symmetrical dialkyl tellurides

3.1.1.1 From alkali tellurides and alkylating agents

$$Te^{2-} + 2RX \longrightarrow RTeR$$

Alkali tellurides, among which sodium telluride is the most widely employed, are powerful nucleophilic reagents and therefore react easily with alkylating agents.

(a) From sodium telluride prepared from the elements

(i) Sodium/liquid ammonia method

$$Te + 2Na \xrightarrow{NH_3 \text{ liquid}} Na_2Te \xrightarrow{RX} RTeR$$

This method has been applied successfully to n- and s-alkyl halides [1–4].

Dialkyl tellurides (general procedure) [1] Elemental Te is added in approximately 0.5 g portions to a well stirred solution of Na in liquid NH_3 until the solution decolorizes, forming a colourless suspension (2 g-atom Na:1 g-atom Te). The quantities of the materials are chosen to give a suspension of about 0.7 M. The alkylating agent is added dropwise in 10% excess to the suspension of Na_2Te. The reaction mixture is stirred until the NH_3 evaporates. H_2O is then added, the mixture extracted with ether and the ethereal solution worked up in the usual manner.

Diethyltelluride: yield 80%; b.p. 38°C/14 torr
Diisopropyltelluride: yield 80%; b.p. 49°C/14 torr

When the alkylating agent is insoluble in liquid ammonia, as in the case of long-chain compounds, an organic solvent is added to the sodium telluride residue after evaporation of the ammonia. Some cyclic and steroidal tellurides have been prepared from sodium telluride in ethanol and the appropriate dihalides [5–7].

(ii) Sodium/DMF method [8, 9]

$$Te + 2Na \xrightarrow[\triangle]{DMF} Na_2Te \xrightarrow{RX} RTeR$$

Bis(2-phenyl-1,2-dioxolan-2-yl)methyl telluride ($R = PhCH_2$) (*typical procedure*) [8]. Finely powdered Te (0.30 g, 2.3 mmol) and Na (0.22 g, 9.6 mmol) are stirred under N_2 in dry DMF (15 ml) at 110°C until all the Te has disappeared (15 min). 2-(Bromomethyl)-2-phenyl-1,3-dioxolane (1.15 g, 4.7 mmol) in dry THF (10 ml) is added to the resulting yellowish suspension. After 2.5 h at room temperature the reaction mixture is poured into H_2O and extracted with ether. Chromatography on SiO_2 (eluent: CH_2Cl_2/petroleum ether 40–60°C 1:1) furnishes the telluride (0.65 g (60%); m.p. 67–68°C (EtOH).

The last method has been used to prepare the following tellurides: R = n-$C_{12}H_{25}$ (66%), *trans*-Ph (41%) [9].

(iii) Sodium/naphthalene method

Sodium telluride can be prepared under mild conditions from the elements in the presence of catalytic amounts of naphthalene [10] or by treating tellurium with sodium naphthalide [11].

$$Te \xrightarrow[THF]{Na/naphthalene} Na_2Te \xrightarrow[(60\text{–}90\%)]{RX} RTeR$$

$$R = Et, n\text{-}Pr, n\text{-}Bu, MeOCH_2CH_2$$

Dialkyl tellurides (general procedure) [10] Te powder (3.58 g, 28 mmol), Na chips (1.29 g, 56 mmol) and naphthalene (0.72 g, 5.6 mmol) in THF (25 ml) are refluxed under N_2 and stirred for 3 h. The heterogeneous white mixture is cooled at 0°C and the alkyl halide (56 mmol) is added slowly, stirring for 30 min. After

30 min of additional stirring, the mixture is filtered, the solution evaporated and the residue distilled under vacuum, giving the telluride.

Diallyl telluride (typical procedure) [11]. In a 25 ml Schlenk flask, Te pieces (6.1 g, 48 mmol, from a Te ingot), Na (2.2 g, 96 mmol) and naphthalene (0.1 g, 0.8 mmol) in THF (20 ml) are stirred under argon at 25°C for 4 days. The mixture is filtered and the filter cake washed with THF (20 ml) and dried under vacuum to give Na_2Te (7.6 g, 92%). Alternatively, Te powder (60 mesh) and sodium naphthalide (2 mol eq) are stirred in THF for 4 h. Na_2Te is isolated as an amber-coloured solid. The Na_2Te (62.9 g, 0.362 mmol) is slurried in absolute EtOH (400 ml). The slurry is cooled at 0°C and allyl bromide (93.3 g, 0.77 mmol) is added dropwise during a 1 h period while stirring. The mixture is stirred for an additional hour at 20°C and then filtered. The grey filter cake is washed with EtOH (400 ml). The combined washing and filtrate are evaporated at atmospheric pressure. The residue is distilled, giving diallyl telluride as a pungent, air-sensitive, yellow liquid (56.7 g (75%); b.p. 70–72°C/13 torr).

(b) From alkali tellurides prepared from tellurium and reducing agents

(i) Rongalite (sodium formaldehyde sulphoxylate) method

This method was first described at the beginning of the century [12], and continues to find a wide application.

$$\mathrm{Te} \xrightarrow[\mathrm{NaOH/H_2O}]{\mathrm{HOCH_2SO_2Na}} \mathrm{Na_2Te} \xrightarrow{\mathrm{RX}} \mathrm{RTeR}$$

n-Alkyl halides, benzyl chloride, ethyl sulphate and s-alkyl bromides give the expected tellurides in medium yields whereas t-butyl chloride is not converted to the telluride [13]. Due to the insolubility of the alkyl halides, ethanol must be added to the reaction mixture.

Di-n-butyl telluride (typical procedure) [14] A mixture of Te powder (40 g, 0.75 mol) $HOCH_2SO_2Na$ (rongalite, 90.0 g, 0.75 mol) and NaOH (70 g, 1.75 mol) is heated under reflux in H_2O (400 ml) under an inert atmosphere for 15–30 min. A solution of n-butyl bromide (90 g, 0.65 mol) in EtOH (350 ml) is added slowly to the above-prepared solution of Na_2Te with stirring. The mixture is heated under reflux for 1 h, then cooled and extracted with ether. The ethereal solution is dried ($CaCl_2$) and evaporated under an inert atmosphere. The oily residue is distilled under vacuum and furnishes the telluride (45 g (57%); b.p. 107–112°C/12 torr).

Dibenzyl telluride (typical procedure) [15] Rongalite (18.0 g, 0.15 mol) is added under N_2 at 80°C to a suspension of Te (2.56 g, 20 mmol) in a solution of NaOH (12 g, 0.3 mol) in 125 ml of H_2O. After stirring for 1 h, a solution of benzyl chloride (1.26 g, 10 mmol) in a small volume of EtOH is added dropwise at room temperature to the almost colourless telluride solution. After stirring for an additional hour, the mixture is extracted with ether, and the ethereal solution dried ($MgSO_4$) and evaporated. The residue is recrystallized from petroleum ether (40–60°C) under red light, giving the telluride as yellow needles (1.09 g (70%); m.p. 47–57°C).

Cyclic tellurides, including some with a steroidal structure [19, 20], have been prepared by the Rongalite method.

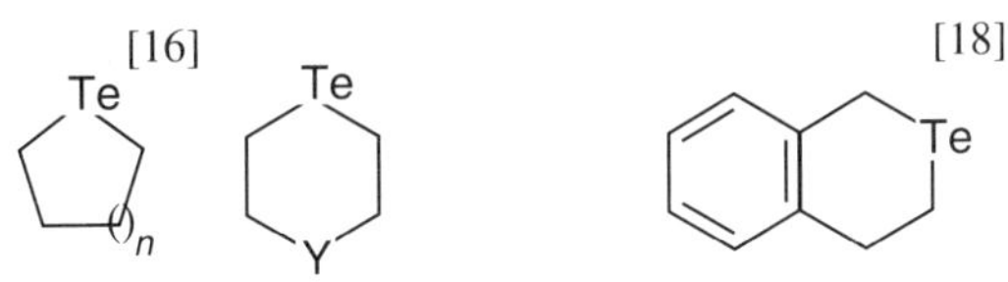

$n = 1, 2$ Y = 0[17], S[18]

(ii) Sodium dithionite and thiourea dioxide method

The use of sodium dithionite ($Na_2S_2O_4$) and thiourea dioxide (TUDO; $HN{=}C(NH_2)S(O)OH$) has been introduced quite recently as a reducing agent for the preparation of sodium telluride in an aqueous medium, followed by reaction with n-alkyl halides to give dialkyl tellurides [21, 22].

The TUDO method has been reinvestigated recently and has found wide use due to the simplicity of the experimental conditions [23].

$$\mathrm{Te} \xrightarrow[\mathrm{NaOH/H_2O/THF}]{\mathrm{TUDO}} \mathrm{Na_2Te} \xrightarrow[\substack{\mathrm{THF\ (CTBA)} \\ (72\text{–}85\%)}]{\mathrm{RX}} \mathrm{RTeR}$$

$$\mathrm{R = n\text{-}C_6H_{13},\ n\text{-}C_8H_{17},\ i\text{-}C_3H_7CH_2CH_2,\ THPO(CH_2)_6}$$

Dialkyl tellurides (general procedure) [23] A mixture of TUDO (0.2 g, 2 mmol), NaOH (0.112 g, 2.6 mmol), Te powder (0.128 g, 1 mmol) in H_2O (0.75 ml) and THF (0.75 ml) is refluxed for 1 h. The alkyl halide (2 mmol) and (cetyl trimethyl ammonium bromide) CTBA (0.004 g, 11×10^{-5} mmol) in THF (0.5 ml) are added to the pale pink solution. After 1 h of additional reflux the mixture is worked up in the usual manner and the residue purified by column chromatography on SiO_2 (elution with petroleum ether 40–60°C).

(iii) Hydride transfer reagent method

A useful method for the reductive conversion of elemental tellurium into Te^{2-} anions employs complex hydrides such as sodium or potassium borohydride and tetraalkyl ammonium borohydride as reducing agents.

$$Te \xrightarrow[\text{solvent/reflux}]{cat^+BH_4^-} Te^{2-} \xrightarrow{RX} RTeR$$

$$cat = Na^+, K^+, Me_4N^+$$

Di-n-octyl telluride (typical procedure) [24] Elemental Te (2.54 g, 20 mmol) is heated at reflux in 20% aqueous NaOH (43 ml) containing KBH_4 (2.7 g, 50 mmol), under argon. After 1.5 h the mixture is deep purple in colour, but after an additional 30 min of reflux turns pale yellow and no Te metal remains. Bromooctane (7.72 g, 40 mmol) in MeOH (50 ml) is added, the solution refluxed for 30 min, cooled, poured into H_2O (100 ml) and extracted with ether (3×). The ether extracts are washed with H_2O, dried (Na_2SO_4) and evaporated. The oily residue is distilled under vacuum, giving the telluride (5.5 g (78%); b.p. 154°C/0.7 torr).

Dibenzyl telluride (typical procedure) [15] Elemental Te (1.27 g, 10 mmol) and Me_4NBH_4 (1.78 g, 20 mmol) are heated in H_2O (100 ml) in a steam bath under N_2 until the initially produced purple colour is discharged. After cooling at room temperature, benzyl chloride (2.53 g, 20 mmol) in EtOH (20 ml) is added slowly with stirring and rigorous exclusion of air. After 2 h of stirring, the mixture is extracted with ether, dried and evaporated. The residue is recrystallized from petroleum ether (under red light), giving the telluride as yellow needles (2.56 g (82%); m.p. 49–57°C).

In an interesting one-pot procedure the borohydride is used as both the reducing and alkylating agent [25].

$$Te + 2R_4NBH_4 \longrightarrow RTeR + 2R_3NBH_3 + H_2$$

Di-n-butyl telluride (typical procedure) [25] Elemental Te (0.63 g, 5 mmol) and n-Bu_4NBH_4 (3.87 g, 15 mmol, 50% excess) are refluxed in toluene (80 ml) under N_2 and stirring. The initial red solution gradually turns colourless, giving a clear solution after 3 h. The cooled solution is washed with H_2O (50 ml) and 1 M HCl (50 ml) (evolution of H_2). The organic phase is dried ($CaCl_2$) and evaporated, giving an oil (mixture of the telluride and Bu_3BNH_3). The telluride is separated by column chromatography on SiO_2 (eluent: CH_2Cl_2/petroleum ether 40–60°C 1:1) (1.20 g (95%); $n_D^{20} = 1.5165$).

(iv) Tin(II)/potassium hydroxide method [26]

Dimethyl telluride (typical procedure) [26] KOH (100 g, 1.78 mol) in DMSO

$$\mathrm{Te} \xrightarrow[\mathrm{DMSO,120^{\circ},10\ h}]{\mathrm{SnCl_2/KOH}} \mathrm{K_2Te} \xrightarrow{\mathrm{MeI}} \mathrm{MeTeMe}$$

(180 ml) is heated on a steam bath. $SnCl_2$ (10 g, 53 mmol) and Te powder (60 g, 0.47 mmol) are added and the mixture heated at 120°C for 20 h. After cooling at 40°C, methyl iodide (157 g, 1.1 mol) is added dropwise during 2 h. The mixture is then heated at 70°C for 2 h and then distilled until the condensing vapour reaches a temperature of 100°C. The organic layer of the residue is separated, dried and distilled (71.6 g (97%); b.p. 93–94°C).

(v) Hydrazine hydrate method [27]

$$\mathrm{Te} \xrightarrow[\mathrm{DMF}]{\mathrm{N_2H_4/NaOH}} \mathrm{Na_2Te} \xrightarrow[(55\text{–}73\%)]{\mathrm{RX}} \mathrm{RTeR}$$

R = Et, n-Bu, n-pentyl, s-Bu, i-pr-CH_2CH_2, n-octyl

Dibutyl telluride (typical procedure) [27] Hydrazine hydrate 80% (0.50 ml, 7.1 mmol) is added dropwise using a syringe to a stirred mixture of finely ground Te (0.64 g, 5 mmol) and powdered NaOH (0.40 g, 10 mmol) in DMF (10 ml) at 50–60°C. After 3 h of stirring, n-butyl bromide (1.4 g, 10 mmol) in DMF (2 ml) is added, the mixture is heated at 60°C for a further 30 min, cooled at room temperature and extracted with petroleum ether (40–60°C). The organic phase is separated, washed with H_2O, dried ($CaCl_2$) and evaporated, giving the pure telluride (0.69 g (57%); b.p. 111–114°C/13 torr).

3.1.1.2 From bis(triphenylstannyl) telluride and alkylating reagents

The title reagent* reacts easily with the more active halides such as benzyl bromides whereas common halides need to be activated by caesium fluoride [29].

Alkyl iodides and bromides react satisfactorily under these conditions whereas alkyl chlorides and aryl halides are unreactive.

*Prepared by the reaction of sodium hydrogen telluride with chlorotriphenylstannane [28].

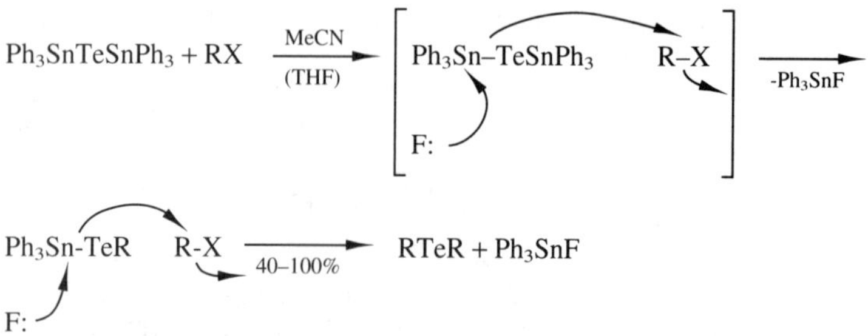

RX = BzBr, n-$C_{10}H_{21}I$, n-$C_{10}H_{21}Br$, i-PrI, EtO_2CCH_2Br, $PhCOCH_2Br$

$Cl(CH_2)_6Br$ (giving $[Cl(CH_2)_6]_2Te$),

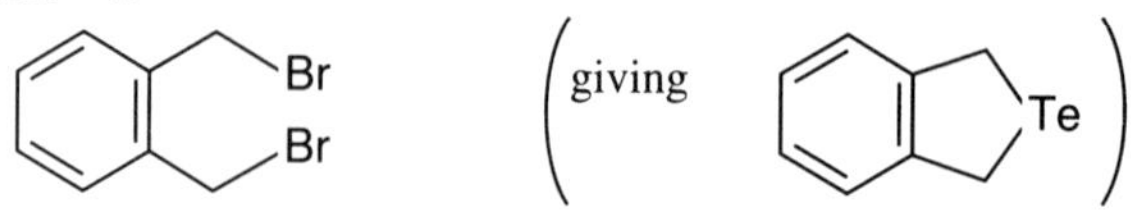

Dialkyl tellurides (general procedure) [29] Bis(triphenylstannyl) telluride (1 eq), alkyl halide (2 eq), excess CsF (4 eq) and the solvent (MeCN or MeCN/THF) are mixed, kept under N_2 and monitored by TLC (or ^1H NMR). Usual work-up of the mixture yields the telluride.

REFERENCES

1. L. Brandsma and H. E. Wijers, *Recl. Trav. Chim. Pays-Bas.* **82**, 68 (1963).
2. R. S. Sukhai, R. A. Jong, H. D. Verkruijsse and L. Brandsma, *Recl. J.R. Neth. Chem. Soc.* **100**, 368 (1981); R. S. Sukhai, R. A. Jong, H. D. Verkruijsse and L. Brandsma, *Chem. Abstr.* **96**, 104204 (1982).
3. US 2398414 (1946), Cal Research Corporation; G. H. Denison and P. C. Condit, *Chem. Abstr.* **40**, 3598 (1946).
4. Y. N. Shlyk, G. M. Bogolyubov and A. A. Petrov, *J. Gen. Chem. USSR* **38**, 1156 (1968) (in English).
5. E. Buchta and K. Greiner, *Chem. Ber.* **94**, 1311 (1961).
6. G. Zanati and M. E. Wolf, *J. Med. Chem.* **15**, 368 (1972).
7. G. Zanati, G. Gaare and M. E. Wolf, *J. Med. Chem.* **17**, 561.
8. L. Engman, *Organometallics* **5**, 427 (1986).
9. K. Ohe, H. Takahashi, S. Uemura and N. Sugika, *Nippon Kagaku Kaishi*, 1469 (1987).
10. K. K. Bhasin, V. Gupta, A. Gautam and R. P. Sharma, *Synthetic Commun.* **20**, 2191 (1990).
11. K. Higa and D. C. Harris, *Organometallics* **8**, 1674 (1989).
12. L. Tschugaeff and W. Chlopin, *Ber. Dtsch. Chem. Ges.* **47**, 1269 (1914).
13. M. D. Balfi and K. M. Nandi, *J. Chem. Soc.*, 70 (1941).
14. M. P. Balfi, C. A. Chaplin and H. Phillips, *J. Chem. Soc.*, 344 (1938).
15. H. K. Spencer and M. P. Cava, *J. Org. Chem.* **42**, 2937 (1977).
16. W. V. Farrar and J. M. Gulland, *J. Chem. Soc.*, 11 (1945).
17. J. B. McCullough, *Inorg. Chem.* **4**, 862 (1965).

18. F. G. Holliman and F. G. Mann, *J. Chem. Soc.*, 37 (1945).
19. F. F. Knapp Jr., *J. Labelled Compd Radiopharm.* **17**, 81 (1980).
20. H. Suginome, S. Yamada and J. B. Wang, *J. Org. Chem.* **55**, 2170 (1990).
21. B. G. Brigov, V. L. Bregadze, L. M. Golubinskaya, L. G. Tonoyan and B. I. Kozyzkin, *USSR* **541**, 851 (C1.CC7 F11/00) 5 Jan. 1977, Appl 2, 182, 164 20 Oct. 1975 *Chem. Abstr.* **87**, 5394 (1977).
22. B. I. Kozyzkin, B. A. Salantin, L. L. Ivanov, I. A. Kuzovlev, B. G. Gribov and V. A. Federov, *Poluch. Anal. Veshchestv Osoboi Chist [Dokl. Vses. Konf.]* **5**, 142 (1976), *Chem. Abstr.* **91**, 140278 (1979).
23. J. T. B. Ferreira, A. R. Oliveira and J. V. Comasseto, *Synthetic Commun.* **19**, 239 (1989).
24. G. Kirsch, M. M. Goodman and F. F. Knapp Jr, *Organometallics* **2**, 357 (1983).
25. J. Bergman and C. Engman, *Synthesis*, 569 (1980).
26. M. G. Voronkov, V. K. Stankevich, P. A. Rodkniko, N. A. Korchevin, E. N. Deryagina and B. A. Trofimov, *J. Gen. Chem USSR* **57**, 2398 (1987).
27. P. Lue, B. Chen, X. Yu, J. Chen and X. Zhou, *Synthetic Commun.* **16**, 1849.
28. F. W. Einstein, C. H. W. Jones, T. Jones and R. D. Sharma, *Can. J. Chem.* **61**, 2611 (1983).
29. C. J. Li and D. N. Harpp, *Tetrahedron Lett.* **31**, 6291 (1990).

3.1.2 Symmetrical diaryl tellurides

3.1.2.1 From alkali tellurides and non-activated aryl halides

The normal low reactivity of aryl halides towards nucleophilic reagents is not generally observed in their reaction with alkali tellurides. The observed reactivity seems to be dependent on the method employed to prepare the alkali telluride as well as on the experimental conditions and solvents.

Non-activated aryl halides react only moderately with sodium telluride prepared from the elements in inert solvents (DMF, *N*-methyl-2-pyrrolidone (NMP), HMPA) [1–4].

$$\text{Te} + 2\text{Na} \xrightarrow[\Delta]{\text{solvent}} \text{Na}_2\text{Te} \xrightarrow[\text{solvent } 130\text{–}170°\text{C, } 16\text{–}24\text{ h}]{\text{ArX}} \text{ArTeAr}$$

(35, 50%)

solvent = DMF, NMP, HMPA

ArX = PhI, 2-bromonaphthyl

Better results are achieved with tellurium and sodium hydride in DMF [5].

$$\text{Te} \xrightarrow[\text{DMF, }140°\text{C}]{\text{NaH}} \text{Na}_2\text{Te} \xrightarrow[\text{DMF, }130°\text{C, }24\text{ h}]{\text{ArI}} \text{ArTeAr}$$

(41–70%)

Ar = 1-naphthyl and derivatives, 2-fluorenyl, and 1,2-diiodoacenaphthylene, giving

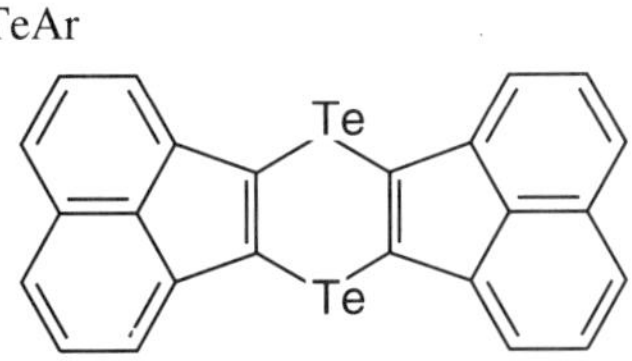

(35%)

Diaryl tellurides (general procedure) [5] A mixture of powdered Te (0.128 g, 1.0 mmol), NaH (0.053 g, 2.2 mmol, 60% suspension in oil, washed with hexane) and dry DMF (3 ml) is heated at 140°C for 1 h. Within 0.5 h the initial deep red colour is lost, and a pale yellow suspension is obtained. After cooling at room temperature, the aryl iodide (2.0 mmol) in dry DMF (3 ml) is added and the mixture heated at 130°C for 24 h. After cooling at room temperature, the mixture is quenched with 10% aqueous $(NH_4)_2SO_4$ (10 ml) and extracted with ether (10 ml). The ethereal extract is washed with H_2O, dried (Na_2SO_4) and evaporated, giving the telluride, which is purified by column chromatography on SiO_2 (eluent: hexane).

The Rongalite method (see Section 3.1.1.1 (b, i)) can be successfully applied to the preparation of diaryl tellurides from aryl iodides [5, 6]. This method seems to be advantageous with respect to the preceding one in view of the higher yields and milder experimental conditions.

$$\mathrm{Te} \xrightarrow[\mathrm{NaOH/H_2O}]{\mathrm{HOCH_2SO_2Na}} \mathrm{Na_2Te} \xrightarrow[\mathrm{DMF, 60°C, 10\ h}]{\mathrm{ArI}} \mathrm{ArTeAr}$$

(55–94%)

Ar = Ph and (alkyl and methoxy) derivatives, 1-naphthyl

2-naphthyl and derivatives, 9-anthryl, 9-phenanthryl, 1-pyrenyl

Some concurrent reactions sometimes observed in the above methods (dehydrohalogenation with the tellurium/Rongalite method and formation of tellurocarbamates $RTeC(O)NMe_2$ with the tellurium/NaH/DMF method) that decrease the yields of the desired tellurides can be avoided by a recently described modified procedure, where sodium telluride is generated by heating elemental tellurium and NaH in NMP [7]. The deep purple solution of the reagent prepared under these conditions can be stored for many days (in contrast to the use of DMF as the solvent) without appreciable degradation.

$$\mathrm{Te} \xrightarrow[\mathrm{100–110°C, 1\ h}]{\mathrm{NaH/NMP}} \mathrm{Na_2Te} \xrightarrow[\mathrm{NMP, 50–110°C, 2–24\ h}]{\mathrm{ArI}} \mathrm{ArTeAr}$$

(42–77%)

Ar = Ph and (alkyl and methoxy) derivatives, 1-naphthyl, 2-naphthyl

1-(2-methyl)naphthyl

Di(1-naphthyl) telluride (typical procedure) [7] Finely ground Te (0.140 g, 1.1 mmol) and NaH (2.2 mmol, 60% suspension in oil, washed with hexane) are heated in NMP (3.5 ml) for 1 h. 1-Iodonaphthalene (0.310 g, 1.2 mmol) is added in small portions over 1.5 h to the obtained solution of Na_2Te at 100°C under argon. The resulting mixture is kept at this temperature for 19 h. During this period the colour of the solution gradually changes from deep purple to black. The progress of the reaction is monitored by TLC. After complete disappearance of the starting material the reaction is quenched by addition of saturated aqueous NH_4Cl (1 ml) followed by benzene (5 ml). The Te deposit is filtered off and the filtrate is partitioned between EtOAc (30 ml) and H_2O (20 ml). The organic phase is separated and dried (Na_2SO_4), the solvent is evaporated and the residue chromatographed over SiO_2 (eluent: hexane), giving the telluride. The product is recrystallized from hexane/$CHCl_3$ (1:1) as yellow crystals (0.160 g (76%); m.p. 123–126°C).

3.1.2.2 From sodium telluride or sodium *O,O*-diethyl phosphorotellurolate and arenediazonium fluoroborates

Sodium telluride and sodium *O,O*-diethyl phosphorotellurolate, prepared respectively by the Te/$NaBH_4$/DMF method and by the reaction of elemental tellurium with NaH and *O,O*-diethyl phosphate in ethanol, react with arenediazonium fluoroborates, giving diaryl tellurides in good yields [8].

*Di(*p*-tolyl) telluride (typical procedure)* [8] *Method A.* Elemental Te (0.65 g, 5 mmol) and $NaBH_4$ (0.45 g, 12 mmol) in DMF (15 ml) are heated in a Schlenk

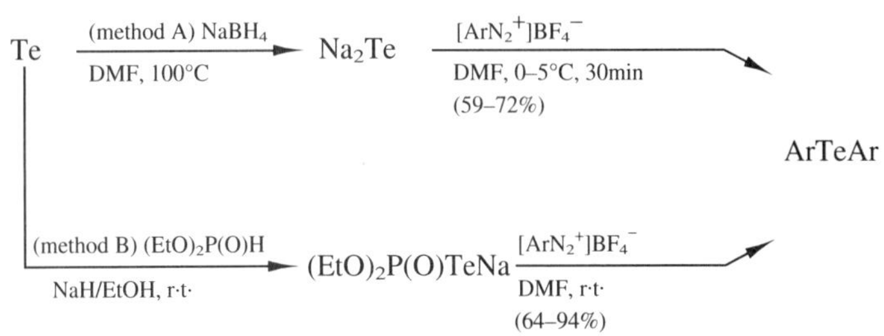

A= Ph and (alkyl, methoxy, halo and acyl) derivatives

reactor at 100°C with stirring and under N_2 until disappearance of the Te powder. The solution is cooled at 0°C and *p*-toluenediazonium fluoroborate* (2.06 g, 10 mmol) in DMF (5 ml) is added dropwise. The mixture is stirred at 0°C for 30 min, quenched with H_2O (20 ml) and then stirred continuously for 30 min. The mixture is filtered, the filtrate extracted with ether (3 × 15 ml), and the extracts washed with H_2O (3 × 15 ml) and dried ($MgSO_4$). The solvent is evaporated and the residue recrystallized from MeOH, giving the telluride (1.05 g (68%); m.p. 67°C).

*For preparation of this reagent, see ref. [8].

Method B. Elemental Te (0.64 g, 5 mmol), NaH (0.30 g, 10 mmol, 80% oil suspension, washed with hexane) and $(EtO)_2P(O)H$ (1.41 g, 10 mmol, 97%) in anhydrous EtOH (20 ml) are stirred in a Schlenk apparatus at room temperature under N_2 until disappearance of the Te powder. *p*-Toluenediazonium fluoroborate (2.27 g, 11 mmol) in DMF (5 ml) is added slowly and the whole stirred for 30 min. The solution is diluted with H_2O, extracted with ether (3 × 15 ml), and the extract dried ($MgSO_4$) and evaporated. The residue is recrystallized from MeOH, giving the telluride (1.38 g (89%); m.p. 67°C).

3.1.2.3 From potassium tellurocyanate and arenediazonium fluoroborates

Another method which uses arenediazonium fluoroborates to prepare diaryl tellurides is the reaction with potassium tellurocyanate [9]. Aryl tellurocyanides are postulated as intermediates.

$$[2KTeCN + 2ArN_2^+]BF_4 \xrightarrow[-N_2,\ -2KBF_4]{\text{DMSO, r.t.}} \underset{(41\text{–}47\%)}{[2ArTeCN]} \longrightarrow ArTeAr + Te(CN)_2$$

$$Te(CN)_2 \longrightarrow Te + (CN)_2$$

Ar = alkyl, methoxy, nitro, cyano, halophenyl derivatives, 2-biphenyl

Diaryl tellurides (general procedure) [9] Finely ground Te (1.6 g, 12.5 mmol) and KCN (0.82 g, 12.5 mmol) in dry DMSO (20 ml) are heated at 100°C for 1 h under N_2. After all the Te has dissolved, the mixture is cooled in an ice bath until most of the solvent has solidified. The diazonium fluoroborate (12.5 mmol) is added rapidly while a brisk stream of N_2 is passed into the system. When the initial violent gas evolution has ceased, the ice bath is removed and stirring is continued at room temperature for 3 h. The mixture is diluted with CH_2Cl_2 (250 ml), filtered from the dark insoluble material (Te) and washed several times with H_2O. The solution is dried ($CaCl_2$) and evaporated, giving an oil or a semi-solid, which is dissolved in CH_2Cl_2 and filtered through a short SiO_2 column. (In the few cases where aryl tellurocyanides are formed as by-products, more careful chromatography is required with CH_2Cl_2/hexane 1:1 as the eluent.)

The small amounts of diaryl ditellurides formed frequently as by-products can be converted to tellurides by treatment with copper powder in refluxing dioxane [10]. For Ar = $2HO_2CC_6H_4$, the crude product is extracted with aqueous Na_2CO_3 and precipitated with acid (yield 8%).

3.1.2.4. From tellurium(IV) halides and arylmagnesium halides

In this method, the electrophilic tellurium tetrachloride is employed as starting material to introduce tellurium in organic substrates, in contrast to the preceding methods using nucleophilic tellurium species.

Tellurium tetrachloride (as well as tellurium tetrabromide and tellurium tetraiodide) reacts with 4 mol eq of arylmagnesium bromide, giving rise to diaryl tellurides in high yields [11, 12].

$$TeX_4 + 4ArMgBr \xrightarrow[0°C,\ reflux\ (>90\%)]{ether/benzene} ArTeAr + ArAr + 4MgBrX$$

Ar = Ph; X = Cl, Br, Y

Ar = 1-naphthyl, o-MeC_6H_4, p-MeC_6H_4; X = Br

The intermediates in these reactions are the corresponding triaryltelluronium halides (which react with an additional equivalent of Grignard reagent) or the unstable tetraaryltellurium derivatives.

$$TeX_4 + 3ArMgX \longrightarrow [Ar_3T^+e]X^- \xrightarrow{ArMgX} ArTeAr + ArAr$$

$$TeX_4 + 4ArMgX \longrightarrow [Ar_4Te] \nearrow$$

This method is an efficient modification of the early method of Lederer [13], employing tellurium dihalides (unstable compounds which disproportionate in elemental tellurium and tellurium(IV) halides).

To make the isolation and purification of the products easier, the obtained tellurides are converted *in situ* into the corresponding dichlorides, dibromides or diiodides by treating, respectively, with sulphuryl chloride, bromine or iodine [11, 12].

Diphenyl telluride (typical procedure) [12] To an ethereal solution of a Grignard reagent (100 ml), prepared from bromobenzene (39.3 g, 0.25 mol) and Mg (6.1 g, 0.25 mol), benzene (100 ml) is added. The solution is cooled at 0°C and $TeCl_4$ (13.5 g, 0.05 mol) in benzene (200 ml) is added slowly with vigorous stirring. The reaction mixture is refluxed for 2 h, then cooled at 0°C and quenched with saturated aqueous NH_4Cl (300 ml). The organic layer is separated, washed with H_2O (3×), dried (Na_2SO_4) and evaporated in a rotatory evaporator, giving the crude telluride. Treatment of the crude telluride with Br_2 gives diphenyltellurium dibromide (20 g (91%); m.p. 197°C).

3.1.2.5 From elemental tellurium and diarylmercury compounds

The reaction of diarylmercurials with elemental tellurium under heating, one of the oldest methods for the preparation of diaryl tellurides [14–16], has been employed more recently in some specific cases [17–19].

$$Ar_2Hg + 2Te \xrightarrow[>200^\circ C]{\Delta} ArTeAr + TeHg$$

3.1.2.6 From diaryl ditellurides by extrusion of a tellurium atom

Diaryl ditellurides are relatively thermolable compounds and eliminate one tellurium atom by heating at about 300°C [20, 21]. In the presence of copper metal, however, the extrusion of tellurium is achieved during reflux with toluene or dioxane [10, 22–24].

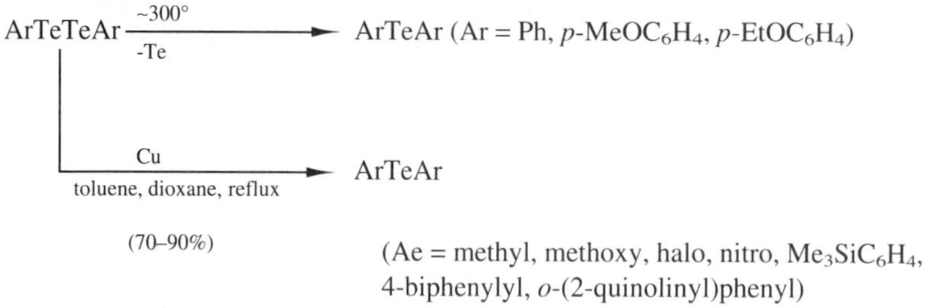

Bis(2,4,6-trimethylphenyl) telluride (typical procedure) [24] A solution of bis(2,4,6-trimethylphenyl) ditelluride (4.94 g, 10 mmol) in toluene (150 ml) is refluxed in the presence of electrolytic Cu (1.40 g, 22 mmol) for 12 h. The mixture is filtered and the filtrate evaporated to give the pure telluride (m.p. 123–125°C).

REFERENCES

1. D. J. Sandman, J. C. Stark, L. A. Acampora and P. Gagne, *Organometallics* **2**, 549 (1983).
2. D. J. Sandman, J. C. Stark, M. Rubner, L. A. Acampora and L. A. Samuelson, *Mol. Cryst. Liq. Cryst.* **93**, 293 (1983).
3. L. A. Acampora, D. L. Dugger, T. Emma, J. Mohammed, M. F. Rubner, L. Samuelson, D. J. Sandman and S. K. Tripathy, *Polymers Electronics* **36**, 461 (1984).
4. D. L. Sandman, J. C. Stark, C. A. Acampora, C. A. Samuelson and G. W. Allen, *Mol. Cryst. Liq. Cryst.* **107**, 1 (1984).
5. S. Suzuki, S. Padmanabhan, M. Inouye and T. Ogawa, *Synthesis*, 468 (1989).
6. H. Suzuki and M. Inouye, *Chem. Letters*, 389 (1985).

7. H. Suzuki and T. Nakamura, *Synthesis*, 549 (1992).
8. J. Li and P. Lue, X. Zhan, *Synthesis*, 281 (1992).
9. L. Engman, *J. Org. Chem.* **48**, 2920 (1983).
10. T. D. Sadekov, A. Y. Bushkov and V. T. Minkin, *J. Gen. Chem. USSR* **43**, 815 (1973).
11. H. Rheinboldt and N. Petragnani, *Chem. Ber.* **89**, 1270 (1956).
12. W. R. McWhinnie and M. G. Patel, *J. Chem. Soc. Dalton Trans.*, 199 (1972).
13. K. Lederer, *Ber.* **48**, 1345, 2049 (1915); K. Lederer, *Ber.* **49**, 334, 345, 1071, 1076, 2002, 2532, 2663 (1916); K. Lederer, *Ber.* **50** 238 (1917); K. Lederer, *Ber. Dtsch. Chem. Ges.* **52**, 1989 (1919); K. Lederer, *Ber.* **53**, 712 (1920).
14. F. Kraft and R. E. Lyons, *Ber.* **27**, 1768 (1896).
15. F. Zeiser, *Ber.* **28**, 1760 (1895).
16. R. E. Lyons and G. H. Bush, *J. Am. Chem. Soc.* **30**, 834 (1908).
17. D. Hellwinkel and G. Farbach, *Tetrahedron Lett.*, 1823 (1965).
18. S. C. Cohen, M. L. N. Reddy and A. G. Massey, *J. Organomet. Chem.* **11**, 563 (1968).
19. C. W. H. Jones, R. D. Sharma and D. Naumann, *Can. J. Chem.* **64**, 987 (1986).
20. W. V. Farrar, *Research* **4**, 177 (1951).
21. N. Petragnani and M. Moura Campos, *Chem. Ber.* **94**, 1759 (1961).
22. W. S. Haller and K. J. Irgolic, *J. Organomet. Chem.* **38**, 97 (1972).
23. I. D. Sadekov, A. Y. Bushkov and V. I. Minkin, *J. Gen. Chem. USSR* **47**, 576 (1977).
24. M. Akiba, M. V. Lakshmikantham, K. Y. Jen and M. C. Cava, *J. Org. Chem.* **49**, 4819 (1984).

3.1.3 Unsymmetrical tellurides

Unsymmetrical tellurides constitute a large class of organotellurium compounds characterized by two different alkyl groups, one alkyl and one aryl group, or two different aryl groups linked to a tellurium atom. These compounds find only a minor use as reagents in organic synthesis, but due to structural, and in some cases biological, interest their preparation will be discussed in the following sections.

3.1.3.1 From sodium telluride and two different alkyl halides

$$\mathrm{Te} \xrightarrow[\mathrm{H_2O}]{\mathrm{NaBH_4}} \mathrm{Na_2Te} \xrightarrow[\mathrm{EtOH/THF}]{(1)\ \mathrm{RX},\ (2)\ \mathrm{R^1X}} \underset{(50\%)}{\mathrm{RTeR^1}}\ (+\ \underset{(25\%)}{\mathrm{RTeR}} + \underset{(25\%)}{\mathrm{R^1TeR^1}})$$

R = n-Bu, n-octyl, $HC{\equiv}C(CH_2)_3$, *p*-$IC_6H_4(CH_2)_9$, *trans* $ICH = CH(CH_2)_3$

$R^1 = (CH_2)_nCO_2Me$

Unsymmetrical telluro-substituted fatty esters (of biological interest) are obtained in about 40% yield after chromatographic separation from the symmetrical tellurides [1].

n-Octyl (7-carbomethoxy)heptyl telluride (typical procedure) [1] To a suspension of elemental Te (0.127 g, 1 mmol) in H_2O (3 ml) heated at 80°C is added a solution of $NaBH_4$ (0.100 g, 2.64 mmol) in H_2O (1 ml), under red light, after careful deaeration and argon purge. The obtained clear solution of Na_2Te is cooled at room temperature and a solution of methyl 8-bromooctanoate (0.223 g, 0.95 mmol) and 1-bromooctane (0.202 g, 1.05 mmol) in THF/EtOH (1:1, 15 ml) is added under argon. The solution is stirred for 1 h, poured into H_2O (150 ml) and then extracted with ether (3 × 50 ml). The ether extracts are washed with H_2O (3 × 50 ml), dried (Na_2SO_4) and evaporated under vacuum. The crude product is dissolved in petroleum ether (30–60°C) and purified by column chromatography on SiO_2 slurried with petroleum ether (elution with 10 × 25 ml aliquots of petroleum ether followed by 10 × 25 ml aliquots of benzene, monitoring the fractions by TLC). By combination of aliquots 13–16 the telluride is obtained as an oil (0.155 g (39%)).

Aryl alkyl tellurides are prepared in medium yields by sequential arylation/alkylation of Na_2Te generated by the recently described Te/NaH/NMP system (see Section 3.1.2.1) [2].

$$\mathrm{Te} \xrightarrow[\text{NMP, 100–110°C, 1 h}]{\text{NaH}} \mathrm{Na_2Te} \xrightarrow[\text{NMP, 120°C, 2–4 h}]{\text{ArI (0·5) eq)}} [\mathrm{ArTe^-}] \xrightarrow[\substack{\text{NMP, 50°C, 2 h}\\ \text{(52–67\%)}}]{\text{RX}} \mathrm{ArTeR}$$

Ar = *o*-MeC_6H_4, 1-naphthyl; R = n-heptyl

Ar = 2-naphthyl; R = Me

Methyl 2-naphthyl telluride (typical procedure) [2] 2-Iodonaphthalene (0.150 g, 0.59 mmol) is added dropwise to a stirred solution of Na_2Te (1.3 mmol) in NMP (4 ml) and the mixture is heated at 120°C for 5 h. An excess of MeI (0.40 ml, 4.3 mmol) is then added and the resulting mixture is stirred at 50°C for 2 h. Usual work-up gives the telluride as pale yellow crystals (0.110 g (67%); m.p. 59–60°C).

3.1.3.2 From organyl tellurolates and alkylating agents

(a) From organyl tellurolates generated by tellurium insertion in organomagnesium or organolithium reagents

The magnesium aryl tellurolates described earlier [3] have seldom been employed for the preparation of unsymmetrical tellurides [4–6].

$$\mathrm{Te + ArMgBr \longrightarrow ArTeMgBr \xrightarrow{RX} ArTeR}$$

Ar = Ph; R = Et, $H_2C\text{-}C{\equiv}CH$

Ar = *o*-MeS-, *o*-Me_2N-, *o*-ClC_6H_4; R = Me

Elemental tellurium also inserts easily into a C(sp^2)—Mg bond of a vinyl or styryl Grignard reagent (see Section 3.11.1.2), but is unaffected by alkylmagnesium halides [7].

In contrast, tellurium insertion in alkyl- or aryllithium compounds followed by alkylation is a useful method for the synthesis of unsymmetrical tellurides. (For a tabulation of unsymmetrical tellurides prepared by alkylation of organyl tellurolates, see ref. [8].)

$$\text{Te} + \text{RLi} \longrightarrow \text{RTeLi} \xrightarrow{\text{R}^1\text{X}} \text{RTeR}^1$$

$$\text{Te} + \text{ArLi} \longrightarrow \text{ArTeLi} \xrightarrow{\text{RX}} \text{ArTeR}$$

Methyl (3-hydroxy)propyl telluride (typical procedure) [9] *MeTeLi.* A suspension of Te powder (12.8 g, 0.1 mol) in THF (100 ml) is frozen in a liquid N_2 bath, and then MeLi (66.7 ml of a 1.5 M solution in ether, 0.1 mol) is injected into the flask. The mixture is allowed to thaw and is stirred magnetically for a further 30 min at room temperature. The resulting yellow-orange solution is stored under N_2 and used within 2 h.

$MeTe(CH_2)_3OH$. 3-Bromopropan-1-ol (6.95 g, 50 mmol) is injected into a frozen (by liquid N_2) solution of MeTeLi (50 mmol). The mixture is allowed to thaw and is stirred magnetically at room temperature for about 1 h. Aqueous NaCl (100 ml) is then added and the aqueous layer extracted several times with ether. The combined organic phases are dried ($MgSO_4$) for 16 h, and evaporated. The residue is chromatographed on an SiO_2 column (elution with petroleum ether 40–60°C), giving the telluride (8.3 g (83%)) in acceptable purity.

Methyl phenyl telluride (typical procedure) [10] To a well-stirred suspension of Te powder (3.0 g, 24 mmol) in anhydrous THF (36 ml) under argon is added dropwise an equimolar amount of PhLi (1.5–2.0 M in ether). The mixture is then stirred for 2 h at room temperature, and for 1 h under reflux, and then MeI (3.4 g, 24 mmol) is added. After refluxing for 0.5 h the mixture is poured into three volumes of H_2O, the aqueous layer is extracted with CH_2Cl_2, and the combined organic phases are washed with H_2O, dried (Na_2SO_4) and evaporated. The oily brown residue is distilled under vacuum, giving the telluride (2.1 g (43%); b.p. 57–58°C/0.7 torr).

(b) From organyl tellurolates generated by the cleavage of ditellurides with alkali metals or reducing agents

The reductive fission of diaryl ditellurides with sodium in liquid ammonia reported earlier [11] has seldom been employed [12, 13]. More recently, aprotic solvents have been substituted for ammonia [14–16].

A practical and successful method employs lithium in THF [17].

$$\mathrm{ArTeTeAr} \xrightarrow[\mathrm{THF}]{\mathrm{Li}} \mathrm{ArTeLi} \xrightarrow[(62\text{–}84\%)]{\mathrm{RX}} \mathrm{ArTeR}$$

Ar = Ph; R = Me, Et, n-Pr, n-Bu, n-$C_{14}H_{29}$, i-Pr*

Alkyl phenyl tellurides (general procedure) [17] Li metal (1.4 g, 0.2 mol) in small pieces is added under N_2 to a solution of diphenyl ditelluride (4.1 g, 10 mmol) in dry THF (100 ml). The mixture is stirred at room temperature for 6 h, unreacted lithium is removed using a spatula, and the alkyl halide (20 mmol) neat or in THF is added dropwise to the stirred yellowish brown solution. The solution is stirred at room temperature for 30 min, and under reflux for an additional 30 min. The solvent is evaporated, and H_2O (10 ml) and ether (25 ml) are added to the residue, mixing thoroughly. The ethereal phase is separated and evaporated. The residue is fractionally distilled under vacuum.

Actually, the method of choice for the preparation of organyl tellurolate anions is the reduction of diorganyl ditellurides with reducing agents.

$NaBH_4$ is the more widely used reagent due to the mild experimental conditions. The choice of solvent, or solvent mixture, and of neutral or basic media is governed by the further use of the tellurolate solution. Typical media are benzene/ethanol/aqueous NaOH [18], MeOH or EtOH [19–21].

$$\mathrm{ArTeTeAr\ (RTeTeR)} \xrightarrow[\mathrm{EtOH\ (MeOH)}]{\mathrm{NaBH_4}} \mathrm{ArTeNa\ (RTeNa)} \xrightarrow{\mathrm{R^1X}} \mathrm{ArTeR^1\ (RTeR^1)}$$

Medium-to-high yields of the expected tellurides are obtained from diaryl or dialkyl ditellurides and n- and s-alkyl halides (and steroidal tosylates).

Reactions with epoxides are also successful, giving the corresponding hydroxy tellurides [19].

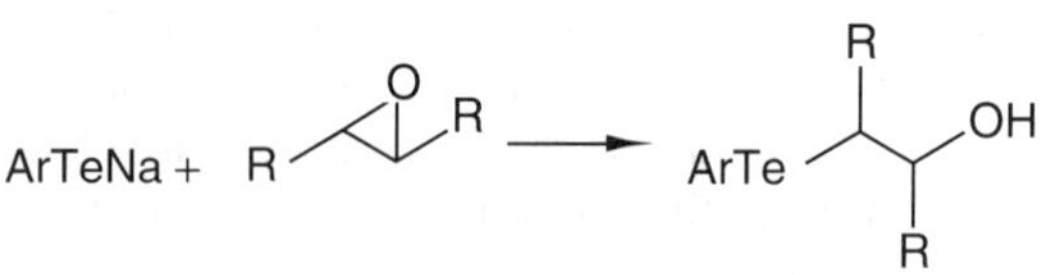

Phenyl dodecyl telluride (typical procedure) [19] To a solution of diphenyl ditelluride (1.21 g, 2.96 mmol) in absolute EtOH (100 ml) is slowly added

*Yield 46%.

powdered $NaBH_4$ (0.23 g, 6.08 mmol) under N_2 at room temperature. At the end of the addition the red colour is discharged and 1-bromododecane (1.5 g, 6.02 mmol) is added by a syringe. The mixture is refluxed for 3.5 h (during this period the progress of the reaction is monitored by TLC; the plates are developed in the dark) and then evaporated. The residue is purified by chromatography on SiO_2 (eluting with CCl_4), giving the telluride (1.73 g (76%)).

Diaryl ditellurides, in analogy to diaryl disulphides and diselenides [22], undergo disproportionation into arene tellurolates and tellurinates by alkaline treatment.

$$2ArTeTeAr + 4OH^- \longrightarrow 3ArTe^- + ArTeO_2^- + 2H_2O$$

The tellurolate anion formed can be alkylated *in situ*, giving alkyl aryl tellurides, if the above reaction is effected in the presence of a phase transfer catalyst [23]. Otherwise, performing the reaction in the presence of the reducing agent TUDO gives a telluride yield near to quantitative due to the reduction of the aryl tellurinate anion to aryl tellurolate [24, 25].

$$2ArTeTeAr + 4OH^- \xrightarrow{PTC} 3ArTe^- + ArTeO_2^- + 2H_2O$$

($ArTeO_2^- \xrightarrow{TUDO} ArTe^-$)

$$ArTe^- + RX \xrightarrow[(81\text{–}96\%)]{} ArTeR$$

Ar = Ph, 2-naphthyl, *p*-$MeOC_6H_4$, *m*-$F_3CC_6H_4$

R = n- and s-alkyl, benzyl

Alkyl aryl tellurides (general procedure) [24] To a solution of diaryl ditelluride (1.0 mmol) in THF (7.5 ml) are added TUDO (0.108 g, 1.0 mmol), 2HT (0.030 g) and the alkyl halide (2.0 mmol). Then, 50% NaOH (7.5 ml) is added and the mixture is stirred vigorously for several hours at room temperature. The phases are separated, and the organic phase is extracted with EtOAc (3 × 20 ml). The combined organic phases are washed with H_2O, dried ($MgSO_4$) and evaporated. The residue is purified by column chromatography on SiO_2 (eluting with EtOAc).

3.1.3.3 By addition of aryl tellurolates to electrophilic alkenes

Aryl tellurolates add to alkenes bearing electron-withdrawing groups in a typical 1,4 Michael addition, giving β-telluro-substituted derivatives [26].

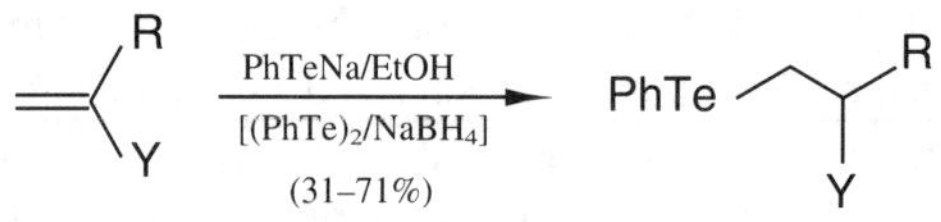

R = H, Me; Y = CO_2Et, CN

3.1.3.4 From organyl tellurolates and arylating agents

The arylation of organyl tellurolates, restricted first to aryl halides activated by electron-withdrawing groups [24], or requiring special conditions such as heating in HMPA in the presence of CuI [27], or photostimulation in liquid ammonia [12, 13], has recently been achieved successfully with non-activated aryl halides under normal conditions [6].

$$\text{PhTeTePh} \xrightarrow[\text{(2) CuI (3) ArI (80–90\%)}]{\text{(1) NaBH}_4\text{/HMPA, 70–80°C}} \text{PhTeAr} \qquad [26]$$

Ar = nitroaryl

$$\text{PhTeTePh} \xrightarrow{\text{Na/NH}_3} \text{PhTeNa} \xrightarrow[\text{UV}]{\text{ArX}} \text{PhTeAr} \qquad [12,13]$$

$$\text{RLi} + \text{Te} \longrightarrow \text{RTeLi} \xrightarrow{\text{ArX}} \text{RTeAr (R = Me, Ph)} \qquad [6]$$

Methyl o-*methoxyphenyl telluride (typical procedure)* [6] To a frozen solution of MeTeLi (50 mmol) is added *o*-bromoanisole (9.3 g, 50 mmol). The mixture is allowed to warm at room temperature, stirred for 1.5 h and then quenched with deoxygenated H_2O (100 ml). The organic product is extracted with ether (3 × 50 ml), the organic extracts dried ($MgSO_4$) for 16 h and evaporated, and the residue is distilled under vacuum, giving the telluride as a pale yellow oil (9.3 g (74%)).

Arenediazonium salts [28, 29] as well as arenesulphonylazo compounds [30] have also found a use as arylating agents for arene tellurolates.

$$\text{ArTeTeAr} \xrightarrow[\text{EtOH/THF/NaOH aq.}]{\text{NaBH}_4} \text{ArTeNa} \xrightarrow{[\text{Ar}^1\text{N}_2]\text{X}^-} \text{ArTeAr}^1$$

$$\text{PhTeLi} + \text{ArN=NSO}_2\text{C}_6\text{H}_4\text{CH}_3 \xrightarrow[\text{MeCN, r.t.}]{\text{18-crown-6}} \text{PhTeAr}$$

Trimethylsylylmethyl telluride generated *in situ* from methyllithium and trimethylchlorosilane can be used instead of the tellurolate in the latter reaction [30].

$$\text{MeTeLi} + \text{Me}_3\text{SiCl} \longrightarrow \text{MeTeSiMe}_3 \xrightarrow{\text{ArN=NSO}_2\text{C}_6\text{H}_4\text{CH}_3} \text{MeTeAr}$$

3.1.3.5 From diorganyl ditellurides or arenetellurenyl halides and organometallic reagents

A different approach to unsymmetrical diorganyl tellurides, in which an electrophilic tellurium species is used, involves the nucleophilic attack of diorganyl ditellurides by organomagnesium or organolithium reagents.

$$\text{RTeTeR} + \text{R}^1\text{Met} \longrightarrow \text{RTeR}^1 + \text{RTeMet}$$

R	R^1	Met	
aryl	aryl	MgX, Li	[31, 32]
aryl, alkyl	alkyl	MgX, Li	[33]
alkyl	aryl	Li	[6, 34]

Only half of the tellurium of the starting ditelluride is converted into telluride. The other half, converted into tellurolate, can be oxidatively recovered as the starting ditelluride.

If the diaryl ditelluride is previously converted *in situ* to the corresponding tellurenyl halide, both the electrophilic tellurium moieties are used [35, 36]. (The stable crystalline 2-naphthyltellurenyl iodide (see Section 3.7) reacts similarly.) The reaction is of general application, giving high yields of the expected tellurides with aromatic, aliphatic, cycloaliphatic, vinylic [37] and acetylenic Grignard reagents, under mild conditions.

$$\text{ArTeTeAr} \xrightarrow[\text{THF/benzene}]{X_2} [\text{ArTeX}] \xrightarrow[(>90\%)]{\text{RMgX}} \text{ArTeR}$$

Ar = Ph, *p*-Me, *p*-MeO, *p*-$EtOC_6H_4$

R = Ph, *n*-Bu, cyclohexyl, PhC = CH, PhC≡C; X = Br, I

Unsymmetrical diorganyl tellurides (general procedure) [35] A solution of the ditelluride (2 mmol) in THF (30 ml) is treated dropwise at 0°C under N_2 with Br_2 (0.32 g, 3 mmol) in benzene (4 ml). The Grignard reagent is then added dropwise by means of a syringe. Gradual disappearance of the red colour of the solution is observed. Finally, the solution becomes almost colourless after the addition of a 10% excess of the Grignard reagent. After stirring for 30 min at room temperature the solution is diluted with petroleum ether 30–60°C (30 ml) and treated successively with aqueous NH_4Cl and brine. The organic layer is dried ($MgSO_4$) and evaporated, giving the telluride.

3.1.3.6 Additional methods

(a) From dialkyl ditellurides and arenediazonium fluoroborates [38]

$$\text{RTeTeR} \xrightarrow[\text{CHCl}_3(\text{CH}_2\text{Cl}_2),\ \text{KOAc},\ 18\text{-crown-6}\ (25\text{–}48\%)]{[\text{ArN}_2^+]\text{BF}_4^-} \text{RTeAr}$$

(b) From diaryl ditellurides and dialkylmercury reagents [39, 40]

$$\text{ArTeTeAr} + \text{R}_2\text{Hg} \xrightarrow[-\text{Hg}\ (70\text{–}80\%)]{\text{dioxane/reflux}} 2\text{ArTeR}$$

(c) From phenyl tellurocyanate and alcohols [41]

$$\text{PhTeTePh} \xrightarrow[(2)\ \text{BrCN}]{(1)\ \text{NaBH}_4} \text{PhTeCN} \xrightarrow[\text{Bu}_3\text{P/CH}_2\text{Cl}_2]{\text{ROH}} \text{PhTeR}$$

(d) From diphenyl ditellurides and trialkyl boranes [42]

$$\text{PhTePh} \xrightarrow[\text{O}_2/\text{THF}]{(\text{RCH}_2\text{CH}_2)_3\text{B}} \text{RCH}_2\text{CH}_2\text{TePh}$$

REFERENCES

1. M. M. Godman and F. F. Knapp Jr, *Organometallics* **2**, 1106 (1983).
2. H. Suzuki and T. Nakamura, *Synthesis*, 549 (1992).
3. M. Giua and F. Cherchi, *Gazz. Chim. It.* **50**, 362 (1920).
4. K. Bowden and A. E. Braude, *J. Chem. Soc.*, 1076 (1952).
5. G. Pourcelot, M. Lequen, M. P. Simmonin and P. Cadiot, *Bull. Soc. Chim. Fr.*, 1278 (1962).
6. T. Kemmitt and W. Levanson, *Organometallics* **8**, 1303 (1989).
7. W. S. Haller and K. J. Irgolic, *J. Organomet. Chem.* **38**, 97 (1972).
8. K. Y. Irgolic, in *Houben–Weyl—Methoden der Organischen Chemie*, 4th edn, Vol. E12b, p. 389. Georg Thieme, Stuttgart, 1990.
9. E. G. Hope, T. Kemmitt and W. Levanson, *Organometallics* **7**, 78 (1988).
10. D. Seebach and A. K. Beck, *Chem. Ber.* **108**, 314 (1975).
11. K. Lederer, *Ber.* **48**, 1345 (1915); see also J. Liesk, P. Schultz and G. Klar, *Z. Anorg. Allg. Chem.* **435**, 98 (1977).
12. A. B. Pierini and R. A. Rossi, *J. Organomet. Chem.* **168**, 163 (1979).
13. A. B. Pierini and R. A. Rossi, *J. Org. Chem.* **44**, 4667.
14. S. Uemura, S. I. Fukuzawa, T. Yamauchi, K. Hattori, S. Mizutaki and K. Tamaki, *J. Chem. Soc. Chem. Commun.*, 426 (1984).

15. L. Engman, *Organometallics* **5**, 427 (1986).
16. K. Ohe, H. Takahashi, S. Uemura and N. Sugita, *J. Organomet. Chem.* **326**, 35 (1987).
17. K. J. Irgolic, P. J. Bush, R. A. Grigsby and M.R. Smith, *J. Organomet. Chem.* **88**, 175 (1975).
18. J. L. Piette and M. Renson, *Bull. Soc. Chim. Belg.* **79**, 383 (1970).
19. D. L. Clive, G. J. Chittattu, V. Farina, W. A. Kiel, S. M. Menchen, C. G. Russel, A. Sing, C. K. Wong and N. J. Curtis, *J. Am. Chem. Soc.* **102**, 4438 (1980).
20. F. F. Knapp, K. R. Ambrose and A. P. Callahan, *J. Med. Chem.* **24**, 794 (1981).
21. S. Uemura and S. Fukuzawa, *J. Am. Chem. Soc.* **105**, 2748 (1983).
22. H. Rheinboldt, in *Houben–Weyl—Methoden der Organischen Chemie* (ed. E. Muller), 4th edn, Vol. IX, p. 1102. Georg Thieme, Stuttgart, 1955.
23. J. V. Comasseto, J. T. B. Ferreira and J. A. Fontanellas, *J. Organomet. Chem.* **277**, 261 (1984).
24. J. V. Comasseto, E. S. Lang, J. T. B. Ferreira, F. Simonelli and V. R. Correia, *J. Organomet. Chem.* **334**, 329 (1987).
25. J. T. Ferreira, A. P. M. Oliveira and J. V. Comasseto, *Tetrahedron Lett.* **33**, 915 (1992).
26. K. Ohe, H. Takahashi, S. Uemura and M. Sugita, *Nippon Kagaku Kaishi*, 1469 (1987).
27. H. Suzuki, H. Abe, N. Ohmasa and A. Osuka, *Chem. Lett.*, 1115 (1981).
28. J. L. Piette, P. Thibaut and M. Renson, *Tetrahedron* **34**, 655 (1978).
29. J. D. Sadekov, A. A. Ladatko and U. I. Minkin, *J. Gen. Chem. USSR*, 2194 (1977) (in English).
30. M. J. Evers, L. E. Christiaens and M. J. Renson, *J. Org. Chem.* **51**, 5196 (1986).
31. N. Petragnani, *Chem. Ber.* **86**, 247 (1963).
32. M. Herberhold and P. Leitner, *J. Organomet. Chem.* **336**, 153 (1987).
33. D. H. O'Brien, N. Dereu, C. K. Huang and K. J. Irgolic, *Organometallics* **2**, 305 (1983).
34. N. Dereu and J. L., Piette, *Bull. Soc. Chim. Fr.*, 623 (1979).
35. N. Petragnani, L. Torres and K. J. Wynne, *J. Organomet. Chem.* **92**, 175 (1975).
36. G. Vicentini, E. Giesbrecht and L. R. M. Pitombo. *Chem. Ber.* **92**, 40 (1959).
37. M. Moura Campos and N. Petragnani, *Tetrahedron* **18**, 527 (1962).
38. A. Luxen and L. Christiaens, *Tetrahedron Lett.*, 3905 (1982).
39. Y. Okamoto and T. Yano, *J. Organomet. Chem.* **29**, 99 (1971).
40. T. K. Vychkova, G. A. Kalabin and D. F. Kushnarev, *J. Org. Chem. USSR* **17**, 1179 (1981) (in English).
41. F. Ogura, H. Yamaguchi, T. Otsuba and K. Chikamatsu, *Synth. Commun.* **12**, 131 (1982).
42. T. Abe, Y. Aso, T. Otsuba and F. Ogura, *Chem. Lett.*, 1671 (1990).

3.1.4 Diorganyl tellurides by reduction of diorganyltellurium dihalides

Diorganyltellurium dihalides are often the primary products in the synthesis of organic derivatives of tellurium and are therefore immediate precursors of the corresponding tellurides.

The most employed reducing agents in the "old" tellurium chemistry were alkali sulphites or hydrogen sulphites, methylmagnesium iodide and sodium sulphide hydrate. The last reagent is still used, together with the more recently introduced sodium borohydride, TUDO and hydrazine.*

*For an extensive literature coverage on the reduction of diorganyltellurium dihalides, see ref. [1].

$$\underset{X\ \ X}{\mathrm{RTeR}} \xrightarrow{\text{red}} \mathrm{RTeR}$$

Often the sequence

$$\mathrm{R_2Te} \xrightarrow{X_2} \mathrm{R_2TeX_2} \xrightarrow{\text{red}} \mathrm{R_2Te}$$

is employed for isolation and purification purposes, since the dihalides are generally solid, easily recrystallizable compounds, and the two steps cause only a little loss in the telluride yield.

Reduction of diorganyltellurium dihalides

With sodium sulphide hydrate (general procedure) [2] The diorganyltellurium dihalide is mixed with a 15 times molar excess of $Na_2S{\cdot}9H_2O$ and the mixture heated at 95–100°C for 10 min or more, until all the solid has melted. Sufficient H_2O is added to dissolve the sulphide and then the mixture is filtered if the obtained telluride is a solid, or extracted with a solvent (ether or petroleum ether) if the telluride is a liquid. The products are purified by crystallization or distillation. Yields are high or quantitative (except for diphenyl telluride or di-*p*-tolyl telluride).

With sodium borohydride–tetrahydrotellurophene (typical procedure) [3] $NaBH_4$ is added to a boiling methanolic solution of diiodotetrahydrotellurophene until the orange colour disappears. The solution is filtered and the filtrate is poured into H_2O (1 litre). Extraction with ether, followed by drying ($CaCl_2$) and evaporation, gives the product as a yellow oil with a persistent odour (b.p. 165–167/760 mmHg; no yield was given).

With TUDO (general procedure) [4] A mixture of diorganyltellurium dihalide (2 mmol) and 2 N NaOH (10 ml) is stirred at room temperature for 15 min. TUDO (0.432 g, 4 mmol) and petroleum ether (30–60°C, 10 ml) are then added and the two-phase system is stirred at room temperature until the mixture turns clear. The phases are separated and the aqueous phase is extracted with ether (3×). The combined organic phases are dried ($MgSO_4$) and the solvent is evaporated. The residue is placed on an SiO_2 column and eluted with petroleum ether giving the telluride (dibutyl, didodecyl, butyl phenyl, di-4-methoxyphenyl telluride; yields 77–90%).

*With hydrazine – di-*p*-anisyl telluride (typical procedure)* [5] *p*-Anisyltellurium dichloride (8.3 g, 0.02 mol) suspended in $EtOH/H_2O$ (150:15) is heated under reflux, and hydrazine (3.2 g, 0.1 mol) is added dropwise (vigorous evolution of N_2). When further addition of hydrazine no longer causes evolution of N_2, the mixture is poured into H_2O (700 ml), and extracted with ether (2 × 300 ml). The extracts are washed with H_2O, dried and evaporated, giving the telluride. The product is recrystallized from MeOH (at −25°C) (5.2 g (77%); m.p. 57°C).

REFERENCES

1. H. Rheinboldt, in *Houben–Weyl—Methoden der Organischen Chemie* (ed. E. Muller), 4th edn, Vol. IX, p. 1068. Georg Thieme, Stuttgart, 1955; K. Y. Irgolic, in *Houben–Weyl—Methoden der Organischen Chemie*, 4th edn, Vol. E12b, p. 426. Georg Thieme, Stuttgart, 1990.
2. L. Reichel and E. Kirschbaum, *Chem. Ber.* **76**, 1105 (1943).
3. A. Z. Al-Rubaie, H. A. Y. Alshirayada, *J. Organomet. Chem.* **287**, 321 (1985).
4. E. S. Lang, J. V. Comasseto, *Synth. Commun.* **18**, 301 (1988).
5. J. Bergman, *Tetrahedron*, 3323 (1972).

3.2 DIORGANYL DITELLURIDES

Three main routes have been well established for the preparation of diorganyl ditellurides:

1. The reaction of sodium ditelluride with alkylating or arylating agents.

$$\text{NaTeTeNa} + \text{RX} \longrightarrow \text{RTeTeR}$$

2. The oxidation of tellurols or tellurolate anions.

$$\text{RTeH} \ (\text{RTe}^-) \xrightarrow{\text{ox.}} \text{RTeTeR}$$

3. The reduction of organyltellurium trichlorides

$$\text{RTeCl}_3 \xrightarrow{\text{red.}} \text{RTeTeR}$$

In 1,2,3: R = alkyl, aryl

3.2.1 From sodium ditelluride

3.2.1.1 From sodium ditelluride and alkylating agents

This is the most direct route to diorganyl ditellurides and therefore parallels the route leading to diorganyl tellurides, substituting sodium telluride for sodium ditelluride. Sodium ditelluride is prepared employing—with the appropriate ratio of the elements—methods analogous to those described for sodium telluride.

(a) Liquid ammonia method

Dimethyl ditelluride (typical procedure) [1] Clean Na metal (3.2 g, 0.14 mol) is added to 100 ml of liquid NH_3 at –78°C. After stirring for 1 h, high-purity powdered Te (18.2 g, 0.14 mol) is added in 0.5 g portions. Methyl iodide (24 g, 0.17 mol) is then added dropwise during 20 min, with stirring, to the dark green solution. The NH_3 is evaporated, H_2O is added to the residue and the mixture extracted with ether (4 × 50 ml). The combined deep red extracts are dried overnight ($CaCl_2$), and evaporated under vacuum. The residue is then distilled (7.6 g (38%); b.p. 97°C/9 torr).

Diethyl ditelluride and dibenzyl ditelluride are prepared similarly [2] to dimethyl ditelluride in yields of 71 and 82%, respectively.

(b) Sodium naphthalene method

Dialkyl ditellurides (general procedure) [3] A mixture of powdered Te (3.58 g, 28 mmol), Na chips (0.65 g, 28 mmol) and naphthalene (0.36 g, 2.8 mmol) in anhdyrous THF (25 ml) is refluxed under N_2 and stirred for 1 h. During this time all the Na is consumed and the mixture turns a light brown colour. The solution is stirred for an additional 3 h to ensure the complete reduction of Te, the temperature is then lowered to 10°C and the alkyl halide (28 mmol) is added dropwise during 30 min with stirring. After an additional hour of stirring at room temperature, the reaction mixture is filtered, the solvent evaporated and the residue distilled under vacuum, giving the pure ditelluride (R = Et (82%), n-Pr (90%), n-Bu (90%), $MeOCH_2CH_2$ (60%)).

(c) Sodium hydride/DMF method

Bis(4-carbomethoxy)butyl ditelluride (typical procedure) [4] Powdered Te (1.27 g, 10 mmol), NaH (0.44 g, 11 mmol) and dry DMF (50 ml) are stirred at 70°C under argon for 3 h. The purple Na_2Te_2 solution is cooled at room temperature, and methyl 5-bromovalerate (2.15 g, 11 mmol) in dry argon-purged DMF (10 ml) is added. The mixture is stirred at room temperature for 1 h, cooled and poured into H_2O (100 ml) and extracted several times with ether. The combined extracts are washed thoroughly with H_2O, dried (Na_2SO_4) and evaporated under vacuum, giving a dark-orange oil. The crude ditelluride is chromatographed on SiO_2 (basic, 125 g eluting with $CHCl_3$). It is a dark orange oil (1.38 g (53%); single component on TLC (R_f 0.56 in $CHCl_3$).

Diisopropyl ditelluride is prepared in the same manner [5].

(d) Rongalite method

Dialkyl ditellurides (general procedure) [6] Rongalite (9.24 g, 0.08 mol) and NaOH (9.0 g, 0.23 mol) are dissolved in distilled H_2O (200 ml). The apparatus is flushed with N_2 and finely powdered Te (15.3 g, 0.12 mol) is added. The mixture is stirred for 5 h and then the alkyl bromide (0.12 mol) is added dropwise during 30 min with cooling and stirring. The solution is extracted with CCl_4 (3 × 100 ml), and the extract is dried ($CaCl_2$) and distilled under vacuum, giving the dark red, foul-smelling ditelluride (R = Et, n-Bu, n-pentyl, n-dodecyl (53%)) [7].

(e) TUDO method

Dialkyl ditellurides (general procedure) [8] A mixture of powdered Te (0.128 g, 1 mmol) and cetyltrimethylammonium bromide (CTAB, 0.004 g, 1.1×10^{-5} mmol) in THF (0.75 ml) and DMSO (0.5 ml) is purged with a flux of deoxygenated N_2 for 15 min and then heated at 80°C. TUDO (0.1 g, 1 mmol) and NaOH (0.112 g, 2.6 mmol) in H_2O (0.75 ml) are added. The mixture is refluxed for 1 h and then the purple solution is cooled at 15°C. The alkyl halide (2 mmol) is added and the mixture is stirred at room temperature for 1 h. Normal work-up and filtration through a pad of Celite (using CH_2Cl_2 as the mobile phase) give the ditellurides as dark red oils (R = n-$C_{11}H_{23}$, n-C_8H_{17}, $Me_2CHCH_2CH_2$, 2-heptyl, $THPO(CH_2)_6$ (93–98%)).

(f) Sodium borohydride method

Dioctyl ditelluride (typical procedure) [5] Na (0.10 g, 4 mmol) in small pieces is added to a suspension of Te (0.50 g, 4 mmol) in absolute alcohol (50 ml), under N_2 and with stirring. When the Na is dissolved, $NaBH_4$ (0.075 g, 2 mmol) is added and the mixture heated under N_2 for 2 h, using a heating mantle. The mixture is cooled at room temperature and neat octyl bromide (0.72 g, 4 mmol) is slowly added while stirring and cooling in an ice bath. After stirring for 30 min, the mixture is poured into distilled H_2O (100 ml), extracted with ether (2 × 40 ml) and the combined extracts are dried. The solvent is evaporated under vacuum and the residue is recrystallized from ether (by cooling at –78°C) (0.77 g (85%); m.p. 9°C).

(g) Hydrazine hydrate method

Dialkyl ditellurides (general procedure) [9] Powdered Te (6.35 g, 50 mmol) is added to a stirred solution of NaOH (3.0 g, 75 mmol) in deoxygenated H_2O (20 ml). The mixture is cooled in a water bath, and 100% hydrazine hydrate (2.5 g, 200 mmol) is added over a period of 30 min and stirring is continued for an additional hour at room temperature. The alkyl halide (50 mmol) is then added dropwise over a period of 2–3 h. During the addition, the temperature is maintained at 15–20°C. The end-point of the alkylation is indicated by a sharp colour change from dark brown to nearly colourless. The mixture is then extracted with ether, the organic layer is washed with H_2O, dried (Na_2SO_4) and the solvent removed by slow distillation. The residue is distilled under vacuum.

3.2.1.2. From sodium ditelluride and aryl halides

The low reactivity of aryl halides towards nucleophilic reagents makes their reaction with sodium ditelluride unattractive for the preparation of diaryl ditellurides. Low yields are obtained, in analogy to the similar reaction with sodium telluride (see Section 3.1.2.1).

$$Na_2Te_2 + 2ArX \xrightarrow[\text{DMF}]{\text{HMPA}} ArTeTeAr$$

(5–40%)

Ar = Ph, 1-naphthyl, 2-naphthyl, 9-anthracenyl

Di-2-naphthyl diltelluride (typical procedure) [10] 2-chloronaphthalene (2 mol eq.) is added to Na_2Te_2 (prepared by heating equimolar amounts of Te and Na in HMPA under N_2). The mixture is heated at 130–170°C for 16–24 h (yield 20%; m.p. 116–118°C, recrystallized from hexane).

3.2.2 By oxidation of organotellurols or organotellurolates

Organyl tellurolates, obtained by the insertion of elemental tellurium in organyllithium (or magnesium) reagents (used as starting materials to prepare unsymmetrical tellurides as shown in Section 3.1.3.2), can be oxidatively converted into the corresponding ditellurides. This general method can be performed by direct oxidation of the tellurolates or through a previous hydrolysis to the tellurols.

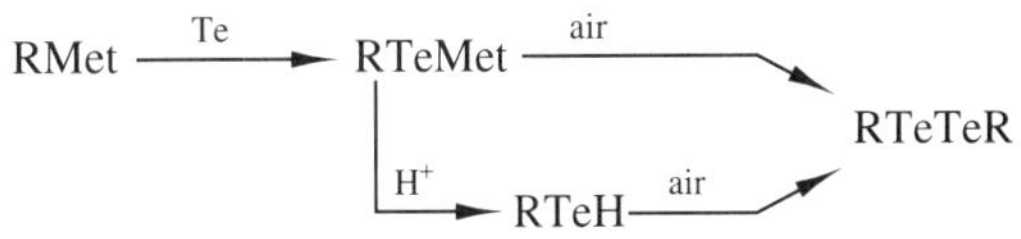

R = alkyl, aryl

Met = Li(MgX)

The Grignard route to diorganyl ditellurides suffers from lack of generality and the mechanism of the oxidation seems to be uncertain. Alkylmagnesium halides demonstrate lack of reactivity towards elemental tellurium [11], whereas arylmagnesium halides in ether as the solvent furnish a mixture of ditellurides and tellurides [12]. Satisfactory results are obtained by tellurium insertion in arylmagnesium halides in THF followed by oxidation before or after aqueous work-up [11, 13, 14].

*Bis(*m*-methoxyphenyl) ditelluride (typical procedure)* [13] To a stirred solution of *m*-bromoanisole (3.74 g, 20 mmol) in dry THF (80 ml) under N_2 are added Mg turnings (0.49 g, 20 mmol) at room temperature. The mixture is heated under reflux until most of the Mg is dissolved. The Grignard reagent solution is cooled to room temperature and well-powdered Te metal is added. After 0.5 h most of the metal will have dissolved. The reaction mixture is then saturated with dry air at room temperature. The *m*-methoxyphenyl ditelluride is obtained after routine work-up as a red oil (2.96 g (63.0%)).

*Bis(*p*-methoxyphenyl) ditelluride (typical procedure)* [14] To a solution of *p*-methoxyphenylmagnesium bromide (prepared from *p*-bromoanisole (5.82 g, 0.0311 mol) and Mg (1.0 g, 0.042 mol) in THF (20 ml)) is added Te shot (3.81 g, 0.0300 mol) at room temperature. The mixture is stirred under reflux for 3 h and then cooled to 0°C and treated with a saturated solution of NH_4Cl (20 ml; vigorous evolution of gas). The mixture is filtered through Celite and the solids washed with saturated solution of NH_4Cl and ether. The organic phase is washed with brine and dried with Na_2SO_4. Evaporation of the solvent and recrystallization from $CHCl_3$/petroleum ether affords the pure product (5.16 g (75%); m.p. 57–59°C).

Early experiments to prepare diaryl ditellurides, employing aryllithium reagents generated by lithium–halogen exchange using ether as the solvent, gave only modest yields [14, 15].

At present the most successful and general route to dialkyl and diaryl ditellurides employs the telluration of organyllithium compounds in THF.

Di-n-butyl ditelluride (typical procedure) [16] n-Butyllithium (8.0 ml, 2.15 M in hexane; 17.2 mmol) is added dropwise at room temperature under N_2 to a stirred suspension of finely ground elemental Te (1.9 g, 14.9 mmol) in dry THF (50 ml). After 15 min all the Te will have been consumed, and the resulting yellowish solution is diluted with H_2O (200 ml). Air is then bubbled through it to oxidize the lithium tellurolate. The red oily ditelluride that separates is extracted with ethyl ether (200 ml). The organic phase is washed with H_2O (2 × 50 ml), dried ($CaCl_2$) and evaporated to give 2.83 g of a viscous red oil. Distillation of 5.66 g of the crude material prepared in this manner affords 4.87 g of pure di-n-butyl ditelluride (89% based on Te; b.p. 103–105°C/0.8 torr).

Di-t-butyl ditelluride is prepared in the same manner as in the preceding synthesis.

Diphenyl ditelluride (typical procedure) [17] To a suspension of powdered Te (25.5 g, 0.2 mol) in anhydrous THF (300 ml), 2.0 M phenyllithium (100 ml, 0.2 mol) is added dropwise, with stirring and under argon. The resultant mixture is stirred at room temperature for 2 h and under reflux for 1 h. The reaction mixture, which should contain only a small amount of unreacted Te, is allowed to cool to 20°C, and is poured into 1000 ml of H_2O. Oxygen (or air) is bubbled for 1 h through the mixture, which is then extracted with benzene (100 ml). The benzene solution is washed three times with H_2O, dried (Na_2SO_4) and the benzene is evaporated completely. The residue is recrystallized from EtOH (77%; m.p. 63.8–65.0°C).

Di-2-thienyl and di-2-furyl ditelluride are prepared from thiophene and furan by lithiation with n-buthyllithium, Te insertion and oxidative work-up [18].

Di-2-thienyl ditelluride (typical procedure) [18] n-Butyllithium (11.0 ml, 2.2 M, 24.2 mmol) is added to an ice-cooled stirred solution of thiophene (2.0 g, 23.8 mmol) in dry THF (50 ml). After 10 min at 0°C and 50 min at room temperature, elemental Te (2.9 g, 22.7 mmol) is added rapidly. All Te is completely dissolved after 30 min, when the yellowish solution is poured into a beaker containing H_2O (300 ml). CH_2Cl_2 (200 ml) is then added and air passed through the two-phase system for 1 h. To effect complete oxidation, the beaker is left overnight in the open air. The organic phase is separated and the aqueous phase extracted several times with CH_2Cl_2. The combined organic extracts are dried ($CaCl_2$) and evaporated to give a red solid. Recrystallization from EtOH affords 3.54 g (74%) of di-2-thienyl ditelluride (m.p. 89–90°C).

The preparation of aryllithium reagents can also be performed by using t-butyllithium in a halogen–metal exchange, and aqueous potassium ferricyanide as an oxidant [18, 19].

*Bis(*p*-methoxyphenyl) ditelluride (typical procedure)* [19] t-Butyllithium (10 ml, 1.7 M, 17.0 mmol) is added dropwise to a stirred solution of *p*-bromoanisole (1.59 g, 8.5 mmol) in THF (40 ml) at –78°C under argon. After 1 h the cooling bath is removed and the mixture allowed to warm to room temperature during 30 min. Finely ground Te (1.08 g, 8.5 mmol) is then added rapidly while a brisk stream of argon is passed through the open system to prevent admission of air. After 1 h, when only traces of Te remain, the mixture is poured into a separating funnel containing $K_3Fe(CN)_6$ (2.80 g, 8.5 mmol) in H_2O (150 ml). The ditelluride product is extracted with CH_2Cl_2 (100 ml + 2 × 50 ml). After drying of the combined extracts ($CaCl_2$), evaporation on SiO_2 (10 g) and flash chromatography (SiO_2; hexane/CH_2Cl_2 8:2) give the ditelluride (0.99 g (50%); m.p. 58–59°C). The material is recrystallized from EtOH.

The following diaryl ditelluride can be prepared in the same manner as in the foregoing procedure: Ar = 2-pyridyl, Ph and (halogen and Me) derivatives, *p*-$(Me_2N)C_6H_4$, 1-naphthyl, 4-biphenylyl (60–84%).

The direct oxidation of lithium aryl tellurolates to the ditellurides without previous aqueous work-up is exemplified by the preparation of di(2,4,6-tri-t-butylphenyl) ditelluride [20].

The recently described arylation of sodium telluride in NMP (see Section 3.1.2.1) followed by air oxidation gives rise to diaryl ditellurides in medium-to-good yields [21].

$$\mathrm{Te} \xrightarrow[\text{NMP, 100–110°C, 1 h}]{\text{NaH}} \mathrm{Na_2Te} \xrightarrow[\text{NMP, 120°C, 5 h}]{\text{ArI (0·5 eq)}} \mathrm{ArTe^-} \xrightarrow[\substack{\text{benzene, r·t·} \\ (51\text{–}71\%)}]{\text{air}} \mathrm{ArTeTeAr}$$

Ar = Ph and (methyl and methoxy) derivatives, 1-naphthyl, 2-naphthyl

Diphenyl ditelluride (typical procedure) [21] Iodobenzene (130 mg, 0.64 mmol) is added dropwise to a stirred solution of Na_2Te (1.3 mmol) in NMP (4 ml) and the mixture is heated at 120°C for 5 h. After cooling, the mixture is diluted with benzene (5 ml), and a stream of air is bubbled into the dark solution. Free Te which separates is removed by filtration and the filtrate is worked up as usual to give the crude ditelluride, which is purified by recrystallization from a mixture of hexane and EtOAc (3:1) to give the pure product (90 mg (66%)).

3.2.3 By reduction of organotellurium trichlorides

The reduction of organyltellurium trichlorides, which are the primary products of several reactions involving tellurium tetrachloride (see Section 3.5.1), is a useful and general method for the preparation of diorganyl ditellurides.

$$TeCl_4 \xrightarrow[\text{or metallo-organic reagent}]{\text{organic substrate}} RTeCl_3 \xrightarrow{\text{red}} RTeTeR$$

Some examples of these reductions are given below.

3.2.3.1 Reduction of β-carboxyalkyltellurium trichlorides

β-Carboxylalkyltellurium trichlorides, obtained by the reaction of tellurium tetrachloride with carboxylic acid anhydrides, are reduced to the corresponding ditellurides with potassium hydrogen sulphite [22, 23].

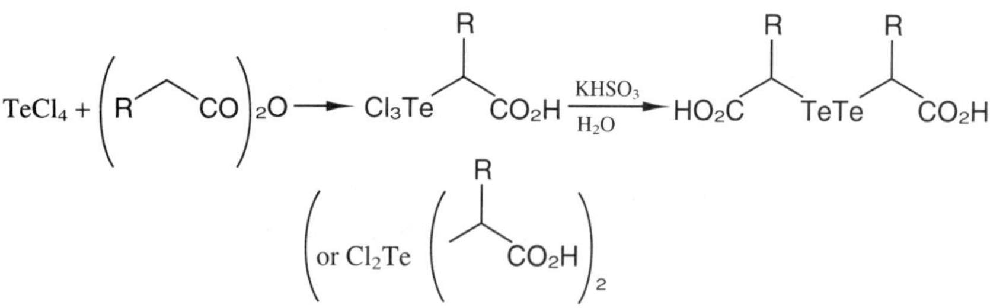

3.2.3.2 Reduction of β-alkoxyalkyltellurium trichlorides

The trichlorides obtained by the alkoxytellurination of alkenes (see Section 3.5.1.2) or by the cyclotellurination of hydroxyalkenes (see Section 4.6.2.4) are reduced *in situ*, without isolation, with sodium metabisulphite [24].

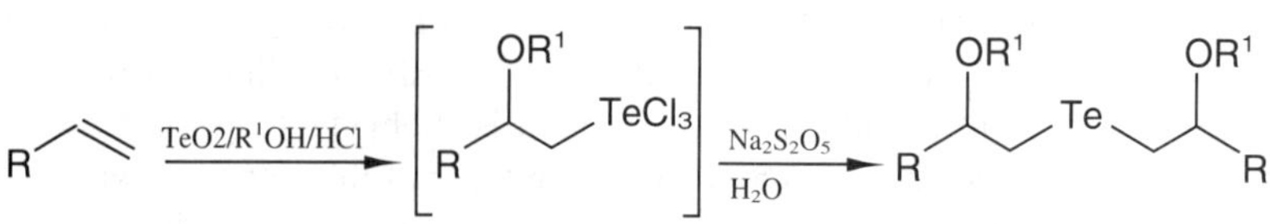

R (yield %): n-Bu (58), s-Bu (22), i-Bu (29), n-hexyl (72), n-decyl (72). n-$C_{14}H_{29}$ (50), HOC_9H_{15} (45), $HOC_{10}H_{20}$ (41),

Bz (80), Ph (Bz); R^1 = Me
n-decyl (55), Bz (65); R^1 = Et
n-decyl (23); R^1 = i-Pr

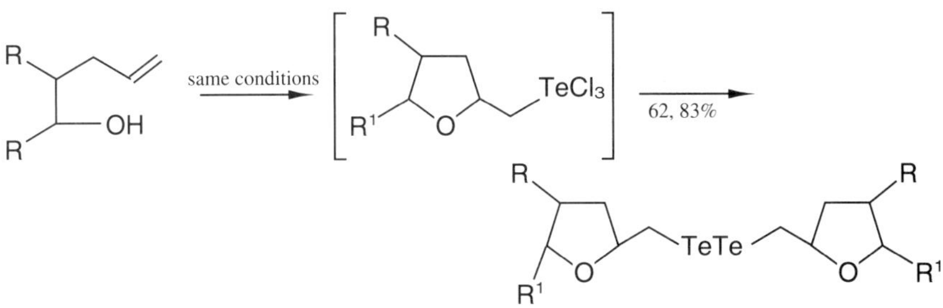

$R^1 = R = H$,

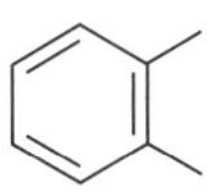

Bis(2-methoxy-3-phenylpropyl) ditelluride (typical procedure) [24] TeO_2 (2.0 g, 12.5 mmol) is dissolved in concentrated HCl (10 ml) and the yellow solution diluted with MeOH (40 ml). Allybenzene (1.75 g, 14.8 mmol) is added and the mixture heated under reflux for 24 h and then allowed to cool to room temperature. The yellowish solution (containing a trace of elemental Te) is shaken in a separating funnel with H_2O/CH_2Cl_2 (100 ml/100 ml) containing $Na_2S_2O_5$ (5 g). The red CH_2Cl_2 phase is separated, dried and evaporated, giving a red oil which is purified by flash chromatography (SiO_2/CH_2Cl_2) to give the ditelluride (2.78 g (80%)).

3.2.3.3 Reduction of aryltellurium trichlorides

Aryltellurium trichlorides (see Sections 3.5.1.3 and 3.5.1.4) are easily reduced by means of several reducing agents. The most popular reduction procedures are described in sequence.

(a) By hydrated sodium sulphide

This reagent, together with potassium hydrogen sulphite [25], has been used to reduce aryltellurium trichlorides [26] since the early period of tellurium chemistry, and continues to find wide application.

Diaryl ditellurides (general procedure) [26] The aryltellurium trichloride is poured into 15 mol eq of $Na_2S{\cdot}9H_2O$ heated (and melted) at 95–100°C in a beaker, with strong manual stirring. An exothermic reaction occurs. After 10 min at the same temperature, excess H_2O is added. By cooling the ditelluride solidifies, and is collected by decantation or filtration. The yields are near to quantitative.

The following ditellurides can be prepared by the foregoing method [27–30]: Ar = Ph, *p*-MeO, *p*-EtO, *p*-PrO, *p*-BuO, *p*-PhO, *p*-PhS, *p*-Cl, *p*-Br, *m*-F, *p*-MeO_2C, *p*-$EtO_2CC_6H_4$, 3,4-$(MeO)_2C_6H_3$, 1-naphthyl, 2-naphthyl.

(b) By TUDO

Diaryl ditellurides (general procedure) [31] A mixture of the aryltellurium trichloride (2.0 mmol) and 2 N NaOH (5 ml) is vigorously stirred at room temperature for 15 min.* TUDO (0.43 g, 4.0 mmol) and petroleum ether 30–60°C (5 ml) are added to the mixture, and stirring at room temperature is continued for 15 min. The phases are separated and the organic layer is dried ($MgSO_4$). The solvent is evaporated, giving the ditelluride (Ar = Ph (92%), *p*-$MeOC_6H_4$ (94%)).

The corresponding aryltellurium tribromides are reduced similarly.

(c) By hydrazine hydrate

*Di-*p*-tolyl ditelluride (typical procedure)* [32] A solution of hydrazine hydrate (3.2 g, 0.1 mol) in EtOH (20 ml) is added slowly dropwise to a refluxing solution of *p*-tolyltellurium trichloride (3.25 g, 10 mmol) in EtOH until evolution of N_2 ceases. The hot mixture is filtered, and the ditelluride crystallizes by cooling (0.96 g (44%; m.p. 52°C).

In the same manner the following ditellurides can be prepared: Ar = Ph, *p*-Me, *p*-EtO, *p*-Br-C_6H_4, *m*-Me, *p*-$MeOC_6H_3$ (45–84%).

REFERENCES

1. M. T. Chen and J. W. George, *J. Organomet. Chem.* **12**, 401 (1968).
2. G. M. Bagolyubov, Y. N. Slyk and A. A. Petrov, *J. Gen. Chem. USSR* **39**, 1768 (1969) (in English).
3. K. K. Bhasin, V. Gupta, A. Gautam and R. P. Sharma, *Synth. Commun.* **20**, 2191 (1990).

*At that time the trichloride has been converted into the corresponding sodium tellurinate.

4. M. M. Goodman and F. F. Knapp, *J. Org. Chem.* **47**, 3004 (1982).
5. W. Lee, *M.S. Thesis*, Texas A&M University (1986); K. Y. Irgolic, in *Houben–Weyl—Methoden der Organischen Chemie*, 4th edn, Vol. E12b, p. 259. Georg Thieme, Stuttgart, 1990.
6. J. Davies, W. R. McWhinnie, N. S. Dance and C. H. W. Jones, *Inorg. Chim. Acta* **29**, L217 (1978).
7. S. Uemura, M. Takahashi, K. Ohe and N. Sugita, *J. Organomet. Chem.* **361**, 63 (1989).
8. J. T. B. Ferreira, A. R. M. Oliveira and J. V. Comasseto, *Synth. Commun.* **19**, 239 (1989).
9. K. H. Bhasin and A. Gautam, *Phosphorus Sulfur* **38**, 211 (1988).
10. D. J. Sandman, J. C. Stark, L. A. Acampora and P. Gagne, *Organometallics* **2**, 549 (1983).
11. W. S. Haller and K. J. Irgolic, *J. Organomet. Chem.* **38**, 97 (1972).
12. N. Petragnani and M. Moura Campos, *Chem. Ber.* **96**, 249 (1963).
13. M. Akiba, M. V. Lakshmikantham, K. Jen and M. P. Cava, *J. Org. Chem.* **49**, 4819 (1984).
14. M. R. Detty, B. J. Murray, D. J. Smith and N. Zumbulyadis, *J. Am. Chem. Soc.* **105**, 875 (1983).
15. J. L. Piette and M. Renson, *Bull. Soc. Chim. Belg.* **79**, 353 (1970).
16. L. Engman and M. P. Cava, *Synth. Commun.* **12**, 163 (1982).
17. D. Seebach and A. K. Beck, *Chem. Ber.* **108**, 314 (1975).
18. L. Engman and P. Cava, *Organometallics*, **1**, 470 (1982).
19. L. Engman and J. Persson, *J. Organomet. Chem.* **388**, 71 (1990).
20. L. Lange and W. W. Du Mont, *J. Organomet. Chem.* **286**, C1 (1985).
21. S. Suzuki and T. Nakamura, *Synthesis*, 549 (1992).
22. G. T. Morgan and H. D. K. Drew, *J. Chem. Soc.*, 531 (1925).
23. G. T. Morgan and R. E. Kellett, *J. Chem. Soc.*, 1080 (1926).
24. L. Engman, *Organometallics* **8**, 1997 (1989).
25. L. Reichel and E. Kirschbaum, *Ann.* **523**, 211 (1936).
26. L. Reichel and E. Kirschbaum, *Ber. Dtsch. Chem. Ges.* **76**, 1105 (1943).
27. N. Petragnani and G. Vicentini, Unpublished results.
28. N. Petragnani, *Tetrahedron* **11**, 15 (1960).
29. G. Vicentini, E. Giesbrecht and L. R. M. Pitombo, *Chem. Ber.* **92**, 40 (1959).
30. I. D. Sadekov, L. M. Sayapina, A. Y. Bushkov and V. I. Minkin, *J. Gen. Chem. USSR* **41**, 2747 (1971) (in English).
31. E. S. Lang and J. V. Comasseto, *Synth. Commun.* **18**, 301 (1988).
32. J. Bergman, *Tetrahedron* **28**, 3323 (1972).

3.3 ORGANYLTELLUROLS

Organyltellurols are very unstable compounds due to their extreme sensitivity to oxygen, giving the corresponding ditellurides. The first short-chain alkyltellurols (C_1–C_4) have been isolated, as yellow liquids with an obnoxious odour, from the reaction of aluminium telluride and hydrogen telluride, respectively, with alcohols and alkyl bromides [1]. Aryltellurols seem not to have been isolated. As shown in Sections 3.1.3.2 and 3.2.2, aryl tellurolates are starting materials for the preparation of aryl tellurides and diaryl ditellurides and, subsequently, aryltellurols find wide use in the manipulation of several organic functionalities.

Due to their intractability, these reagents are generated *in situ*, the most appropriate methods being described in several sections in the text.

REFERENCE

1. K. Y. Irgolic, in *Houben–Weyl—Methoden der Organischen Chemie*, 4th edn, Vol. E12b, p. 152. Georg Thieme, Stuttgart.

3.4 BIS-ORGANYL TELLUROMETHANES

Bis-organyl telluromethanes, or organyl telluroketals, can be prepared by the conventional method involving the reaction of 2 eq of organyl tellurolates with dihalomethanes.

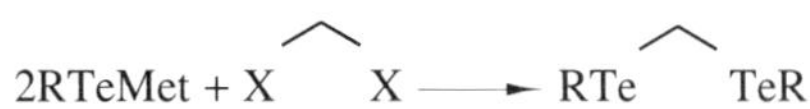

R = Me; Met = Li; X = Cl [1]

R = Ph, Met = Li; X = I [2]

R = *p*-$EtOC_6H_4$; Met = Na; X = I [3]

R =

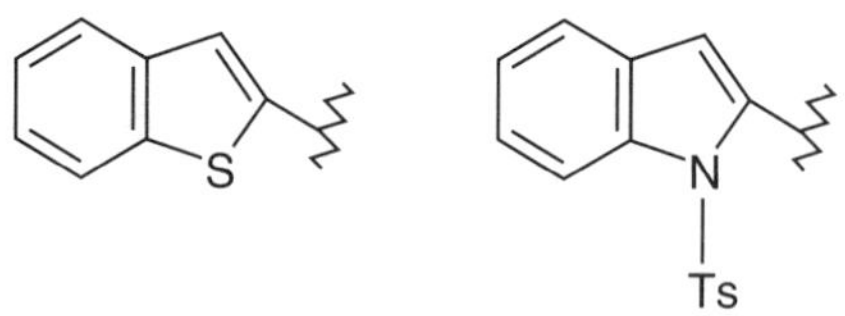

[4]

A more convenient method due to the mild experimental conditions and high yields involves the reaction of diaryl ditellurides with diazomethane [5].*

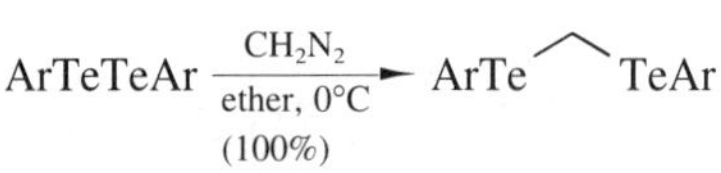

Ar = *p*-$MeOC_6H_4$, *p*-$EtOC_6H_4$

Reaction of diaryl ditellurides with diazomethane—formation of diaryl telluroketals (typical procedure) [5] Di-*p*-methoxyphenyl ditelluride (2.34 g, 5 mmol) in ether (30 ml) is treated dropwise, at 0°C and while stirring, with an ethereal solution of diazomethane until disappearance of the red colour of the ditelluride. Evolution of N_2 ensues. The product is obtained by evaporation of the solvent. Recrystallization from petroleum ether 60–70°C or EtOH gives colourless crystals (2.41 g (100%); m.p. 98–99°C).

*Diaryl diselenide and diaryl disulphide react analogously.

Telluroketals exhibit synthetic interest since they easily generate methyllithium derivatives by treatment with alkyllithium compounds (see Section 4.9.1.3).

REFERENCES

1. E. G. Hope, T. Kemmitt and W. Levanson, *Organometallics* **7**, 78 (1988).
2. D. Seebach and A. B. Beck, *Chem. Ber.* **108**, 314 (1975).
3. H. M. K. K. Pathirane and W. R. McWhinnie, *J. Chem. Soc. Dalton Trans.*, 2003 (1986).
4. L. Engman and M. P. Cava, *Organometallics* **1**, 470 (1982).
5. N. Petragnani and G. Schill, *Chem. Ber.* **103**, 2271 (1970).

3.5 ORGANYLTELLURIUM TRIHALIDES

Organyltellurium trichlorides, tribromides and triiodides are well-known compounds, the trichlorides being the most familiar and useful for synthetic purposes.

Organyltellurium trichlorides are easily accessible, exploiting the electrophilic character of tellurium tetrachloride by the direct introduction of the $—TeCl_3$ moiety in organic substrates. Otherwise, organyltellurium trichlorides as well as the tribromides and triiodides, seldom generated by the preceding method, can be obtained by the halogenolysis of the corresponding ditellurides.

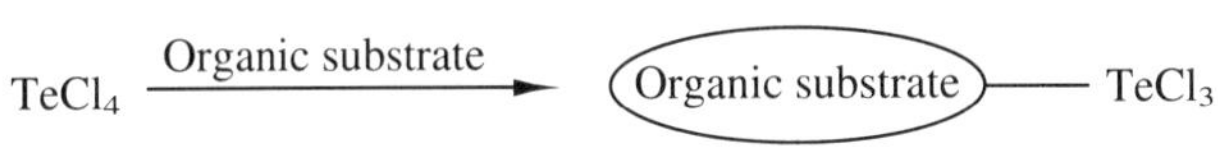

$$RTeTeR \xrightarrow{X_2} RTeX_3 \quad X = Cl, Br, I$$

3.5.1 Organyltellurium trichlorides from tellurium tetrachloride and organic substrates

3.5.1.1 From tellurium tetrachloride and ketones and carboxylic anhydrides

Reactions involving the attack of a nucleophilic carbon atom activated by an α-carbonyl group at the electrophilic tellurium atom of tellurium tetrachloride belong to the oldest methods of preparative organic tellurium chemistry.

Depending on the structure of the ketone, different regiosubstituted trichlorides are obtained, in addition to the corresponding dichlorides [1].

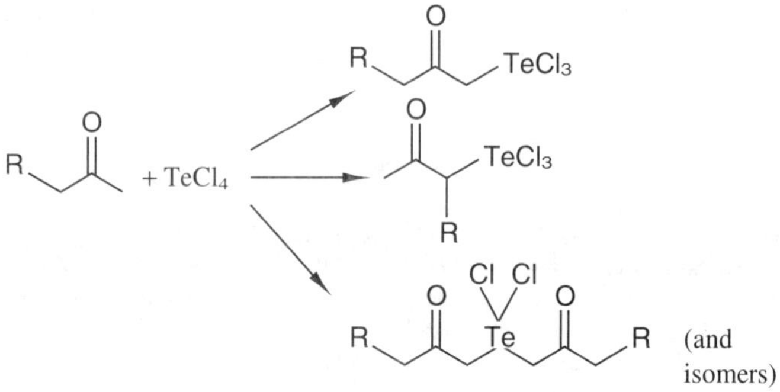

3.5.1.2 From tellurium tetrachloride and alkenes

The addition of tellurium tetrachloride to alkenes produces chloroalkyltellurium trichlorides. Butenes, 1-decene, cycloalkenes and 3-substituted 1-propenes give mixtures of *syn* and *anti* additions [2–6].

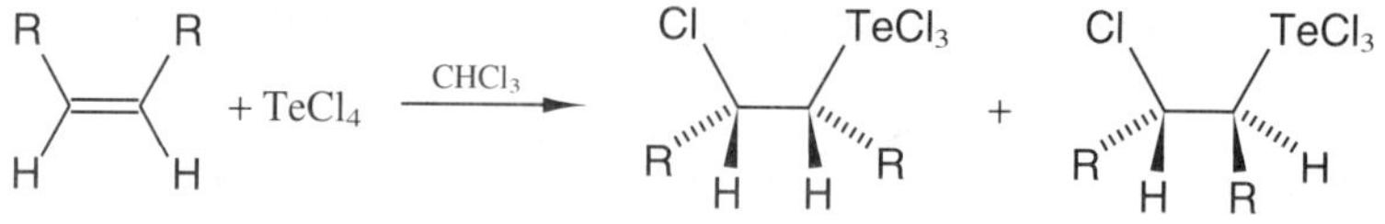

β-Chloroalkyltellurium trichlorides (general procedure) [5] Freshly sublimed $TeCl_4$ (1.67 g, 6.2 mmol) and the olefin (6.4 mmol) are stirred in an ice bath for 3 h in dry, ethanol-free $CHCl_3$ (20 ml), until almost all the $TeCl_4$ has disappeared. Filtration and evaporation yield the adduct as a mixture of olefin isomers (percentage yield, *sin*/*anti* ratio): (*Z*)-2-butene (67, 20/80), (*E*)-2-butene (94, 42/58; after an additional overnight stirring at room temperature), cyclopentene (98, 50/50), cyclohexene (99, > 97% *anti*), cyclooctene (83, > 97% *syn*), 1-decene (88), (*E*)-1-D-1-decene (96).

Depending on the $TeCl_4$/alkene ratio and the polarity of the solvent, bis(chloroalkyl)tellurium dichlorides can be formed (see Section 3.9.2.2).

The alkoxytrihalotellurination of alkenes or cycloalkenes is afforded by treatment with tellurium dioxide/hydrochloric acid/alcohol [7] or with tellurium tetrachloride (or tetrabromide)/alcohol in carbon tetrachloride [8].

The addition is regiospecific, the tellurium moiety bonding to the less hindered carbon atom, and stereospecific as demonstrated by the *anti* addition to cycloalkenes. By the first procedure the trichloride is reduced without isolation to the corresponding ditelluride (see Section 3.2.3.2).

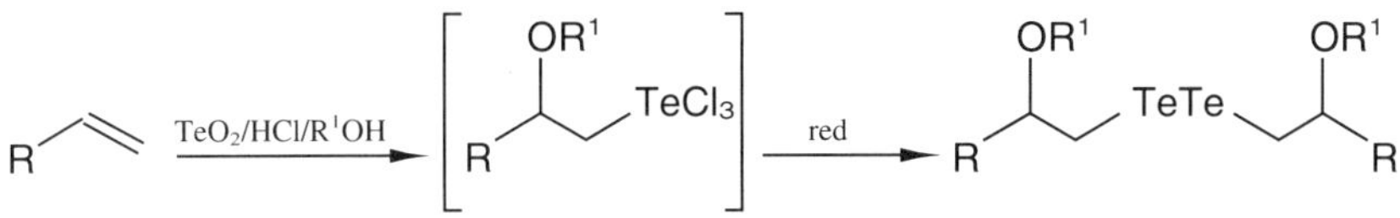

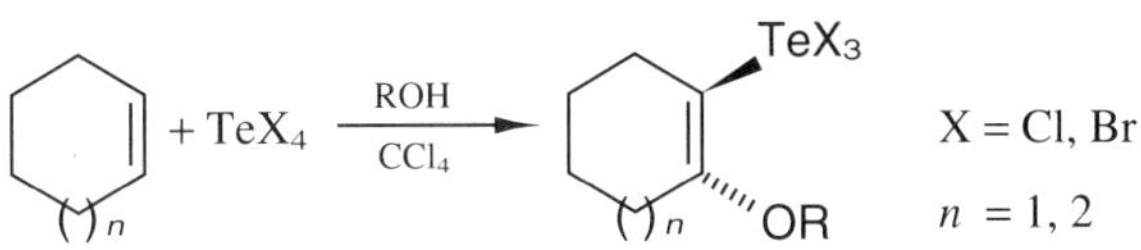

trans-2-Ethoxycyclohexyltellurium trichloride (typical procedure) [8] Cyclohexene (1.6 g, 19.5 mmol) is added to $TeCl_4$ (5.00 g, 18.6 mmol) in CCl_4 (50 ml) and EtOH (2.76 g, 60 mmol). The mixture is heated under reflux for 2 h, cooled, filtered to remove traces of elemental Te, shaken with charcoal, and refiltered. Evaporation under vacuum and recrystallization from petroleum ether 30–60°C give the pure product (4.23 g (63%); m.p. 97–98°C).

The following *trans*-2-alkoxycycloalkyltellurium trihalides are prepared similarly: $n = 1.2$; X = Cl, Br; R = MeO, n-Pr, i-Pr.

It must be pointed out, however, that the addition of tellurium tetrachloride to olefins lacks generality, and undefined results have been mentioned such as formation of tarry products, reduction to elementary tellurium or no reaction at all [9].

3.5.1.3 From tellurium tetrachloride and arenes

The condensation of tellurium tetrachloride with aromatic compounds is of general application and has long been used as an efficient route for the preparation of aryltellurium trichlorides.

In accordance with a typical electrophilic aromatic substitution, this reaction is successfully accomplished by simply treating aromatic compounds activated by electron-donating groups, such as alkoxy, phenoxy, hydroxy and thiophenoxy moieties, with tellurium tetrachloride in refluxing chloroform, carbon tetrachloride or toluene [10–15].

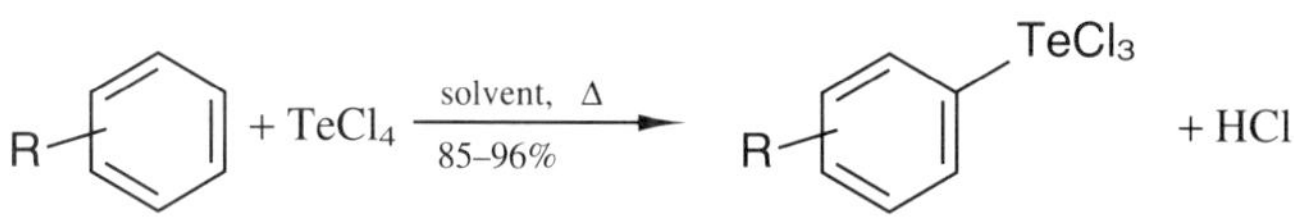

representative examples: R = *p*-MeO, *p*-EtO, *p*-PhO, *p*-HO

Non-activated arenes react only under more severe conditions [16, 17], or in the presence of a Lewis acid [18]. In the last case, however, the reaction does not stop at the trichloride step, and is more appropriate for the preparation of diaryltellurium dichlorides (see Section 3.9.2.3).

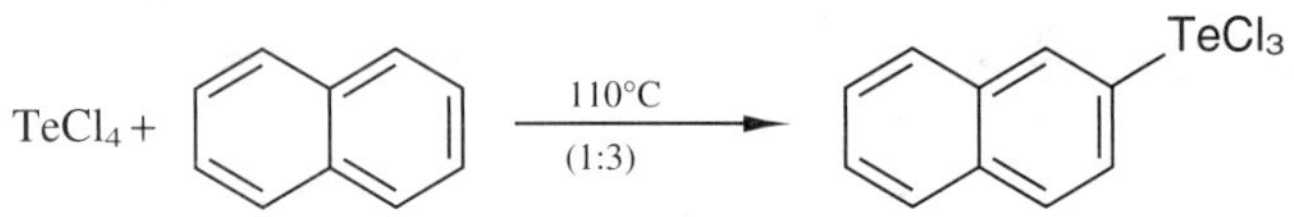

p-*Methoxyphenyltellurium trichloride (typical procedure)* [13] Anisol (5.4 g, 50 mmol) and $TeCl_4$ (13.5 g, 50 mmol) in CCl_4 (40 ml) are heated under reflux for 2 h. Evolution of HCl occurs and the heavy, crystalline $TeCl_4$ is progressively converted into yellow flakes of trichloride. The product is filtered off and washed with CCl_4 (15.5 g (91%)). The crude product is recrystallized from glacial HOAc as yellow needles (m.p. 182°C).

2-Naphthyltellurium trichloride (typical procedure) [16] $TeCl_4$ (7.5 g, 27.8 mmol) is added to molten naphthalene (11.0 g, 86 mmol) kept at 110°C under N_2. After 12 h at this temperature the greenish solid obtained is pulverized in a mortar and washed thoroughly with hot toluene (15.4 g (54%)). The product is recrystallized from benzene (m.p. 200–202°C).

3.5.1.4 From tellurium tetrachloride and arylmercury chlorides

Arylmercury chlorides are valuable reagents for the preparation of aryltellurium trichlorides since they are employed when the aromatic substrate is not sufficiently reactive for an easy direct condensation with tellurium tetrachloride [14, 19, 20].

$$TeCl_4 + ArHgCl \xrightarrow[-HgCl_2 \cdot \text{dioxane}]{\text{dioxane}} ArTeCl_3$$

representative examples [21]:

Ar = Ph [19], 2-naphthyl [22], *o*-$PhSC_6H_4$ [14], *p*-Cl, *p*-Br, *m*-F, *p*-NO_2, *m*-$NO_2C_6H_4$ [15], 2-thienyl [23]; yields 56–96%.

The formed mercury dichloride is separated as a crystalline complex with dioxane.

Since the starting arylmercury chlorides are easily prepared from diazonium salts, this method allows the conversion of anilines into aryltellurium trichlorides.

1-Naphthyltellurium trichloride (typical procedure) [14] A solution of 1-naphthyl mercury chloride [24] (7.2 g, 20 mmol) and $TeCl_4$ (5.4 g, 20 mmol) in pure dioxane (40 ml) is heated under reflux for 2 h. The crystalline $HgCl_2$·dioxane that precipitates on cooling the solution at 10°C is removed by filtration. The dark yellow solution is evaporated, giving a viscous oil. By cooling and adding petroleum ether 30–60°C the oil crystallizes. The crude trichloride is crystallized from glacial HOAc as yellow crystals (m.p. 175–180°C; 6.9 g (96%)). The crude product can also be purified by reduction to the corresponding detelluride with $Na_2S \cdot 9H_2O$ (see Section 3.2.3) and subsequent chlorinolysis with SO_2Cl_2 (see next section).

3.5.2 By chlorinolysis of diorganyl ditellurides

Dialkyl and diaryl ditellurides are easily converted to the corresponding tellurium trichlorides by means of the fission of the Te—Te bond by a chlorinating reagent. This route is very convenient for the preparation of the alkyl derivatives in view of the easy accessibility of dialkyl ditellurides via the alkylation of sodium ditelluride (see Section 3.2.1).

The chlorinolysis can be effected by employing chlorine [25] in solvents (chloroform, carbon tetrachloride, dichloromethane, benzene) or, more conveniently, avoiding the use of gaseous chlorine, by treatment with thionyl chloride [26–28] or sulphuryl chloride [29–31].

$$\mathrm{RTeTeR} \xrightarrow[\text{or } SOCl_2 \text{ or } SO_2Cl_2]{Cl_2} \mathrm{RTeCl_3} \quad R = \text{alkyl, aryl}$$

Organyltellurium trichlorides (general procedure) [29, 31] A solution of the ditelluride in benzene or CCl_4 is treated dropwise at 0°C with a slight excess of SO_2Cl_2 in the same solvent. The crude precipitate of the trichloride is separated by filtration. The yield is close to 100%.

3.5.3 Organyltellurium tribromides and triiodides by halogenolysis of the corresponding ditellurides

In analogy to chlorinolysis, the brominolysis and iodinolysis of diorganyl ditellurides offer a facile route to tellurium tribromides and triiodides [14, 32].

$$\mathrm{RTeTeR} \xrightarrow[\text{solvent}]{X_2} \mathrm{2RTeX_3}$$

$X = Br, I$

The aryl derivatives which are not easily accessible from tellurium tetrabromide and tetraiodide are therefore prepared by this route.

p-*Methoxyphenyltellurium tribromide (typical procedure)* [14] Bromine (0.48 g, 3 mmol) dissolved in CCl_4 (5 ml) is added dropwise, with stirring, to a solution of di-*p*-methoxyphenyl ditelluride (0.47 g, 1 mmol) in CCl_4 (5 ml) at 0°C. The tribromide precipitates as an orange crystalline product and is separated by filtration (0.95 g (100%)). It is recrystallized from glacial HOAc (m.p. 188–190°C (dec.)).

p-*Methoxyphenyltellurium triiodide (typical procedure)* [14] The triiodide is prepared following the previous procedure, using the corresponding ditelluride (1 mmol) in CCl_4 and iodine (3 mmol) in benzene. The yield is quantitative. The product is recrystallized from benzene (m.p. 131–133°C (dec.)).

The following tribromides can be prepared using the same procedure as described above: Ar = *p*-EtO; *p*-PhO, *p*-$PhSC_6H_4$ [14], 1-naphthyl [14], 2-naphthyl [30]. The following triiodides can also be prepared: Ar = *p*-Eto, *p*-PhO, *p*-$PHSC_6H_4$, 1-naphthyl [14].* Dibenzyl ditelluride on treatment with bromine undergoes cleavage of the Te–benzyl bond, giving benzyl bromide and tellurium tetrabromide [33].

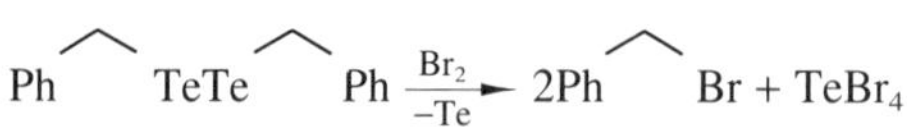

3.5.4 Additional method. Preparation of organyltellurium trichlorides and tribromides by the reaction of tetraorganyltin compounds with tellurium tetrachloride and tetrabromide [34]

$$SnR_4 + TeX_4 \xrightarrow[r\cdot t\cdot]{benzene} RTeX_3 + R_3SnX$$

R = Me, Et, C_3H_7, Ph

X = Cl, Br

*$(2\text{-naphthyl-Te})_2$ forms exclusively 2-naphthyl-TeI, the triiodide being unknown (see Section 3.7).

REFERENCES

1. G. T. Morgan and O. C. Elvins, *J. Chem. Soc.* **127**, 2625 (1925).
2. H. Funk and W. Weiss, *J. Prakt. Chem.* **1**, 33 (1954).
3. M. Moura Campos and N. Petragnani, *Tetrahedron Lett.*, 11 (1959).
4. M. Ogawa and R. Ishioka, *Bull. Chem. Soc. Jpn* **43**, 496 (1970).
5. J. E. Backwall, J. Bergman and L. Engman, *J. Org. Chem.* **48**, 3918 (1983).
6. S. Uemura and S. Fukuzawa, *J. Organomet. Chem.* **268**, 223 (1984).
7. L. Engman, *Organometallics*, **8**, 1997 (1989).
8. M. E. S. Ali, M. A. Malik and S. Smith, *Inorg. Chim. Acta* **162**, 157 (1989).
9. K. Y. Irgolic, in *Houben–Weyl—Methoden der Organischen Chemie*, 4th edn, Vol. E12b, p. 304. Georg Thieme, Stuttgart, 1990.
10. G. T. Morgan and R. E. Kellet, *J. Chem. Soc.*, 1080 (1926).
11. G. T. Morgan and H. D. Drew, *J. Chem. Soc.* **127**, 2307 (1925).
12. H. D. Drew, *J. Chem. Soc.*, 223 (1926).
13. L. Reichel and E. Kirschbaum, *Liebigs Ann. Chem.* **523**, 211 (1936).
14. N. Petragnani, *Tetrahedron*, **11**, 15 (1960).
15. I. D. Sadekov, L. M. Sayapina, A. Y. Bushkov and V. I. Minkin, *J. Gen. Chem. USSR* **41**, 2747 (1971) (in English).
16. J. Bergman and L. Engman, *Tetrahedron* **36**, 1275 (1980).
17. J. Bergman, *Tetrahedron* **28**, 3323 (1972).
18. W. H. H. Gunther, J. Nepywoda and J. Y. C. Chu, *J. Organomet. Chem.* **74**, 79 (1974).
19. W. V. Farrar, *Research* **4**, 179 (1951).
20. I. G. M. Campbell and E. E. Turner, *J. Chem. Soc.*, 39 (1938).
21. K. Y. Irgolic, in *Houben–Weyl—Methoden der Organischen Chemie*, 4th edn, Vol. E12b, p. 311. Georg Thieme, Stuttgart, 1990.
22. H. Rheinboldt and J. Vicentini, *Chem. Ber.* **89**, 624 (1956).
23. L. Y. Chia and W. R. McWhinnie, *J. Organomet. Chem.* **148**, 165 (1978).
24. N. Nesmejanov, *Ber. Dtsch. Chem. Ges.* **62**, 1010 (1929); N. Nesmejanov, *Org. Synth. Coll.*, Vol. II, p. 432. Wiley, New York, 1943.
25. K. J. Wynne and P. S. Pearson, *Inorg. Chem.* **9**, 106 (1970); K. J. Wynne and P. S. Pearson, *Inorg. Chem.* **10**, 1871 (1971).
26. H. Rheinboldt, in *Houben–Weyl—Methoden der Organischen Chemie* (ed. E. Muller), 4th edn, Vol. IX, p. 1156. Georg Thieme, Stuttgart, 1955.
27. I.D. Sadekov, A. Y. Bushkov and V. I. Minkin, *J. Gen. Chem. USSR* **42**, 124 (1972) (in English).
28. I. D. Sadekov, A. Y. Bushkov, U. S. Yur'eva and V. I. Minkin, *J. Gen. Chem. USSR* **47**, 2321 (1977) (in English).
29. N. Petragnani and M. Moura Campos, *Chem. Ber.* **96**, 249 (1963).
30. G. Vicentini, H. Giesbrecht and L. R. M. Pitombo, *Chem. Ber.* **92**, 40 (1959).
31. L. Engman, *Organometallics* **8**, 1997 (1989).
32. K. Y. Irgolic, in *Houben–Weyl—Methoden der Organischen Chemie*, 4th edn, Vol. E12b, p. 315. Georg Thieme, Stuttgart, 1990.
33. H. K. Spencer and M. P. Cava, *J. Org. Chem.* **42**, 2937 (1977).
34. H. Schumann and M. Magerstadt, *J. Organomet. Chem.* **232**, 147 (1982).

3.6 THE PRODUCTS OF THE HYDROLYSIS OF ARYLTELLURIUM TRIHALIDES

Aryltellurium trihalides are sensitive to water and can be converted to products at different stages of hydrolysis. Oxohalides, aryltellurinic acids and aryltellurium anhydrides have been known for a long time, the oxohalides being the products of partial hydrolysis.

$$ArTeX_3 \xrightarrow{H_2O} ArTe(O)X \xrightarrow{H_2O}$$

$$ArTe(O)OH \longrightarrow ArTe(O)\text{-}O\text{-}Te(O)AR$$

The ease of hydrolysis follows the order $ArTeCl_3 > ArTeBr_3 > ArTeI_3$. Aryltellurium trichlorides and tribromides are converted to the oxohalides by simple treatment with cold water. The triiodides are unaffected under these conditions,* and on treatment with boiling water give mixtures or undefined products in most cases [1].

Aryltellurinic acids were described in the early period of tellurium chemistry as the products of the alkaline hydrolysis of the trichlorides [2]. However, the structure of these compounds was later argued about, and the corresponding anhydrides are today accepted as the unequivocal products [1, 3, 4]. The anhydrides are stable, colourless, high-melting-point solids and have recently attained a relevant role as synthetic reagents.

Aryltellurium anhydrides (general procedure) [1] The aryltellurium trichlorides (or tribromides) are treated by heating them with 10% aqueous Na_2CO_3 or NaOH until dissolution (a few minutes are sufficient). The cooled solution is treated with 10% HOAc. The anhydrides precipitate as white solids and are separated by filtration and dried under vacuum (in the presence of $CaCl_2$) (m.p. > 200°C; Ar = Ph, *p*-HO, *p*-MeO, *p*-EtO, *p*-$PhOC_6H_4$, 2-naphthyl (yields 90–100%)).

Mixed anhydrides are prepared by reacting tellurinic anhydrides with carboxylic acids or anhydrides [5].

$$(ArTeO)_2O + 2RCO_2H \longrightarrow 2ArTe(O)OCOR + H_2O$$

$$Ar = Ph;\ R = Me,\ F_3C$$

*Aryltellurium triiodides can be prepared by treating aryltellurium trichlorides, tribromides, oxohalides or aryltellurinic anhydrides with aqueous sodium iodide [1].

Mixed aryltellurium anhydrides (typical procedure) [5] Benzenetellurinic anhydride (0.45 g, 1 mmol) is treated with TFA (0.251 g, 2.2 mmol) or anhydride (0.23 g, 1.1 mmol) in CH_2Cl_2 (10 ml) at room temperature. After 30 min the starting anhydride has dissolved, and the solution is then evaporated, giving benzenetellurinyl trifluoroacetate (quantitative yield). The product is recrystallized from benzene as colourless crystals (m.p. 181–182°C).

REFERENCES

1. N. Petragnani and G. Vicentini, *Universidade de São Paulo, Faculdade de Filos. Cienc. Letras, Biol. Quím.* **5**, 75 (1959). *Chem. Abstr.* **11256**, 58 (1963).
2. L. Reichel and E. Kirschbaum, *Ann.* **523**, 211 (1936).
3. P. Thavornyutikarn and W. R. Whinnie, *J. Organomet. Chem.* **50**, 135 (1973), and references therein.
4. D. H. R. Barton, J. P. Finet and M. Thomas, *Tetrahedron* **42**, 231 (1986).
5. N. X. Hu, Y. Aso, T. Otsubo and F. Ogura, *J. Org. Chem.* **54**, 4398 (1989).

3.7 ARYLTELLURENYL HALIDES

Halogenated derivatives of tellurium in the oxidation state +2, namely tellurenyl halides, are unfamiliar compounds in comparison to the well-known selenium analogues.

These compounds are in principle easily accessible by the controlled halogenolysis of the parent ditellurides. However, the first compound obtained in 1959, 2-naphthyltellurenyl halide [1], remained for a long time the only stable representative of this class until the preparation of other members of this class by a successful utilization of the same method [2].

$$ArTeTeAr + X_2 \longrightarrow 2ArTeX$$

Ar = 2-naphthyl; X = I

Ar = Ph, *p*-MeO, *o*-Ph, *p*-PhC_6H_4, 2-biphenylyl; X = Br, I

Ar = *m*, *p*-$(MeO)_2C_6H_3$; X = I

Several aryltellurenyl halides have been prepared, bearing an ortho carbonyl, nitro or another electron-attracting group, exhibiting a stabilizing effect by coordination to tellurium [3–6].

With the exception of 2-naphthyltellurenyl iodide, the unsubstituted members are, however, thermally unstable in the solid state. Therefore, to exploit their highly electrophilic character, aryltellurenyl halides are not isolated, instead being generated *in situ* and used directly for further conversion (see Section 3.1.3.4) [7].

REFERENCES

1. G. Vicentini, E. Giesbrecht and L. R. M. Pitombo, *Chem. Ber.* **92**, 40 (1959).
2. P. Schulz and G. Klar, *Z. Naturforsch. B* **30**, 40, 43 (1974).
3. J. L. Piette and M. Renson, *Bull. Soc. Chim. Belg.* **80**, 669 (1971).
4. J. L. Piette, R. Lysy and M. Renson, *Bull. Soc. Chim. Fr.*, 3559 (1972).
5. P. Wiriyachitra, S. J. Falcone and M. P. Cava, *J. Org. Chem.* **44**, 3957 (1979).
6. L. Engman, *J. Org. Chem.* **54**, 2964 (1989).
7. N. Petragnani, L. Torres and K. Y. Wynne, *J. Organomet. Chem.* **92**, 175 (1975).

3.8 ARYLTELLURENYL PSEUDOHALIDES. ARYL TELLUROCYANATES

Aryl tellurocyanates are easily prepared by reacting aryltellurenyl bromides (generated *in situ* by reduction of the corresponding tribromides) with potassium cyanide [1].

$$\mathrm{ArTeBr_3} \xrightarrow{\mathrm{Na_2S_2O_5/H_2O/MeOH}} [\mathrm{ArTeBr}] \xrightarrow[\text{93, 86\%}]{\mathrm{KCN}} \mathrm{ArTeCN}$$

Ar = p-MeOC$_6$H$_4$, Ph

p-*Methoxyphenyl tellurocyanate (typical procedure)* [1] p-Methoxyphenyl tellurium tribromide (5.0 g, 14.2 mmol) is added to a solution of $NaHCO_3$ (2.45 g, 29.2 mmol) in H_2O (50 ml) and MeOH (3 ml). The mixture is stirred for 30 min and then KCl (3.70 g, 58.4 mmol) followed by $Na_2S_2O_5$ (11.0 g, 58.4 mmol) are added. The resultant mixture is stirred for 8 min and KCN (3.70 g, 58.4 mmol) is added. A white precipitate forms. The reaction mixture is diluted with H_2O, extracted with CH_2Cl_2, and the organic phase is dried (Na_2SO_4) and filtered. The filtrate is diluted with hexane and the resultant solution is concentrated by distilling some of the solvent. The formed precipitate is filtered (3.60 g (93%); m.p. 77°C).

REFERENCE

1. S. J. Falcone and M. P. Fava, *J. Org. Chem.* **45**, 1044 (1980).

3.9 DIORGANYLTELLURIUM DIHALIDES

3.9.1 From elemental tellurium

The reaction of elemental tellurium with alkyl iodides giving dialkyltellurium diiodides is a 100-year-old method [1], but still finds some use [2] even if the low yields make this method unattractive.

$$Te + 2RI \longrightarrow R_2TeI_2$$

In an improved procedure, employing alkyl chlorides and bromides in the presence of sodium iodide, tellurium diiodides are obtained in increased yields [3].

$$Te + 2RX \xrightarrow[\text{solvent}]{\text{NaI}} R_2TeI_2$$

$$X = Cl, Br$$

1,3-Dihydro-2-benzotellurophene diiodide (typical procedure) [3] A mixture of α,α′-dichloro-*o*-xylene (3.5 g, 20 mmol), finely powdered Te (2.55 g, 20 g at) and NaI (12.0 g, 80 mmol) in 2-methoxyethanol (100 ml) is stirred and gently boiled in an open beaker. After 1 h the mixture contains a heavy orange precipitate and a small amount of unreacted Te. After the addition of deionized H_2O (200 ml), which causes additional precipitation, the precipitate is filtered off, washed with H_2O, rinsed with acetone and air dried (8.5 g (83%)). The product is recrystallized from 2-methoxyethanol (two crystalline forms) (m.p. 222–225°C).

3.9.2 From tellurium tetrahalides (or tellurium dioxide)

Employing an appropriate ratio of reagents and experimental conditions, some of the reactions which convert tellurium tetrachloride into organyltellurium trichlorides (see Section 3.5.1) can be applied to the synthesis of diorganyltellurium dichlorides.

3.9.2.1 With ketones and carboxylic acid anhydrides

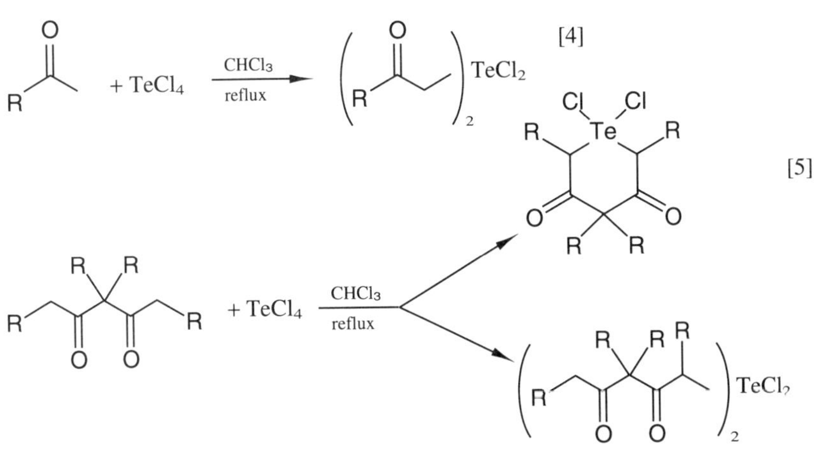

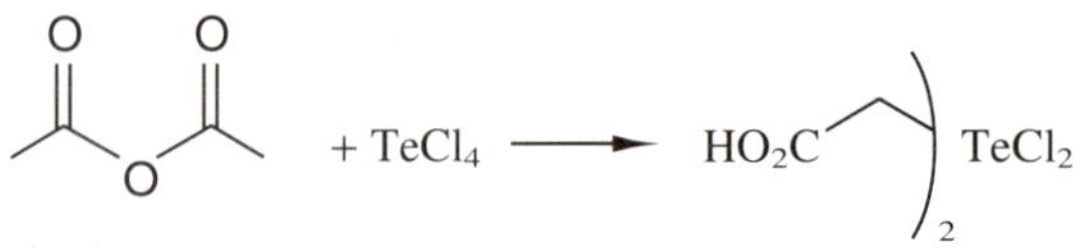

3.9.2.2 With alkenes

[7–10]

Cyclotelluroetherification reaction, such as shown in the accompanying equation, will be discussed in Section 4.6.2.4 [11].

$n = 1,2$

3.9.2.3 With arenes

When the reactions of tellurium tetrachloride with activated arenes (such as alkoxybenzenes) are performed employing a large excess of the organic substrate and/or under prolonged heating at high temperatures, diaryltellurium dichlorides are formed in high yields [12–15].

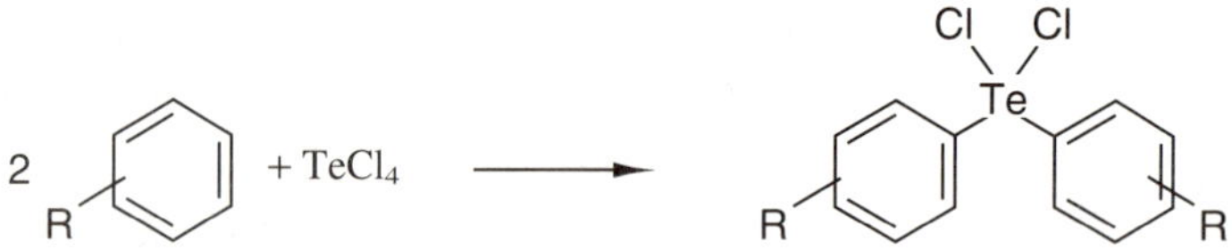

*Di-(*p*-methoxyphenyl)tellurium dichloride (typical procedure)* [14] Anisol (64.8 g, 0.6 mol) and $TeCl_4$ (27.0 g, 0.1 mol) are heated at 160°C for 6 h. The mixture is allowed to crystallize under vacuum (30.6 g (90%)). The product is recrystallized from MeCN as colourless crystals (m.p. 181–182°C).

In the presence of Lewis acids such as aluminium trichloride, diaryltellurium dichlorides are formed even with non-activated arenes [14, 16].

Diphenyltellurium dichloride (typical procedure) [14] $TeCl_4$ (27.0 g, 0.1 mol) and benzene (150 ml) are heated under reflux for 0.5 h, the solution is then cooled at 60°C and $AlCl_3$ (1.0 g) is added. Evolution of HCl ensues. The mixture is then refluxed for 12 h, cooled, filtered to remove the formed trichloride and inorganic impurities. After evaporation, the residue is recrystallized from MeCN (12.6 g (36%); m.p. 160–161°C).

3.9.2.4 With arylmercury chlorides

$$TeCl_4 + 2ArHgCl \xrightarrow[-HgCl_2 \cdot \text{dioxane}]{\text{dioxane}} ArTe(Cl)_2Ar$$

Diaryltellurium dichlorides (general procedure) [17] $TeCl_4$ (10.8 g, 40 mmol) and the arylmercury chloride (80 mmol) are heated under reflux in anhydrous dioxane (100 ml) for 2 h. The solution is cooled at 10°C and the $HgCl_2$·dioxane precipitate filtered off and washed with cold dioxane (15 ml). The filtrate is poured while stirring into 1% ice-cold HCl (400 ml). The dichlorides precipitate as crystalline solids (yields > 90%), and are recrystallized from benzene/petroleum ether 30–60°C (Ar = Ph, m.p. 160–162°C; *p*-MeC_6H_4, m.p. 165–166°C).

It must be pointed out that the reaction of tellurium tetrachloride with aryl Grignard or lithium reagents is inappropriate for the preparation of diorganyltellurium dichlorides due to the difficulty of stopping the reaction at the 1:2 stage.

3.9.2.5 With arenediazonium salts

$TeCl_4$ in concentrated hydrochloric acid reacts with arenediazonium chlorides, giving bis(arenediazonium) hexachlorotellurates that are decomposed by copper powder in acetone to give diaryltellurium dichlorides in low-to-medium yields [18].

$$TeCl_4 \xrightarrow[H_2O]{HCl} H_2TeCl_6 \xrightarrow{[ArN_2^+]Cl^-} [ArN_2^+]_2TeCl_6^{2-} \xrightarrow[\text{acetone}]{Cu} Ar_2TeCl_2$$

(21–55%)

Ar = Ph, *p*-MeO, *p*-Me, *p*-ClC_6H_4

*Bis(*p*-methoxyphenyl)tellurium dichloride (typical procedure)* $TeCl_4$ (27 g, 0.1 mol) or TeO_2 (16 g, 0.1 mol) dissolved in concentrated HCl (75 ml) is added to a stirred solution of $(p\text{-}MeOC_6H_4N_2)^+Cl^-$ (0.2 M, from aminobenzene (24.6 g, 0.23 mol), half-concentrated HCl (150 ml) and $NaNO_2$ (14 g, 0.21 mol) in H_2O

(20 ml)) at −5°C. The mixture is stirred at −5°C for 20 min and then filtered. The filter cake is washed successively with acetone (3 × 50 ml), ether and petroleum ether. The diazonium hexachlorotellurate product is dried in air (52.8 g (86%)).

Of the above product, 18.3 g (30 mmol) is carefully ground and suspended in acetone (95 ml). The suspension is vigorously stirred at −15°C while Cu powder (11.5 g, 180 mmol) is added over 40 min (the temperature must not exceed −10°C). The mixture is stirred for a further 20 min and is then allowed to warm at 20°C. The dark-coloured mixture is filtered to remove CuCl, the filter cake is washed with acetone, and the filtrate and washings are combined. The acetone solution is warmed at 20°C until a grey precipitate forms. The precipitate is collected and extracted repeatedly with boiling toluene. Activated charcoal is added to the toluene extracts, the mixture is heated at reflux and filtered, and most of the toluene is distilled from the filtrate. Petroleum ether is added to precipitate the product, $(p\text{-}MeOC_6H_4)_2TeCl_2$, which is recrystallized from benzene (6.8 g (52%); m.p. 182°C).

3.9.3 From organyltellurium trihalides

Similarly to tellurium tetrachloride, aryltellurium (and alkyltellurium) trichlorides (and also tribromides) undergo condensation reactions with reactive organic substrates or metallo-organic reagents as well as additional reactions to C=C bonds, giving rise to tellurium dichlorides.

3.9.3.1 With ketones

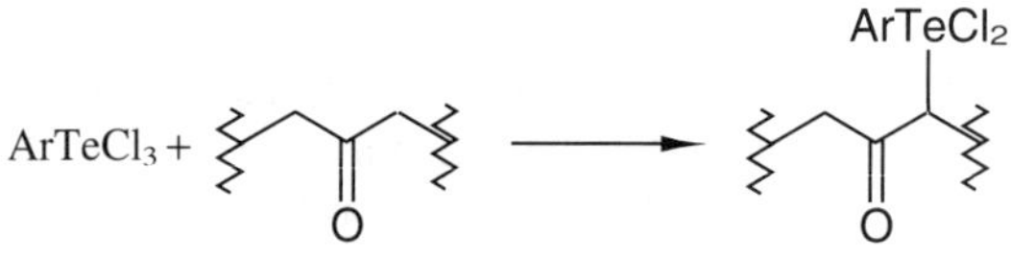

[19–22]*

Ar = p–MeO, p–EtO, p–$PhOC_6H_4$;
ketone = acetone, acetophenone, pinacolone, cycloalkanones

The reaction can be carried out by simply maintaining a mixture of the trichloride and excess ketone (1:3) at room temperature, or by refluxing equimolar amounts of the reagents in benzene. Alkyltellurium trichlorides give only low yields of the products.

*Aryltellurium tribromides are generally less reactive towards ketones. p-$MeOC_6H_4Te(Br_2)CH_2COPh$ was prepared by this method [19].

Improved reaction times and yields are achieved employing ketone silyl enol ether [20–22].

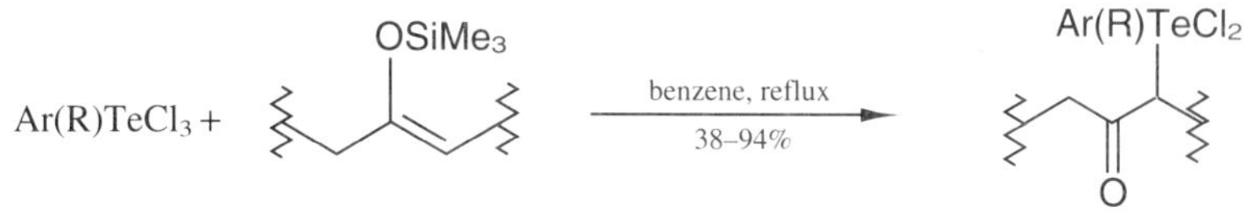

[Ar = p-MeOC$_6$H$_4$;]

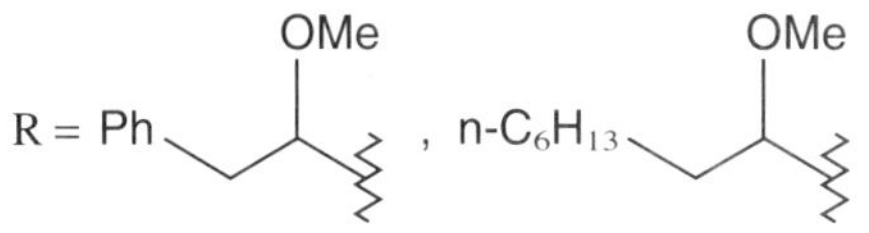

ketone = acetone, pinacolone, cycloalkanones

*(*p*-Methoxyphenyl)phenacyltellurium dichloride (typical procedure)* [19] *p*-Methoxyphenyltellurium trichloride (0.34 g, 1 mmol) and acetophenone (0.36 g, 3 mmol) are intimately mixed and allowed to stand for 48 h. During this time the product crystallizes and is then washed thoroughly with ether and successively with a small volume of cold methanol. The crude dichloride is recrystallized from benzene/petroleum ether 30–60°C (0.25 g (60%); m.p. 136–137°C).

*(*p*-Methoxyphenyl)pinacolyltellurium dichloride (typical procedure)* [21] A mixture of *p*-methoxyphenyltellurium trichloride (1.020 g, 3 mmol) and pinacolone trimethylsilyl enol ether (0.516 g, 3 mmol) in benzene (10 ml) is heated under reflux. The reaction is monitored by TLC. After 10 h the mixture is treated with MeOH/H_2O (1:1, 100 ml) and then extracted with $CHCl_3$, and the organic layer is then dried ($MgSO_4$) and evaporated. The residual oil is filtered on an SiO_2 pad (eluting with $CHCl_3$). The product is recrystallized from $CHCL_3$/petroleum ether 30–60°C (0.97 g (80%); m.p. 90–92°C).

3.9.3.2 With alkenes

The addition of aryltellurium trichlorides to alkenes occurs with a high stereospecificity, giving exclusively *anti*-2-chloroalkyltellurium dichlorides [23–26], in contrast to the addition of tellurium tetrachloride which gives mixtures of *syn* and *anti* adducts (see Section 3.5.1.2).

Ar = 2-naphthyl; olefin = propene, (*E*)- and (*Z*)-2-butenes, 1-decene, (*E*)-10-1-hexane, cyclopentene, cyclohexene

2-Chloroalkyl-2-naphthyltellurium dichloride (general procedure) [24] 2-Naphthyltellurium trichloride (2.98 g, 0.83 mmol) is heated under reflux with the olefin (1.2 mmol) in dry, ethanol-free $CHCl_3$ (15 ml) until the trichloride has dissolved (0.5–1 h). Filtration from a small amount of elemental Te and evaporation gives an oil or a semi-solid that is recrystallized from a large amount of petroleum ether 40–60°C. The reactions with propene and 2-butenes are carried out in a sealed tube at 80°C. The yields are in the range 70–95%.

In the presence of alcohols the addition of aryltellurium trihalides to alkenes gives rise to alkoxytelluration products [27–29]. The addition is highly *anti*-stereospecific and regiospecific for terminal alkenes, tellurium species attacking exclusively the terminal carbon.

The reaction is usually carried out with aryltellurium tribromides, generated *in situ* by the treatment of the corresponding ditellurides with bromine (although trichlorides, ditellurides/$CuCl_2$ and tellurocyanates/$CuCl_2$ are also effective).

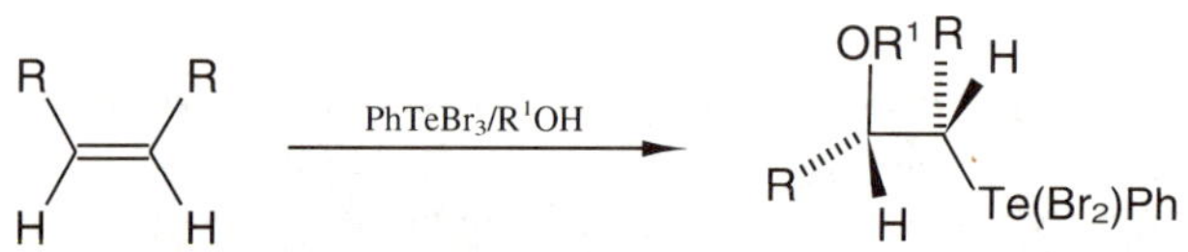

olefin = 1-hexene, 1-octene, 1-decene, styrene, α-methylstyrene, 2-methyl-1-pentene, isobutylene, cyclopentene, cyclohexene, cycloheptene, cyclooctene

(2-Methoxycyclohexyl)phenyltellurium dibromide (typical procedure) [28, 29] To a solution of phenyltellurium tribromide (from diphenyl ditelluride (1.02 g, 2.5 mmol) and Br_2 (1.20 g, 7.5 mmol) in MeOH (5 ml) is added cyclohexene (0.49 g, 6.0 mmol) and the mixture is heated under reflux for 1 h. By cooling the yellow solution at room temperature, the dibromide precipitates (1.04 g (46%); m.p. 180–181°C).

3.9.3.3 With arenes

Aryltellurium trichlorides and tribromides condense with aromatic compounds bearing electron-donating groups, giving unsymmetrical diaryltellurium dihalides [19]. This electrophilic substitution therefore parallels that described for tellurium tetrachloride.

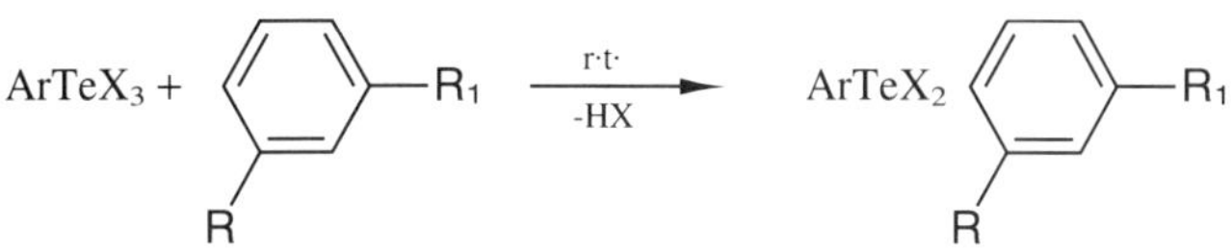

representative examples:

X = Cl, Br; Ar = *p*-MeO, *p*-EtO, *p*-$PhOC_6H_4$;

R = H; R^1 = NMe_2; R = R^1 = OH

p-*Methoxyphenyl*-p-*dimethylaminophenyltellurium dichloride (typical procedure)* [19] *p*-Methoxyphenyltellurium trichloride (0.34 g, 1 mmol) and *N,N*-dimethylaniline (0.36 g, 3 mmol) are intimately mixed. After 24 h at room temperature, the mixture is extracted several times with MeOH leaving a yellow crystalline residue (0.32 g (75%)). Recrystallization from MeOH or benzene/MeOH gives the product as yellow prisms (m.p. 170–172°C (dec.)).

3.9.3.4 With organylmercury chlorides

The reaction of aryltellurium trichlorides with aryl- or alkylmercury chlorides is a well-established and general method for the preparation of unsymmetrical diaryl- or alkylaryltellurium dichlorides [31, 32].

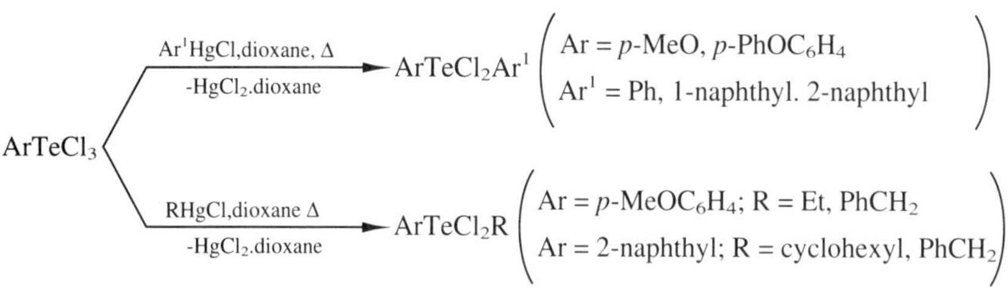

Unsymmetrical diaryltellurium dichlorides (general procedure) [32] The aryltellurium trichloride (20 mmol) and the arylmercury chloride (20 mmol) are heated under reflux in dry dioxane (30 ml) for 1.5 h. The solution is cooled at 10°C and the $HgCl_2$·dioxane precipitate so formed is filtered off and washed with a small volume of cold dioxane. The dioxane solution is then poured into 1% HCl (100 ml) with vigorous stirring. The oily product formed crystallizes by friction and is separated by filtration and then washed with cold EtOH.

The reaction with ethyl- and cyclohexylmercury chlorides requires heating for 24 h.

3.9.4 By addition of halogens to diorganyl tellurides

Diorganyl tellurides, which as shown in Section 3.1 are the primary products of several methods for the synthesis of organotellurium compounds, can easily be converted to the corresponding dihalides by the simple addition of halogens.

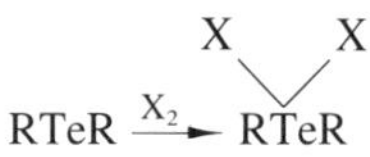

X = Cl, Br, I

This reaction has a general character and can be applied to the different types of tellurides [33]. The halogenation is accomplished in inert solvents (ether, benzene, dichloromethane, chloroform) and the dihalides are normally easily crystallizable products.

The chlorination of tellurides is more conveniently carried out employing, instead of chlorine (gaseous or dissolved in solvents), chlorinating reagents such as sulphuryl or thionyl chloride (in analogy to the chlorinolysis of ditellurides, see Section 3.5.2).

Diaryltellurium dihalides (general procedure) [32] The diaryl telluride dissolved in a small amount of benzene or $CHCl_3$ is treated dropwise, while cooling (ice bath) and stirring, with a solution of an equimolar amount of SO_2Cl_2, Br_2 or I_2 in the same solvent (in the case of SO_2Cl_2, evolution of SO_2 is observed). By addition of a large excess of petroleum ether 30–60°C (50 ml for 1 mmol) the dihalide precipitates as a crystalline solid (the dichlorides are colourless, the dibromides yellow and the diiodides red). The yields are quantitative.

Representative examples of diaryltellurium dihalides prepared by the above method are:

$ArTeX_2Ar'$	Ar = Ph; Ar′ = *p*-MeO, *p*-$EtOC_6H_4$, 1- and 2-naphthyl
X = Cl, Br, I	Ar = *p*-MeO, *p*-$EtOC_6H_4$; Ar′ = 1- and 2-naphthyl
	Ar = 1-naphthyl; Ar′ = 2-naphthyl

Dibenzyl and arylbenzyl tellurides (in analogy to dibenzyl ditelluride, see Section 3.5.3) exhibit an unusual behaviour towards halogens, undergoing cleavage of the tellurium–benzyl bond [33–35].

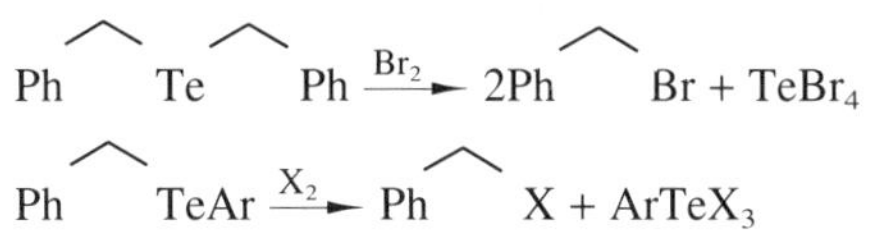

X = Cl, Br, I

3.9.5 Additional methods

3.9.5.1 Reaction of elemental tellurium with arenediazonium salts [36]

$$[2ArN_2^+]Cl^- + Te \xrightarrow[-N_2]{} Ar_2TeCl_2$$

3.9.5.2 Reaction of TeO_2/LiCl with aryl hydrazines [37]

$$2ArNHNH_2 + 2TeO_2 + 2LiCl \longrightarrow ArTeCl_2 + 2N_2 + 2H_2O + 2LiOH + Te$$

3.9.5.3 Reaction of diphenyl ditelluride with arenediazonium salts/$CuCl_2$ [38]

$$(PhTe)_2 + 2[RC_6H_4N_2^+]Cl^- \xrightarrow[-N_2]{} 2PhTeC_6H_4R \xrightarrow{CuCl_2} 2PhTeCl_2C_6H_4R$$

REFERENCES

1. H. Rheinboldt, in *Houben–Weyl—Methoden der Organischen Chemie* (ed. E. Muller), 4th edn, Vol. IX, p. 1066. Georg Thieme, Stuttgart, 1955.
2. J. S. Thayer and K.V. Smith, *Synth. Inorg. Met. Org. Chem.* **3**, 101 (1973); see also K. Y. Irgolic, *Houben–Weyl—Methoden der Organischen Chemie*, 4th edn, Vol. E12b, pp. 525 and 526. Georg Thieme, Stuttgart, 1990.
3. R. F. Ziolo and W. H. H. Gunther, *J. Organomet. Chem.* **146**, 245 (1978).
4. G. T. Morgan and O. C. Elvins, *J. Chem. Soc.* **127**, 2625 (1925).

5. H. Rheinboldt, in *Houben–Weyl—Methoden der Organischen Chemie* (ed. E. Muller), 4th edn, Vol. IX, p. 1061. Georg Thieme, Stuttgart, 1955.
6. G. T. Morgan and H. D. K. Drew, *J. Chem. Soc.* **127**, 531 (1925).
7. H. Funk and W. Weiss, *J. Prakt. Chem.* **1**, 33 (1954).
8. H. J. Arpe and H. Kuckertz, *Angew. Chem. (Engl.)* **10**, 73 (1971).
9. M. Ogawa and R. Ishioka, *Bull. Chem. Soc. Jpn* **43**, 496 (1970).
10. M. Ogawa, *Bull. Chem. Soc. Jpn* **41**, 3031 (1968).
11. J. Bergman and L. Engman, *J. Am. Chem. Soc.* **103**, 5196 (1981).
12. G. T. Morgan and R. E. Kellett, *J. Chem. Soc.*, 1080 (1926).
13. G. T. Morgan and R. D. Drew, *J. Chem. Soc.* **127**, 2307 (1925).
14. J. Bergman, *Tetrahedron* **28**, 3323 (1972).
15. K. Y. Irgolic, *Houben–Weyl—Methoden der Organischen Chemie*, 4th edn, Vol. E12b, p. 529, Table 16. Georg Thieme, Stuttgart, 1990.
16. W. H. H. Gunther, J. Nepywoda and J. Y. C. Chu, *J. Organomet. Chem.* **74**, 79 (1974).
17. N. Petragnani, J. V. Comasseto and N. H. Varella, *J. Organomet. Chem.* **120**, 375 (1976).
18. L. D. Sadekov and V. I. Minkin, *Dokl. Akad. Nauk. SSSR* **197**, 316 (1971) (in English).
19. N. Petragnani, *Tetrahedron* **12**, 219 (1961).
20. I. D. Sadekov, A. A. Maksimenko and B. B. Rivkin, *J. Org. Chem. USSR* **14**, 810 (1978) (in English).
21. H. A. Stefani, J. V. Comasseto and N. Petragnani, *Synth. Commun.* **17**, 443 (1987).
22. H. A. Stefani, A. Chieffi and J. V. Comasseto, *Organometallics* **10**, 1176 (1991).
23. M. Moura Campos and N. Petragnani, *Tetrahedron* **18**, 521 (1962).
24. J. E. Backwall, J. Bergman and C. Engman, *J. Org. Chem.* **48**, 3918 (1983).
25. J. Bergman and L. Engman, *Tetrahedron* **36**, 1275 (1980).
26. S. Uemura and S. Fukuzawa, *J. Organomet. Chem.* **268**, 223 (1984).
27. S. Uemura, S. Fukuzawa, A. Toshimitsu and M. Okano, *Tetrahedron Lett.* **23**, 1177 (1982).
28. S. Uemura, S. Fukuzawa and A. Toshimitsu, *J. Organomet. Chem.* **250**, 203 (1983).
29. S. Uemura and S. Fukuzawa, *J. Am. Chem. Soc.* **105**, 2748 (1983).
30. S. Uemura and S. Fukuzawa, *J. Chem. Soc. Perkin Trans. I*, 471 (1985).
31. N. Petragnani, *Tetrahedron* **12**, 219 (1961).
32. H. Rheinboldt and G. Vicentini, *Chem. Ber.* **89**, 624 (1956).
33. For a large literature coverage upon addition of halogens to organyl tellurides, see K. Y. Irgolic, *Houben–Weyl–Methoden der Organischen Chemie*, 4th edn, Vol. E12b, Section 5.1, p. 553. Georg Thieme, Stuttgart, 1990.
34. G. Vicentini, *Chem. Ber.* **91**, 801 (1958).
35. H. K. Spencer and M. P. Cava, *J. Org. Chem.* **42**, 2937 (1977).
36. H. Taniyama, F. Miyoshi, E. Sakakibara and H. Uchida, *Yakugaku Zasshi* **77**, 191 (1957); H. Taniyama, F. Miyoshi, E. Sakakibara and H. Uchida, *Chem. Abstr.* **51**, 10407 (1957).
37. J. Bergman and L. Engman, *Z. Naturforsch B* **35**, 882 (1980).

3.10 DIORGANYL TELLUROXIDES

Diorganyl tellurides are easily oxidized to the corresponding telluroxides. Dialkyl tellurides are especially sensitive to oxidation, and their exposure to air, either neat or in solution, causes the formation of dialkyl telluroxides. Diaryl tellurides are much more stable and are oxidized only by chemical methods.

Diaryl telluroxides have recently attained outstanding importance as mild and selective oxidizing reagents (see Section 4.5).

Two methods are well established and generally employed to prepare diaryl telluroxides.

3.10.1 Hydrolysis of diaryltellurium dihalides

This method, which dates from the early German tellurium chemistry [1] and is today widely employed [2], uses aqueous sodium hydroxide or ammonia as hydrolytic reagents.

$$Ar_2TeX_2 \xrightarrow[H_2O]{OH^-} Ar_2TeO$$

Di-p-anisyl telluroxide (typical procedure) [2] Di-*p*-anisyltellurium dichloride (8.00 g, 23 mmol) is stirred at 95°C in 5% aqueous NaOH (100 ml) for 1 h. After cooling at 0°C the white solid is separated by filtration, washed with cold H_2O (3 × 10 ml) and dried under vacuum over P_2O_5, giving the telluroxide (5.60 g (81%); m.p. 187–189°C).

3.10.2 Oxidation of diaryl tellurides

Sodium periodate [3] or *N*-haloamides [4] are useful oxidizing reagents.

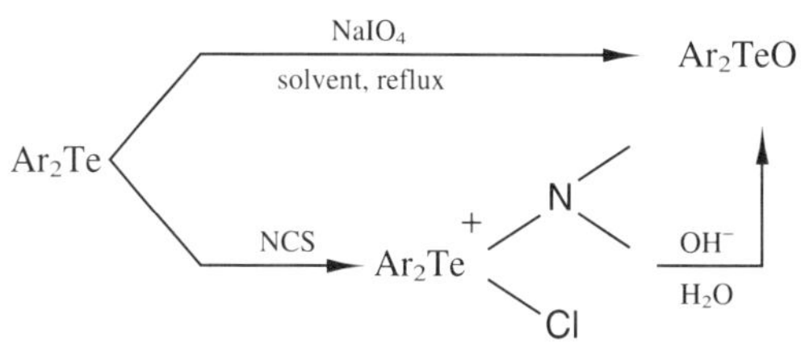

Di-p-anisyl telluroxide (typical procedure) [3] Di-*p*-anisyl telluride (0.80 g, 2.3 mmol) and $NaIO_4$ (0.50 g, 2.3 mmol) in $CH_3CN/MeOH/H_2O$ (100 ml, 1:1:1) are heated under reflux for 2 h. The mixture is then cooled, treated with H_2O (100 ml) and extracted with $CHCl_3$ (3 × 30 ml). The extracts are evaporated and the residue chromatographed on SiO_2 (elution with $CHCl_3$/MeOH 9:1). The crude product is recrystallized from benzene (0.26 g (22%); m.p. 189–191°C).

Di-$(2,4,6\text{-}Me_3C_6H_2)_2TeO$ can also be prepared by the above procedure (yield 95%).

REFERENCES

1. K. Lederer, *Ber. Dtsch. Chem. Ges.* **49**, 345, 2532, 2663 (1916); K. Lederer, *Ber. Dtsch Chem. Ges.* **53B**, 1674 (1920).
2. S. V. Ley, C. A. Meerholz and D. H. R. Barton, *Tetrahedron* **37**, 213 (1981).
3. M. Akiba, M. V. Lakshmikantham, K. Y. Jen and M. P. Cava, *J. Org. Chem.* **49**, 4819 (1984).
4. M. R. Detty, *J. Org. Chem.* **45**, 274 (1980).

3.11 VINYLIC TELLURIDES AND DITELLURIDES

These classes of tellurium compounds are treated separately due to their peculiar behaviour as synthons and intermediates of fundamental organic transformations.

3.11.1 Starting from electrophilic tellurium

3.11.1.1 From organyltellurenyl halides and vinylic Grignard reagents

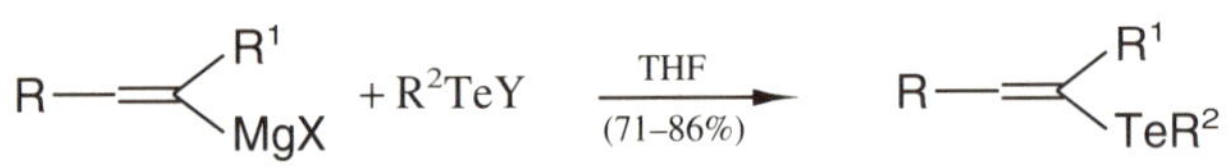

$R = R^1 = H$ { $R^2 = Ph$; $Y = I$ [1]; $R^2 = p\text{-}MeOC_6H_4$; $Y = Br$ [2] }

$R = Ph$; $R^1 = H$; $R = H$; $R^1 = Me$ } $R^2 = p\text{-}MeOC_6H_4$; $Y = Br$ [2]

This reaction is an application of the general method for the preparation of unsymmetrical tellurides (see Section 3.1.3.5). The tellurenyl halide is prepared *in situ* by halogenolysis of the corresponding ditelluride.

Vinylic tellurides (general procedure) [2] A solution of the organotellurenyl bromide, prepared by adding Br_2 (0.40 g, 2.5 mmol) in benzene (5.0 ml) to the corresponding ditelluride (2.5 mmol) in THF (10 ml) at 0°C under N_2, is added dropwise to a solution of the Grignard reagent (7.0 mmol) in THF (10 ml) at 0°C under N_2. A gradual disappearance of the red colour of the reaction mixture is observed. After stirring the mixture for 1 h at 0°C, the solvent is evaporated, and the residue is treated with saturated aqueous NH_4Cl, extracted with ether, dried ($MgSO_4$) and evaporated. The product is purified by SiO_2 column chromatography (elution with benzene/hexane 9:1).

3.11.1.2 From vinyltellurenyl iodides and Grignard reagents

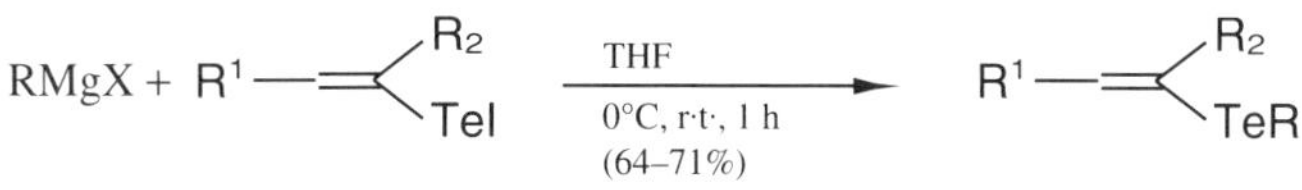

[2]

R = p-MeOC$_6$H$_4$; n-Bu; R^1 = R^2 = H

R = p-MeOC$_6$H$_4$; R^1 = H; R^2 = Me

3.11.2 Starting from nucleophilic tellurium

3.11.2.1 From organyl tellurolate (and telluride) anions and vinyl bromides

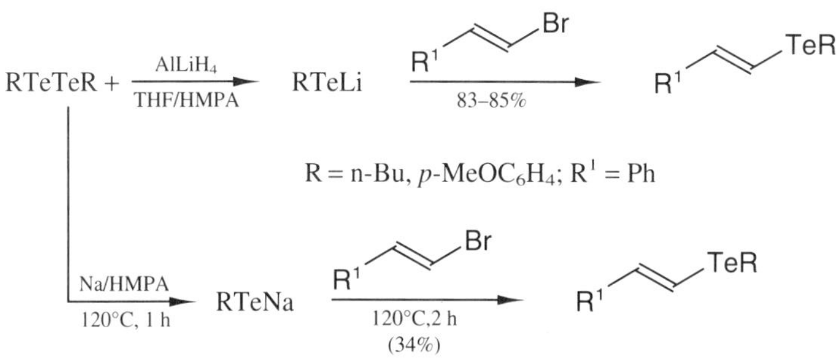

[2]

R = n-Bu, p-MeOC$_6$H$_4$; R^1 = Ph

[3]

R = R^1 = Ph

The reactions of lithium n-butyl and p-methoxyphenyl tellurolate with *trans*-β-bromostyrene require, respectively, a 15 min and a 3 h reflux in THF, indicating a greater nucleophilicity of the aliphatic tellurolate towards the aromatic tellurolate.

Vinylic tellurides (general procedure) [2] A solution of the ditelluride (2.0 mmol) and *trans*-β-bromostyrene (0.73 g, 4.0 mmol) in THF/HMPA (10 ml) is added dropwise at room temperature and under N_2 to $AlLiH_4$ (0.17 g, 4.5 mmol) in THF (2 ml). The dark red colour of the ditelluride disappears at once. The resulting pale green solution is refluxed (R = n-Bu, 15 min; R = p-MeOC$_6$H$_4$, 3 h). After cooling to room temperature the reaction mixture is treated with H_2O (1.0 ml) and 10% NaOH (1.0 ml) and then extracted with ether. The extract is washed with brine, dried ($MgSO_4$) and evaporated. The residue is purified by recrystallization or Kugelrohr distillation.

Sodium telluride, obtained by heating elemental tellurium with sodium hydride in DMF, reacts similarly, leading to symmetrical divinylic tellurides [4] (compare the similar reaction with aryl iodides, see Section 3.1.2.1).

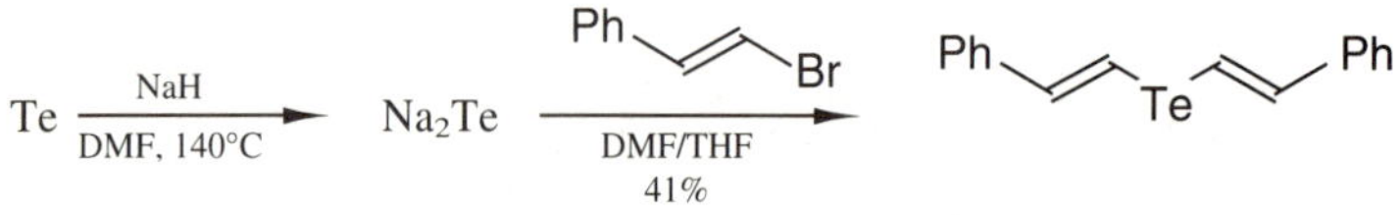

3.11.2.2 From vinylic tellurolate anions and alkyl halides

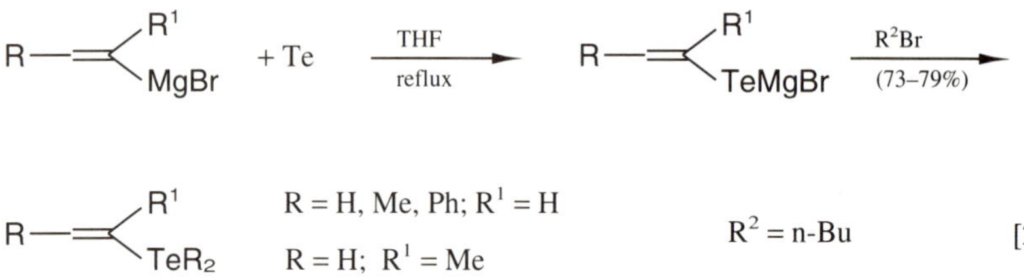

This reaction is an application of the general method for the preparation of unsymmetrical tellurides (see Section 3.1.3).

Vinylic tellurides (general procedure) [2] Elemental Te (0.6 g, 5.0 mmol) is added to a solution of the vinylic magnesium bromide (5.5 mmol) in THF (10 ml) under reflux and an N_2 atmosphere, and the reflux maintained for 20 min. The mixture is allowed to reach room temperature and then treated with n-butyl bromide (0.7 g, 5.0 mmol). After stirring for 10 min, the reaction mixture is cooled at 0°C, treated dropwise with saturated aqueous NH_4Cl, extracted with ether, dried ($MgSO_4$) and then evaporated. Kugelrohr distillation of the residue under vacuum gives the vinyl alkyl tellurides as yellow liquids.

By treating the vinylic magnesium tellurolates with air while stirring, instead of with alkyl bromides, divinylic ditellurides are formed [2, 5].*

*If the vinylic magnesium tellurolate is hydrolysed prior to the oxidation, a complex mixture of products is formed.

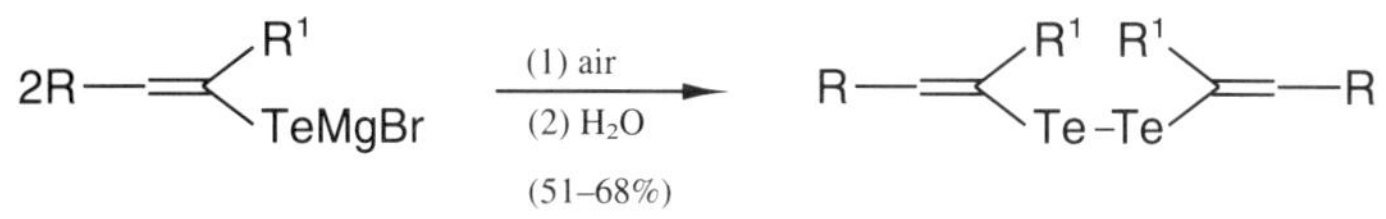

R = Ph, Me, H; R^1 = H

R = H; R^1 = Me

Divinylic ditellurides (general procedure) [5] Elemental Te (12.7 g, 100 mmol) is added at once to a stirred solution of the vinylic magnesium bromide (0.11 mol) in THF (150 ml) under N_2. The mixture is refluxed for 30 min with stirring, allowed to reach room temperature, then stirred for 1 h in the presence of air (by opening the apparatus). Deposition of some Te is observed during the oxidation process. The mixture is treated with brine. The ditelluride is extracted with ether (3 × 50 ml), and the extract dried ($MgSO_4$) and evaporated. The residue is purified by SiO_2 flash chromatography.

3.11.2.3 From divinylic ditellurides

The divinylic ditellurides, obtained as described in the preceding section, are converted into alkyl vinyl tellurides by reduction with $AlLiH_4$ followed by alkylation [5].

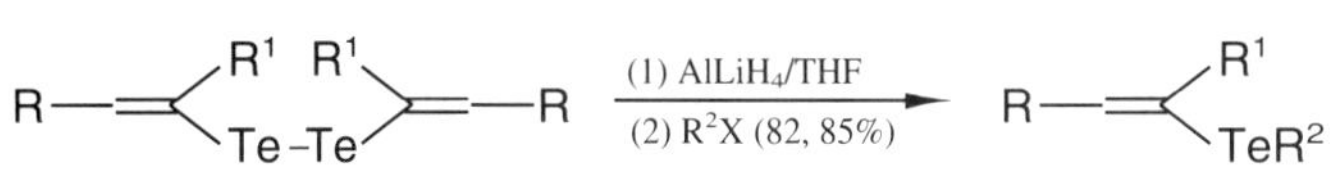

R = H; R^1 = Me
R = Me; R^1 = H
R^2 = n-Bu

Alkyl vinyl tellurides (general procedure) [5] A solution of the divinylic ditelluride (1.0 mmol) in THF (6 ml) is added dropwise, at room temperature under N_2, to $AlLiH_4$ (0.09, 2.5 mmol) in THF (3 ml). The dark red colour of the ditelluride disappears. The resulting mixture is treated with butyl bromide (0.3 g, 2.2 mmol) and stirred for 15 min. Then H_2O (0.5 ml), 10% aqueous NaOH (0.5 ml) and H_2O (0.5 ml) are added in sequence. The product is extracted with ether (3 × 10 ml). The extract is washed with brine, dried ($MgSO_4$) and evaporated, giving the product as a yellow oil.

3.11.2.4 From organyltellurols and terminal acetylenes

Alkyl- and aryltellurols generated *in situ* by the well-established reduction of ditellurides with $NaBH_4$/EtOH add to terminal acetylenes, giving (*Z*)-vinylic tellurides [2, 4, 6–9].

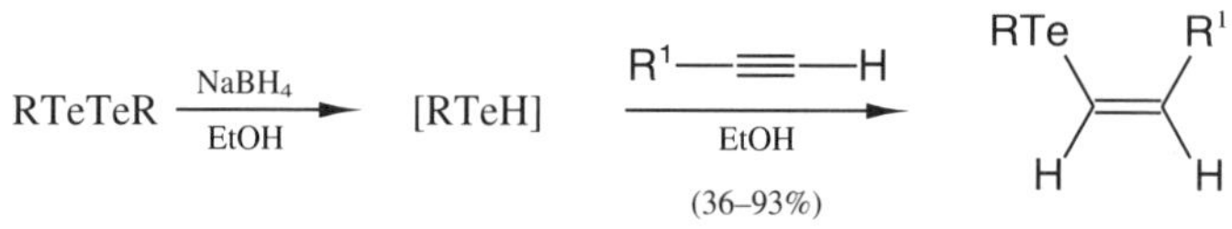

R = R^1 = Ph R = n-Bu; R^1 = Ph, CH_2OTHP,

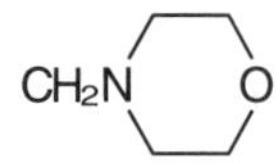

R = *p*-$MeOC_6H_4$; R^1 = n-pentyl

R = Ph; R^1 = n-hexyl, CH_2OH, CO_2Et

R = Ph, *p*-Me, *p*-MeO, *p*-Cl, *m*-$H_2NC_6H_4$; R^1 = Ph

The addition of the tellurium occurs exclusively at the terminal carbon of phenylacetylene, whereas a minor amount (11–28%) of the 2,2-disubstituted adduct ($RTeR^1C{=}CH_2$) is formed in addition to alkylacetylenes.

1-Phenyltelluro-2-phenylethene (typical procedure) [7] To a solution of diphenyl ditelluride (2.05 g, 5 mmol) in EtOH (10 ml) is added a solution of $NaBH_4$ (0.50 g, 13.2 mmol) in EtOH (10 ml) at room temperature under N_2. The orange colour of the solution changes to pale yellow. Phenylacetylene (1.1 g, 10.8 mmol) is then added to the resulting solution and the mixture is stirred under reflux for 20 h. After cooling, the mixture is treated with brine and extracted with $CHCl_3$ (3 × 50 ml) and the extract dried ($MgSO_4$). Evaporation of the solvent gives a yellow-orange oil, which is purified by column chromatography on SiO_2 (elution with hexane/EtOAc 10:1), giving the crude product (R = R^1 = Ph) as a yellow solid (2.15 g (70%)). It is recrystallized from EtOH (m.p. 43–44°C).

The above-described hydrotelluration is successfully applied to conjugated enines and dialkynes, providing a useful method for the synthesis of organotelluro-1,3-butadienes [8] and organotelluro-1,3-enines [10] with the (*Z*) configuration at the newly formed C=C bond.

1-Organyltelluro-1,3-butadienes (general procedure) [8] To a solution of the conjugated enyne (4.0 mmol) and the ditelluride (1.0 mmol) in EtOH (6.0 ml) at room temperature under N_2 is added $NaBH_4$ (0.09 g, 2.5 mmol) in small portions. Towards the end of the addition, when the red colour has disappeared, the mixture is refluxed for 5 h, successively cooled at room temperature, treated with H_2O (0.5 ml) and then with 10% aqueous NaOH (0.5 ml). The resulting mixture is diluted with ether (50 ml) and washed with brine (3 × 30 ml). The organic layer is separated, dried ($MgSO_4$), and evaporated in a rotary evaporator

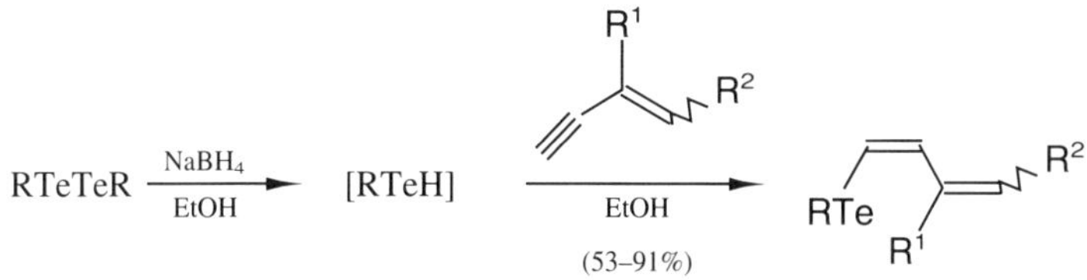

R = n-$C_{12}H_{25}$, *p*-$MeOC_6H_4$; enine =

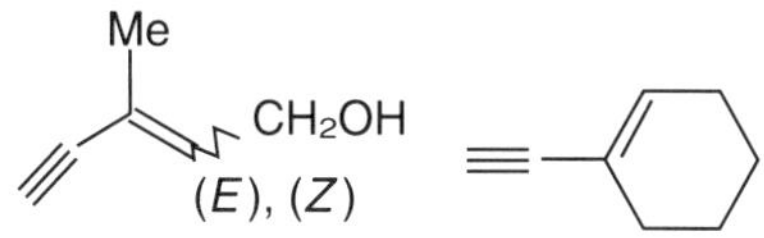

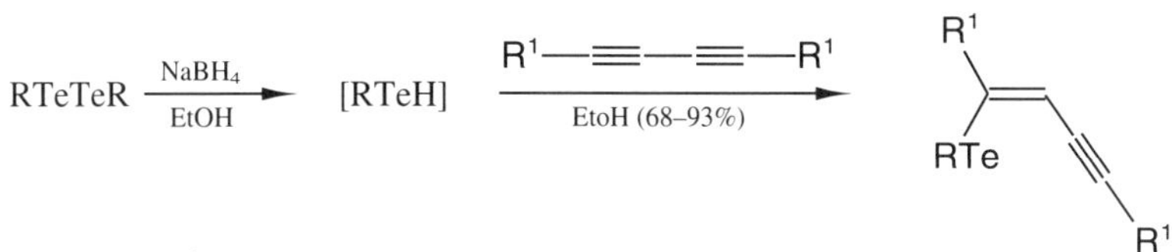

R = n-Bu; R^1 = H, CH_2OH, Me, Ph, *p*-Me, *p*-$MeOC_6H_4$

(20 torr). The residue is purified by SiO_2 flash chromatography (elution with heptane or heptane/EtOAc).

(Z)-1-Butyltelluro-1,4-(bis-hydroxymethyl)but-1-en-3-yne (typical procedure) [10] To a solution of (1,6-dihydroxy)hexa-2,4-diyne (0.11 g, 1.0 mmol) and dibutyl ditelluride (0.184 g, 0.5 mmol) in 95% EtOH (20 ml) under N_2 is added $NaBH_4$ (0.045 g, 1.25 mmol) at room temperature. After the disappearance of the red colour, the yellow solution is refluxed for 15 min, allowed to reach room temperature, diluted with EtOAc (40 ml) and washed with brine (3 × 20 ml). After drying the organic phase ($MgSO_4$), the solvent is evaporated under vacuum and the residue purified by SiO_2 flash chromatography (elution with hexane/EtOAc 7:3) to give the product (0.21 g (71%)).

Another application of the above method involves the addition of sodium vinylic tellurolates, generated by the reduction of the corresponding ditelluride, to terminal acetylenes, giving unsymmetrical divinylic tellurides [8].

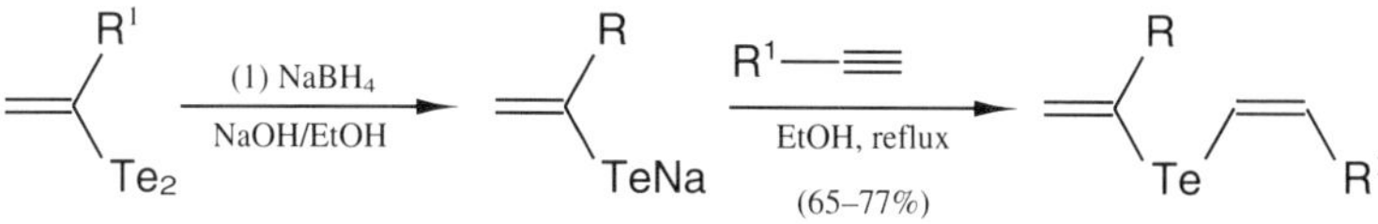

R = H; R^1 = *p*-ClC_6H_4 R = Me; R^1 = Ph, *p*-BrC_6H_4

In neutral conditions, the reduction of the divinylic ditellurides to the corresponding dialkyl ditellurides occurs preferentially to the hydrotelluration reaction [5].

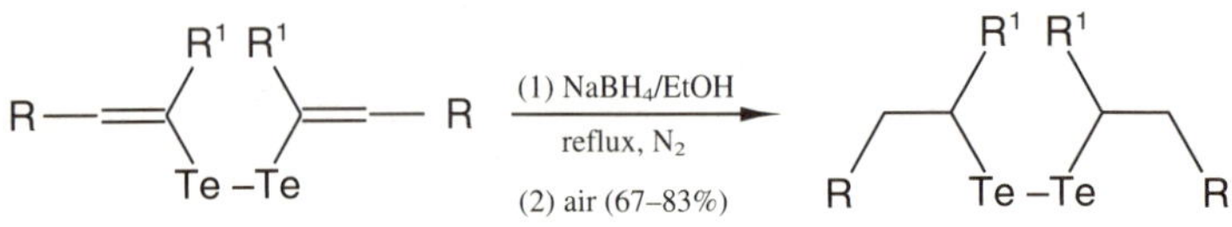

R = R[1] = H

R = H; R[1] = Me

R = Me; R[1] = H

The actual reducing agent of the C=C bond seems to be the intermediate tellurol, present if the hydrogenolysis of the Te—Te bond is effected in a neutral medium. In the presence of OH^- the tellurol is converted to tellurolate, which does not react with the double bond.

Symmetrical divinylic tellurides are prepared by a similar approach starting from alkali tellurides.

By treatment of acetylene with the triad Te/KOH/DMSO or, more effectively, Te/KOH/HMPA/H_2O at 110–120°C, divinyl telluride is obtained in a medium yield [11, 12].

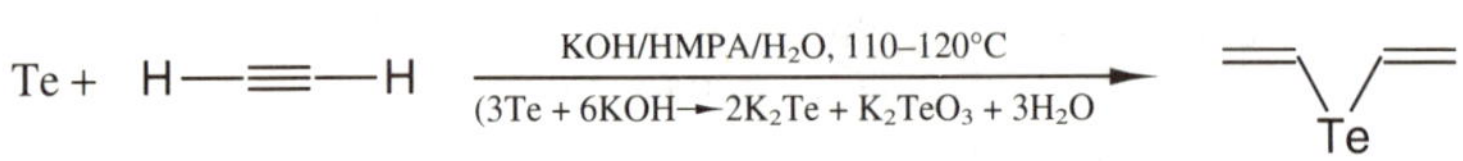

Vinylacetylene reacts similarly, the isomer ratio of the formed bis-1-(1,3-butadienyl) telluride depending on the reaction temperature and isolation procedure.

The yield of divinyl telluride is markedly improved if the reaction is carried out in the presence of the reducing agent $SnCl_2$ [12, 13].*

*A minor amount of more complex hydroxylated by-products are formed such as

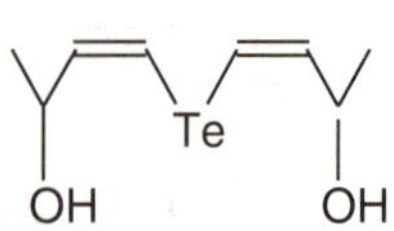

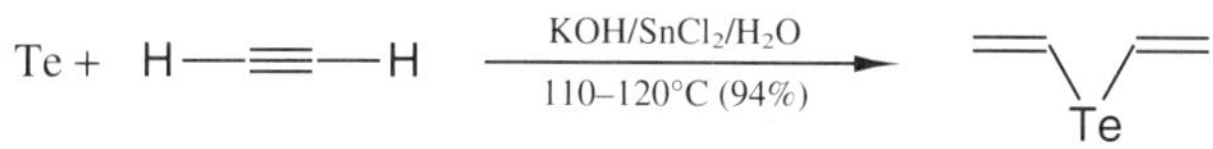

A method of general application employs the familiar system Te/$NaBH_4$/NaOH/EtOH/H_2O to generate the telluride anion [8].

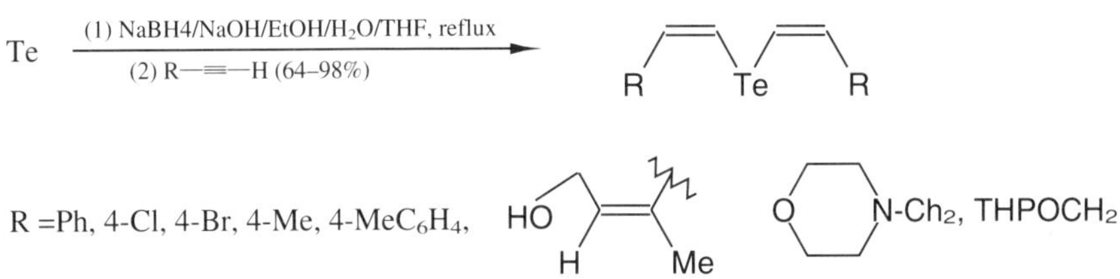

Symmetrical divinylic tellurides (general procedure) [8] $NaBH_4$ (0.34 g, 9 mmol) is added in small portions to a stirred suspension of Te (0.383 g, 3 mmol) in EtOH (15 ml) under N_2. The mixture is heated. NaOH (0.32 g, 8 mmol), H_2O (15 ml) and THF (5 ml) are added to the hot mixture, which is then refluxed. After 15 min the mixture turns dark violet and the reflux is continued until all the Te has disappeared. The heat source is removed and the acetylene (7.8 mmol) added. The mixture is then refluxed and monitored by TLC. When the colour changes to yellow, the mixture is cooled at room temperature, diluted with EtOAc (50 ml) and washed first with saturated aqueous NH_4Cl (3 × 30 ml), and then with brine (3 × 30 ml). The organic phase is separated, dried ($MgSO_4$) and evaporated in a rotary evaporator (20 torr). The residue is purified by SiO_2 flash chromatography (elution with hexane/EtOAc 7:3).

3.11.2.5 From aryl tellurolates and electrophilic acetylenes

Aryl tellurolate anions effect 1,4 additions to acetylenes bearing electron-withdrawing groups such as acetylenic ketones, aldehydes and esters, as well as diacetylenic ketones, giving aryl 2-carbonylvinyl tellurides with the (*Z*)-configuration* [12, 14–16].

*The authors propose the tellurolate anions ArTeNa as the reacting species, although aryltellurols ArTeH are probably operating, since no NaOH is added to the reaction mixture [6, 7].

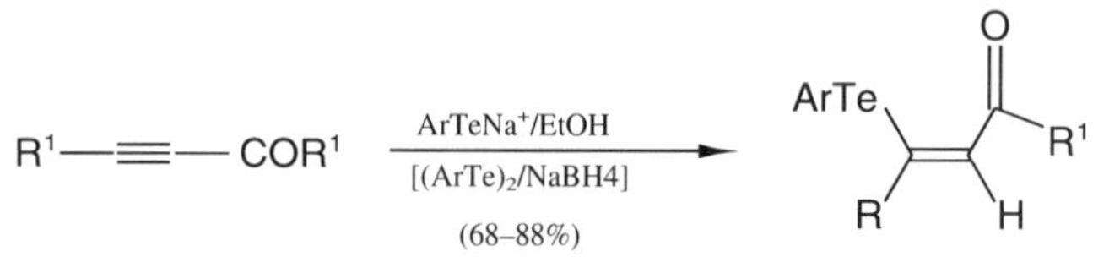

R = H, n-Bu, Ph; R^1 = Ph, H, OEt, OMe; Ar = Ph, *p*-MeO-, *p*-MeC_6H_4

R = Ph; R^1 = ≡—Ph

Addition of aryl tellurolates to acetylenic ketones (general procedure) [14] $NaBH_4$ (0.55 g, 15 mmol) is added in small portions, under N_2, to a solution of the diaryl ditelluride (5 mmol) in EtOH (20 ml) at room temperature. A solution of the acetylenic ketone (10 mmol) in EtOH (10 ml) is then added to the reaction mixture, with stirring. The resulting precipitate of the 2-carbonylvinyl telluride is filtered off and recrystallized from EtOH.

3.11.3 Via radical reactions

Diorganyl tellurides add to alkynes via a radical mechanism in the presence of a catalytic amount of a radical initiator such as 2,2′-azobisisobutyronitrile (AIBN) or 2,2′-azobis(4-methoxy-2,4-dimethylvaleronitrile) (In), as well as under irradiation by visible light (tungsten lamp), affording vinylic tellurides in a variable (*Z*)/(*E*) ratio [17, 18].

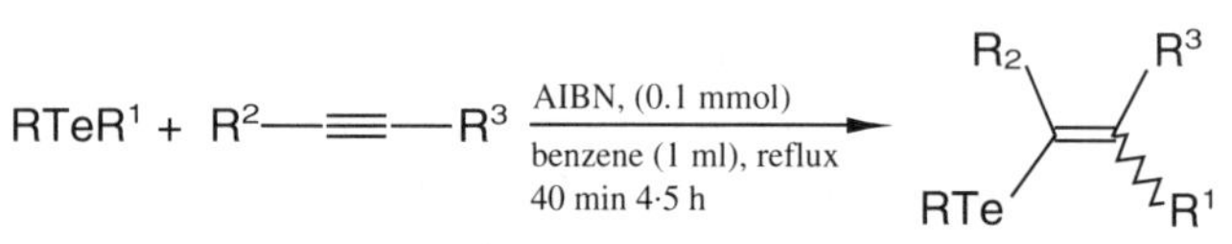

Telluride	Alkyne (percentage yield, (*E*)/(*Z*) ratio)
PhTePr-i	R^3 = H; R^2 = Ph (97, 56/44), Ac (99, 19/81), MeO_2C (93, 79/21); $R^2 = R^3 = CO_2Et$ (95, 37/63)
PhTeBu-n	Ac—≡—H (73, 20/80)
n-BuTeBu-t	R^3 = H; R^2 = Ac (84, 88/12), n-hexyl (42, 80/20), Me_3Si (77, 0/100), Ph (93, 100/0[a]) R^2, R^3 = $(CH_2)_6$ (52)
n-BuTeCH_2Ph	Ac—≡—H (91, 56/44)

[a] Catalyst = In; the adduct is also obtained by irradiation (tungsten lamp), 50°C, 3 h, neat (84%, 100/0).

The examination of the structure of the formed products makes it evident that the fission of the telluride bonds occurs exclusively at $C(sp^3)$—Te rather than $C(sp^2)$—Te bonds, at tertiary C—Te rather than primary C—Te bonds and at $PhCH_2$—Te rather than n-Bu—Te bonds.

Diphenyl ditelluride, on irradiation with visible light, effects the *vic*-phenyltelluration of acetylenes, giving the corresponding bis-phenyltelluroalkenes in good yields [19].

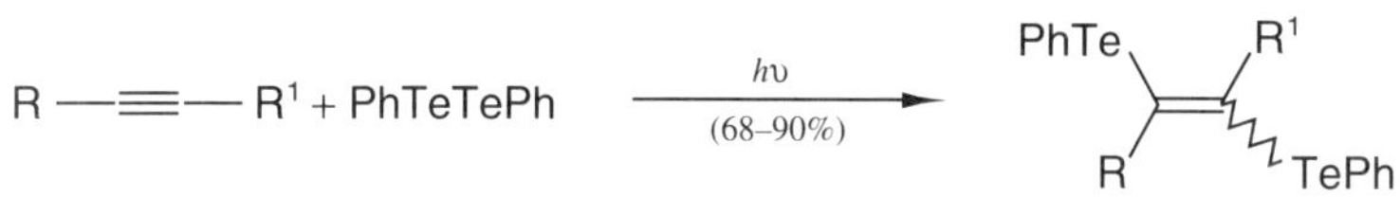

(a) R = n-hexyl, $HOCH_2$, H_2NCH_2; R^1 = H

R = Ph, —⌬; R^1 = H

(b) R = Ph; R^1 = CHO

R = R^1 = CO_2Et

With non-activated acetylenes (R—≡—R^1 (a)) the best results are provided by irradiation with visible light (400 nm) at 40–70°C, giving exclusively the (*E*) isomers. On irradiation in the near-ultraviolet (300 nm), lower yields are obtained, probably owing the reverse reaction to the starting materials.

The addition to activated acetylenes (R—≡R^1 (b)) proceeds even upon irradiation at 300 nm. In this case the intermediate vinyl radical

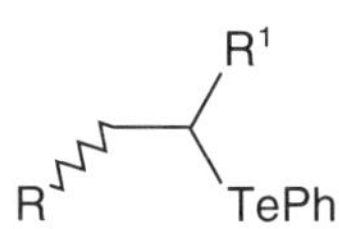

is stabilized by the carbonyl, vinyl or aryl groups, and the reverse reaction to the starting materials is suppressed.

*(*E*)-1,2-Bis(phenyltelluro)-3-hydroxyprop-1-ene (typical procedure)* [19] A mixture of propargyl alcohol (0.11 g, 0.25 mmol) and diphenyl ditelluride (1.02 g, 0.25 mmol), in a Pyrex glass tube sealed under reduced pressure (after filling with argon), is irradiated at 70°C for 24 h through a filter (400 nm) with a tungsten lamp (500 nm) situated approximately 20 cm from the tube. The mixture is purified by preparative TLC on SiO_2 (elution with hexane), giving the pure product (0.90 g (78%)).

3.11.4 By addition of tellurium tetrachloride and aryltellurium trichlorides to acetylenes

Tellurium tetrachloride adds to acetylenes regiospecifically and stereoselectively to produce (*Z*)-2-chlorovinyltellurium trichlorides [20, 21].

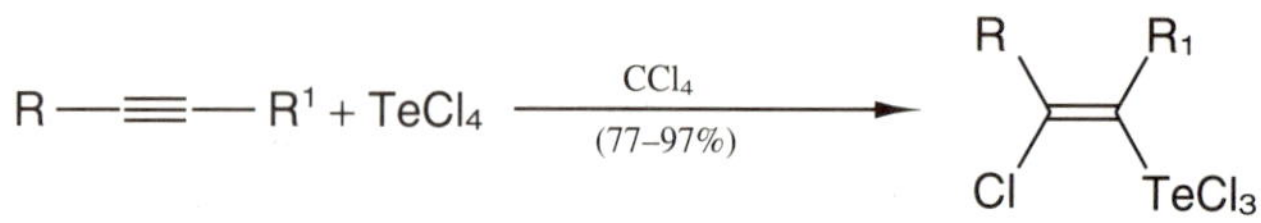

R = Ph*, CH_2OH; R^1 = H

R = Ph; R^1 = Ph, Me, Et

(Z)-2-Chloro-1,2-diphenylethenyltellurium trichloride (typical procedure) [20, 21] A solution of $TeCl_4$ (1.35 g, 5 mmol) and diphenylacetylene (1.00 g, 5 mmol) in CCl_4 (20 ml) is refluxed for 2 h. The solvent is then evaporated under vacuum and the yellow residue recrystallized from glacial HOAc (1.70 g (77%); m.p. 125–128°C).

The trichloride is reduced to the corresponding ditelluride by $Na_2S{\cdot}9H_2O$ [21].

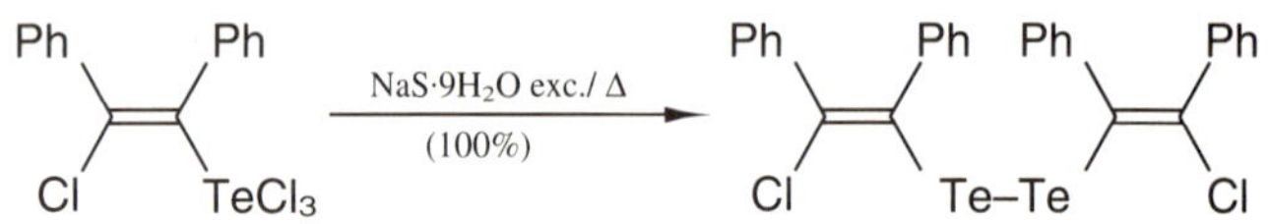

Aryltellurium trichlorides accomplish similar additions to terminal acetylenes, giving adducts with the (*Z*) stereochemistry in good yields. The reduction of these products with $NaBH_4$ furnishes the corresponding tellurides [22].

*The compound exhibits the unusual property of disproportionating into a mixture of $TeCl_4$ and the corresponding dichloride when treated with hydroxylated solvents. Attempts to reduce the trichloride (with $Na_2S{\cdot}9H_2O$) result in decomposition to elemental tellurium. The dichloride is normally reduced to the telluride.

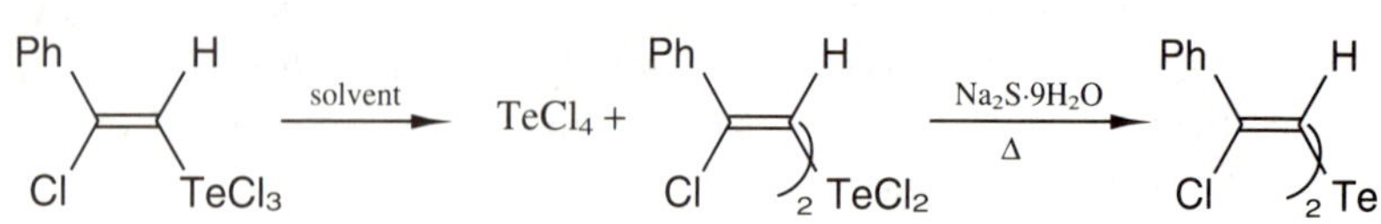

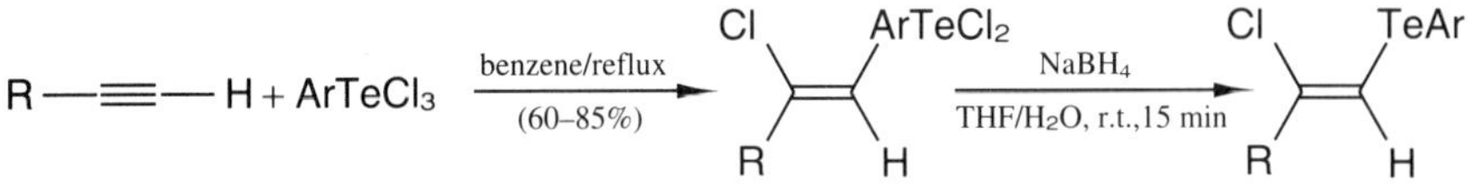

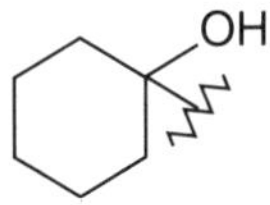

(3),n-pentyl(4)

Ar = p-$MeOC_6H_4$

R (reaction time, h) = Ph (6), p-MeC_6H_4* (4), p-$MeOC_6H_4$ (4), p-$EtOC_6H_4$ (2), p-ClC_6H_4 (3), p-BrC_6H_4 (24), CH_2OH (7),

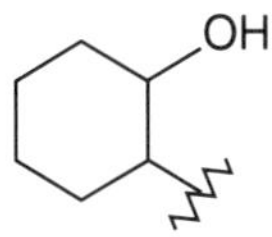

(3),n-pentyl(4)

*2-Chloro-2-phenylethenyl-1-*p*-methoxyphenyltellurium dichloride (typical procedure)* [22] p-Methoxyphenyltellurium trichloride (1.7 g, 5 mmol) is added to phenylacetylene (0.52 g, 5 mmol) in benzene (50 ml). The mixture is refluxed for 6 h and the reaction is monitored by TLC (elution with $CHCl_3$). The mixture is washed with $MeOH/H_2O$ (1:1, 3×30 ml). The organic phase is dried ($MgSO_4$) and the solvent evaporated in a rotary evaporator. The residual oil is filtered through a column of SiO_2, eluting first with CCl_4 and then with $CHCl_3$/MeOH (9:1). The product is recrystallized from $CHCl_3$/petroleum ether 30–60°C (1.7 g (77%); m.p. 134–135°C).

*2-Chloro-2-phenylethenyl-1-*p*-methoxyphenyl telluride (typical procedure)* [22] The dichloride from the preceding preparation (1.55 g, 3.5 mmol) in THF (30 ml) is treated dropwise with $NaBH_4$ (0.14 g, 3.5 mmol) in H_2O (17 ml). An immediate reaction occurs with gas evolution. After stirring for 15 min at room temperature the mixture is diluted with ether (30 ml) and washed in sequence with H_2O, saturated aqueous NH_4Cl and brine. The organic phase is dried ($MgSO_4$), and evaporated in a rotary evaporator. The residual oil is filtered through a column of SiO_2 (elution with petroleum ether 30–60°C) (1.07 g (84%)).

REFERENCES

1. T. Kauffmann and H. Ahlers, *Chem. Ber.* **16**, 1001 (1983).
2. M. Y. Dabdoub, V. B. Dabdoub, J. V. Comasseto and N. Petragnani, *J. Organomet. Chem.* **308**, 211 (1986).

*X ray analysis of this compound confirms the regio- and stereochemistry of the reaction [23].

3. K. Ohe, H. Takahashi, S. Uemura and N. Sugita, *J. Organomet. Chem.* **326**, 35 (1987).
4. K. Ohe, H. Takahashi, S. Uemura and N. Sugita, *Nippon Kagaku Kaishi*, 1469 (1987).
5. M. Y. Dabdoub and J. V. Comasseto, *J. Organomet. Chem.* **344**, 167 (1988).
6. K. Ohe, H. Takayashi, S. Uemura and N. Sugita, *J. Org. Chem.* **52**, 4859 (1987).
7. S. Uemura, S. I. Fukuzawa and S. R. Patil, *J. Organomet. Chem.* **243**, 9 (1989).
8. S. M. Barros, M. J. Dabdoub, V. M. Dabdoub and J. V. Comasseto, *Organometallics* **8**, 1661 (1989).
9. The system DMSO/KOH/$SnCl_2$/H_2O has also been used as reducing agent for diphenylditelluride. V. A. Potapov, A. S. Kashik, N. K. Gusarova, V. I. Minkin, I. D. Sadekov, B. A. Trofimov and S. A. Amosova, *J. Org. Chem. USSR* **23**, 596 (1983) (in English).
10. M. J. Dabdoub, V. B. Dabdoub and J. V. Comasseto, *Tetrahedron Lett.* **23**, 2261 (1992).
11. B. A. Trofimov, S. V. Amosova, N. K. Gusarova and G. K. Musorin, *Tetrahedron* **38**, 713 (1982).
12. N. K. Gusarova, A. A. Tatarinova and L. M. Sinegovskaya, *Sulfur Rep.* **11**, 1 (1991).
13. B. A. Trofimov, S. V. Amosova, N. K. Gusarova, V. A. Potapov and A. A. Tatarinova, *Sulfur Lett.* **1**, 151 (1983).
14. S. R. Buzilova, I. D. Sadekov, T. L. Lipovich, T. M. Filippova and L. I. Vereschchagin, *J. Gen. Chem. USSR* **47**, 1828 (1977) (in English).
15. H. Takahashi, K. Ohe, S. Uemura and N. Sugita, *Nippon Kagaku Kaishi*, 1508 (1987).
16. A. G. Proidakov, G. A. Kalabin, S. R. Buzilova, L. I. Vereschchagin, A. M. Panov and G. V. Ratovskii, *J. Org. Chem. USSR* **17**, 2157 (1981) (in English).
17. L. B. Han, K. I. Ishihara, N. Kambe, A. Ogawa and N. Sonda, *Phosphorus Sulfur Silicon* **67**, 243 (1992).
18. L. B. Han, K. I. Ishihara, N. Kambe, A. Ogawa and N. Sonoda, *J. Am. Chem. Soc.* **114**, 7591 (1992).
19. A. Ogawa, K. Yokoyama, H. Yokoyama, R. Obayashi, N. Kambe and N. Sonoda, *J. Chem. Soc. Chem. Commun.*, 1748 (1991).
20. M. Moura Campos and N. Petragnani, *Tetrahedron* **18**, 527 (1962).
21. S. Uemura, H. Miyoshi and M. Okano, *Chem. Lett.*, 1357 (1979).
22. J. V. Comasseto, H. A. Stefani and A. Chieffi, *Organometallics* **10**, 845 (1991).
23. J. Zuckerman-Schpector, E. E. Castellani, G. Oliva, J. V. Comasseto and H. A. Stefani, *Acta Cryst.* **C47**, 960 (1991).

3.12 ACETYLENIC TELLURIDES

Similarly to vinylic tellurides, acetylenic tellurides can be prepared by two general approaches, starting respectively from nucleophilic (see Section 3.12.1) or electrophilic tellurium species (see Sections 3.12.2 and 3.12.3).

3.12.1 Alkylation of ethynyl tellurolate anions

3.12.1.1 Sodium ethynyl tellurolates

$$\text{R—≡—H} \xrightarrow[\text{(2) Te}]{\text{(1) Na/NH}_3} \text{R—≡—TeNa} \xrightarrow[(21\text{–}65\%)]{\text{R}^1\text{X/NH}_3\text{ liq.}} \text{R—≡—TeR}^1 \quad [1\text{–}4]$$

R = H, Ph, Me, t-Bu, S-Pr, vinyl, -CH=CH-CH=CH_2; R^1 = Me

R = Me; R^1 = Et

R = vinyl; R^1 = n-Bu

Methyl phenylethynyl telluride (typical procedure) [3] To $NaNH_2$ (from 6.0 g, 0.26 mol Na) in liquid NH_3 (250 ml) is added phenylacetylene (25 g, 0.25 mol) dropwise, and then Te powder (30 g, 0.24 mol) in small portions, stirring well for 30 min. Methyl iodide (36 g, 0.25 mol) is added over 20 min to the tellurolate solution. The NH_3 is then evaporated, the residue extracted with ether, and the ether solution washed with H_2O and dried ($MgSO_4$). The residue is distilled under vacuum, giving the product (28 g (46%); b.p. 122–124°C/2 torr).

3.12.1.2 Lithium ethynyl tellurolates

$$\mathrm{R{-}{\equiv}{-}H} \xrightarrow[\text{(2) Te, THF, reflux}]{\text{(1) n—BuLi/THF/0°C}} \mathrm{R{-}{\equiv}{-}TeLi} \xrightarrow[(73\text{–}97\%)]{\mathrm{R^1X}} \mathrm{R{-}{\equiv}{-}TeR^1}$$

R = Me_3Si, Me, n-Bu, Ph; R^1 = CH_2Cl; X = I [5]

R = Ph; R^1 = Et, n-Bu, n-dodecyl, CH_2CH_2Ph, i-Pr, CH(Me)C_3H_7, X = Br [6]

Bis-ethynyl tellurides are accessible from lithium ethynyl tellurolates.

$$\mathrm{Ph{-}{\equiv}{-}TeLi + Ph{-}{\equiv}{-}Br} \xrightarrow[\text{0°C to r·t·}]{\text{THF}} \mathrm{Ph{-}{\equiv}{-}Te{-}{\equiv}{-}Ph}\ [7]$$

Phenyl(alkyltelluro)acetylenes (general procedure) [6] n-BuLi (1.35 M in hexane, 22.2 ml, 30 mmol) is added dropwise to phenylacetylene (3.10 g, 30.0 mmol) in THF (15 ml) at 0°C under N_2. After stirring for 5 min at 0°C, elemental Te (3.90 g, 30.0 mmol) is added and the mixture refluxed until the Te disappears (~ 30 min). The heat source is removed and the alkyl halide (30 mmol) is added. The mixture is stirred for 40 min at room temperature, then diluted with ether (60 ml), washed with brine and the layers separated. The organic phase is dried ($MgSO_4$), evaporated, and the residue purified by SiO_2 flash chromatography (elution with hexane).

3.12.2 From alkynyllithium compounds and aryltellurenyl halides

$$\mathrm{Ph{-}{\equiv}{-}Li + ArTeBr} \xrightarrow[\text{0°C, r·t· (73\%)}]{\text{THF}} \mathrm{Ph{-}{\equiv}{-}TeAr}$$

Ar = p-$MeOC_6H_4$ [6]

*Phenyl(*p*-methoxyphenyltelluro)acetylene (typical procedure)* [6] To a solution of lithium phenylacetylide (20.0 mmol), prepared as described above, is added dropwise a solution of *p*-methoxyphenyltellurenyl bromide (prepared by adding bromine (1.60 g, 10.0 mmol) in benzene (5 ml) to a solution of the corresponding ditelluride (4.69 g, 10 mmol) in THF (10 ml) at 0°C under N_2). The reaction mixture is stirred for 1 h at room temperature and then treated as described above to give the product (4.90 g (73%)), which is recrystallized from EtOH (m.p. 71–72°C).

3.12.3 From tellurium tetrachloride and alkynyllithium compounds

In strict correlation with the early method for the preparation of symmetrical diaryl tellurides by treatment of tellurium(IV) halides with 4 mol eq of aryl Grignard reagent [8] (see Section 3.12.4),* dialkynyl tellurides are prepared employing alkynyllithium reagents [9].

$$4\ \mathrm{R{-}{\equiv}{-}H} \xrightarrow{\text{n—BuLi}} \mathrm{R{-}{\equiv}{-}Li} \xrightarrow[\text{THF/hexane}]{\mathrm{TeCl_4}}$$

$$\mathrm{R{-}{\equiv}{-}Te{-}{\equiv}{-}R + R{-}{\equiv}{-}{\equiv}{-}R}$$

$$\mathrm{R = Me,\ Et,\ n\text{-}Br,\ Ph,\ Me_3Si}$$

Dialkynyl tellurides (general procedure) [9] To a solution of $TeCl_4$ (11 mmol) in THF (15 ml), cooled at −78°C, is added dropwise with vigorous stirring the 1-alkynyllithium reagent (46 mmol) in THF (25 ml). The solution is warmed slowly to room temperature, heated at 55–60°C for 1 h in the dark, and then evaporated under vacuum. The residue is extracted with pentane, the pentane solution evaporated under vacuum, and the obtained crude products are purified by vacuum distillation for R = Et (b.p. 45–46°C/0.075 torr; 56% based on $TeCl_4$), R = n-Pr (b.p. 67–68°C/0.075 torr; 69% based on $TeCl_4$), R = Me_3Si (b.p. 135°C/15 torr; 60% based on $TeCl_4$), and by crystallization (pentane, m.p. 50–52°C) or distillation (110°C/15 torr) for R = Me (42% based on $TeCl_4$). In the case of R = Ph, no pure $(\mathrm{Ph{-}{\equiv}})_2\mathrm{Te}$ is obtained, and the by-product $(\mathrm{Ph{-}{\equiv}{-}})_2$ is sublimed from the crude product under a dynamic vacuum at 100°C (30%).

*Di-phenylethynyl telluride was prepared by this method [8]. Due to difficulties in purification, the product was isolated as the corresponding diiodide $(\mathrm{Ph{-}{\equiv}{-}Te})_2\mathrm{I_2}$ in a yield of 33% [10].

$$\mathrm{TeCl_4 + 4Ph{-}{\equiv}{-}MgBr \longrightarrow Ph{\equiv}{-}Te{-}{\equiv}Ph + Ph{-}{\equiv}{-}{\equiv}{-}Ph}$$

3.12.4 From tellurium tetrachloride/lithium amides and alkynyllithium compounds

By treatment of tellurium tetrachloride with lithium amides, tellurium(II) amides are generated. These species, which provide Te^{2+} electrophiles, react with alkynyllithium reagents, giving dialkynyl tellurides in moderate yields [11].

$$TeCl_4 \xrightarrow{LiNR_2} [Te(NR_2)_2] \xrightarrow[(24\text{–}76\%)]{R^1\text{—}\equiv\text{—}H/LiNR_2}$$

$$R^1\text{—}\equiv\text{—}Te\text{—}\equiv\text{—}R^1$$

R^1 = Ph, *p*-Me, *p*-MeOC_6H_4, n-Bu, Cl$(CH_2)_3$, Me_3Si

HC—≡C$(CH_2)_4$,

R = i-Pr, Me_3Si

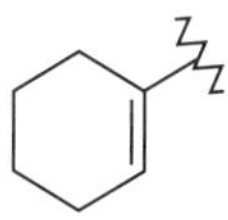

The diynes (R^1—≡—≡—R^1) are formed as by-products, therefore suggesting that the previously described reaction of $TeCl_4$ with 4 mol eq of the lithium alkynyls is partially operative.

Bis-phenylethynyl telluride (typical procedure) [10] To LDA (8 mmol) in THF is added a solution of $TeCl_4$ (0.28 g, 2 mmol) in THF at –78°C, and the mixture stirred at that temperature for 2 h. Phenylacetylene is injected into the resulting orange solution, the mixture is stirred at –78°C for 2 h and then allowed to rise to 0–18°C for an additional 16 h. The mixture is poured onto brine, and the organic layer extracted with ether and dried ($NaSO_4$). The crude material is purified by SiO_2 column chromatography (elution with n-hexane/CH_2Cl_2), giving bis-phenylethynyl telluride and 1,4-diphenylbutadiyne in yields of 40 and 4%, respectively.

3.12.5 Acetylenic tellurides as starting materials for dialkyl ditellurides and tellurides

By treatment of acetylenic tellurides with $AlLiH_4$ (in THF) or $NaBH_4$ (in ethanol), in the presence of air, free acetylenes and diorganyl ditellurides are formed. If the reduction is performed with $NaBH_4$ (in EtOH), in an N_2 atmosphere, vinylic tellurides are produced. Finally, if an alkyl halide is added to the reaction mixture, unsymmetrical tellurides are obtained in good yield [6].

These results are consistent with an attack by a hydride ion at the tellurium atom, generating a tellurol and an acetylide anion which suffer a proton exchange, giving the corresponding acetylene and tellurolate anion.

In the first case, i.e. in the presence of air, the tellurolate is oxidized to the corresponding ditelluride (a). In an N_2 atmosphere a nucleophilic addition of the tellurolate at the acetylene occurs (a protic solvent is necessary to capture the resulting vinyl anion) (b). In the presence of an alkyl halide, the tellurolate is alkylated, giving dialkyl tellurides (c).

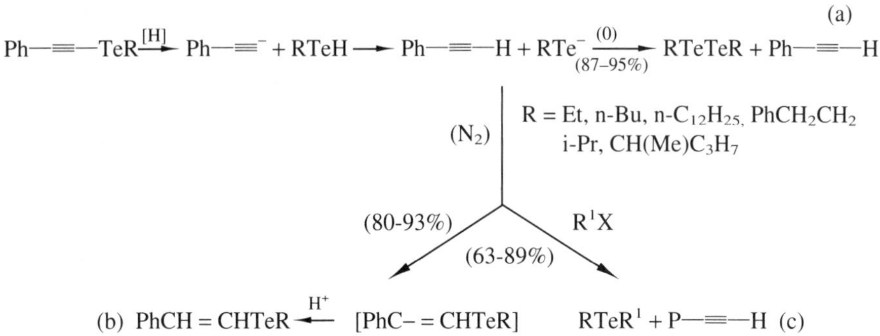

$R = nBu, n\text{-}C_{12}H_{25}$ $\quad R = n\text{-}Bu;\ R^1 = n\text{-}Bu,\ PhCH_2CH_2$

$R = n\text{-}C_{12}H_{25};\ R^1 = n\text{-}C_{12}H_{25},\ i\text{-}Pr$

$R = PhCH_2CH_2;\ R^1 = n\text{-}C_{12}H_{25}$

$R = i\text{-}Pr;\ R^1 = n\text{-}C_{12}H_{25}$

Reaction of acetylenic tellurides with Grignard or organolithium reagents results in a nucleophilic attack at the tellurium atom with formation of the free acetylene and diorganyltellurides [6].

$$Ph{-}{\equiv}{-}TeR \xrightarrow[(2)\ H_2O]{(1)\ R^1Met} Ph{-}{\equiv}{-}H + RTeR^1$$
(67–95%)

Met = MgBr; $R = i\text{-}Pr;\ R^1 = n\text{-}C_{12}H_{25}$

$R = n\text{-}C_{12}H_{25},\ i\text{-}Pr;\ R^1 = n\text{-}Bu$

Met = Li; $R = n\text{-}Bu,\ n\text{-}C_{12}H_{25};\ R^1 = n\text{-}Bu$

Preparation of diorganyl ditellurides from phenyl(organyltelluro)acetylenes (reduction with $NaBH_4$) [6] $NaBH_4$ (0.081 g, 2.20 mmol) in EtOH (3.0 ml) is added to a solution of the acetylenic telluride (2.00 mmol) in EtOH (10 ml) at 0°C (for primary compounds) or at 40–50°C (for secondary compounds). The pale yellow solution turns dark red. After the solution is stirred for 10 min at one of these temperatures, H_2O (0.5 ml) and 10% NaOH (0.5 ml) are added, and the mixture stirred at 0°C for 10 min. The reaction mixture is diluted with ether (30 ml) and washed with brine, and the organic layer is dried ($MgSO_4$). The solvent is evaporated, and the phenylacetylene is cautiously evaporated at reduced pressure (collected in a trap at −78°C). The evaporated residue yields the pure diorganyl ditelluride.

Preparation of diorganyl ditellurides from phenyl(organyltelluro)acetylenes (reduction with $LiAlH_4$) [6] To a suspension of $LiAlH_4$ (0.09 g, 2.2 mmol) in THF (3 ml) at room temperature is added a solution of the acetylenic telluride (2 mmol) in THF (10 ml). The mixture is stirred for 10 min, then H_2O (0.5 ml), 10% NaOH (0.5 ml) and H_2O (0.5 ml) are added, the mixture is diluted with ether, and the red solution is washed with brine. The organic layer is dried ($MgSO_4$), and the solvent and phenylacetylene are evaporated at reduced pressure. The residue remaining after evaporation gives the pure diorganyl ditelluride.

Preparation of vinylic tellurides by the reaction of phenyl(alkyltelluro)acetylenes with $NaBH_4$ under reflux [6] $NaBH_4$ (0.093 g, 2.5 mmol) in EtOH (3.0 ml) is added to a solution of the phenyl(alkyltelluro)acetylene (2.0 mmol) in EtOH (10 ml) under N_2. The solution is stirred under reflux for 3 h (the colour turns a pale yellow). Heating is stopped, and the pale yellow solution is treated with H_2O (0.5 ml) and 10% NaOH (0.5 ml) and diluted with ether (30 ml). The mixture is washed with brine, the organic layer is dried ($MgSO_4$), and the solvent is evaporated. The residue is chromatographed on SiO_2 (eluting with hexane) to give the vinylic telluride.

Preparation of dialkyl tellurides from phenyl(alkyltelluro)acetylenes (reduction with $NaBH_4$ followed by alkylation) [6] $NaBH_4$ (0.081 g, 2.2 mmol) in EtOH (3.0 ml) is added to a solution of the phenyl(alkyltelluro)acetylene (2.0 mmol) and an alkyl halide (2.0 mmol) in EtOH (10 ml) at room temperature under a strong flow of N_2. The mixture is warmed slightly (~ 60–65°C) until the red colour turns pale yellow and gas evolution is observed (25–30 min). The mixture is stirred at room temperature for 10–30 min in the case of primary halides, and for 4 h in the case of secondary halides (i-PrBr). The reaction is quenched with H_2O (0.5 ml) and 10% NaOH (0.5 ml), and diluted with ether (30 ml). The mixture is washed with brine, the organic layer is dried ($MgSO_4$), and the solvent is removed by evaporation. The phenylacetylene is evaporated at reduced pressure (collected at −78°C), and the residue chromatographed on SiO_2 (eluting with hexane) to give the dialkyl telluride.

Preparation of dialkyl tellurides by the reaction between phenyl(alkyltelluro)acetylenes and alkyl Grignard reagents [6] A solution of the phenyl(alkyltelluro)acetylene (2 mmol) in ether (3 ml) is added dropwise to a solution of the Grignard reagent (4 mmol) in ether (5 ml) at room temperature under N_2. After stirring for 30 min at room temperature the mixture is washed with saturated aqueous NH_4Cl and the organic layer is dried ($MgSO_4$). The solvent is evaporated, and the residue is chromatographed on SiO_2 (eluting with heptane) to give the dialkyl telluride.

Preparation of dialkyl tellurides by the reaction of phenyl(alkyltelluro)acetylenes with alkyllithium reagents [6] n-BuLi (1.59 ml, 2.2 mmol, 1.38 M solution in hexane) is added dropwise to a solution of phenyl(alkyltelluro)acetylene (2 mmol) in THF (10 ml) at 0°C under N_2. Following the addition, the solution is stirred for 10 min at this temperature, then brine is added, and the product is extracted with ether. The organic layer is dried ($MgSO_4$), and evaporated under reduced pressure to give the dialkyl telluride.

REFERENCES

1. L. Brandsma, H. E. Wijers and J. F. Arens, *Recl. Trav. Chim. Pays-Bas* **81**, 583 (1962).
2. A. A. Petrov, S. I. Radchenko, K. S. Mingaleva, I. G. Savich and V. B. Lebedov, *J. Gen. Chem. USSR* **34**, 1911 (1964) (in English).
3. Y. A. Boiko, B. S. Kupin and A. A. Petrov, *J. Org. Chem. USSR* **4**, 1307 (1968); Y. A. Boiko, B. S. Kupin and A. A. Petrov, *J. Org. Chem. USSR* **5**, 1516 (1969) (in English).
4. S. A. Radchenko and K. S. M. Mingaleva, *J. Org. Chem. USSR* **13**, 2303 (1977) (in English).
5. S. L. Bender, M. R. Detty and N. F. Haley, *Tetrahedron Lett.* **23**, 1531 (1982).
6. M. G. Dabdoub and J. V. Comasseto, *Organometallics* **7**, 84 (1988).
7. M. G. Dabdoub, J. V. Comasseto and A. L. Braga, *Synth. Commun.* **18**, 1879 (1988).
8. H. Rheinboldt and N. Petragnani, *Chem. Ber.* **89**, 1270 (1956).
9. R. W. Gedridge Jr, L. Brandsma, R. A. Nissan, H. D. Verkruijsse, S. Harder, R. L. P. de Jong and C. J. O'Connor, *Organometallics* **11**, 418 (1992).
10. M. Moura Campos and N. Petragnani, *Tetrahedron* **18**, 527 (1962).
11. T. Murai, K. Imada, S. Kajita, H. Ishibara and S. Kato, *J. Chem. Soc. Chem. Commun.*, 832 (1991).

4

Tellurium in Organic Synthesis

4.1 REDUCTIONS

This section is devoted to the application of tellurium compounds as reducing reagents of organic substrates. Inorganic species such as hydrogen telluride, sodium hydrogen telluride, sodium telluride, and organic derivatives such as aryltellurols (and tellurolates) and diorganyl tellurides are widely employed in several reduction processes such as, for example, reduction of carbonyl compounds, nitro derivatives, epoxides, α,β-unsaturated carbonyl compounds, arylalkenes, arylalkynes, as well as in the reductive cleavage of several types of carbon–heteroatom bonds.

In several cases the use of tellurium reagents is clearly advantageous compared to the usual previously described methods in view of the mildness of the experimental conditions and the great selectivity.

Additional advantages are the possibility of generating the reducing agents *in situ*, as well in catalytic amounts in the presence of an inexpensive co-reductant, and recovery of the tellurium material (for example elemental tellurium from the above inorganic reagents and ditellurides from tellurols).

4.1.1 Reduction of carbonyl compounds

4.1.1.1 With hydrogen telluride

Hydrogen telluride, H_2Te, generated *in situ* by the hydrolysis of aluminium telluride, reduces aldehydes and ketones to the corresponding alcohols [1, 2]. In the presence of deuterium oxide, deuterated derivatives are formed.

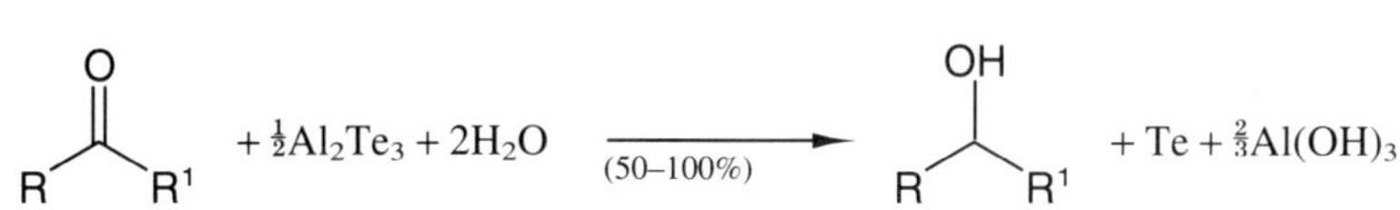

R = H; R^1 = Ph, *o*-MeC_6H_4, n-heptyl

R, R^1 = $(CH_2)_5$

Reduction of benzaldehyde with Al_2Te_3/D_2O (typical procedure) [1] D_2O (0.22 ml, 12 mmol) is added at −78°C to a suspension of Al_2Te_3 (0.876 g, 2 mmol) in THF (10 ml) containing benzaldehyde (0.106 g, 1 mmol) with vigorous stirring. The temperature is raised to 0°C in 30 min and maintained there for another 2 h. Filtration and evaporation of the solvent leaves pure PhCHDOD (0.11 g (100%)).

4.1.1.2 With phenyltellurol

The reagent is prepared by protonation of lithium phenyl tellurolate with trifluoroacetic acid [3] or by methanolysis of phenyltellurotrimethylsilane [4]. Aldehydes and alkyl aryl ketones or diaryl ketones are reduced to the corresponding alcohols in good yields.

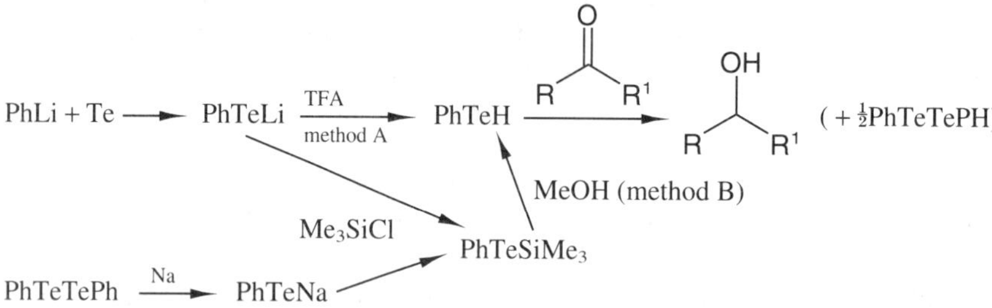

Method A: R = H; R^1 = Ph, *p*-$MeOC_6H_4$ (31, 93%)

Method B: R = H; Ph, *p*-$MeOC_6H_4$, *o*-ClC_6H_4, n-$C_{12}H_{25}$ (73–98%)

R = Ph; R^1 = Me, n-C_3H_7, Ph (80–95%)

Reduction of p-*methoxybenzaldehyde (method A, typical procedure)* [3] To lithium phenyl tellurolate prepared from PhLi (2.1 M, 4.8 ml, 10 mmol) and Te (1.27 g, 10 mmol) in THF (30 ml) under N_2 is added *p*-methoxybenzaldehyde (0.272 g, 2 mmol) at −78°C followed by TFA (0.77 ml, 10 mmol). The solution is stirred for a further 2 h, and allowed to warm at room temperature. H_2O (200 ml) is added and the product is extracted with ether. The extract is dried (Na_2SO_4), evaporated and the residue separated by SiO_2 chromatography from the formed diphenyl ditelluride, giving *p*-methoxybenzyl alcohol (0.23 g (93%)).

Reduction of o-*chlorobenzaldehyde (method B, typical procedure)* [4] Diphenyl ditelluride (20.5 g, 50 mmol) in THF (50 ml) is treated with Na metal (3.5 g, 0.15 mol) for 18 h under ultrasound. After cooling to room temperature, Me_3SiCl (14 ml, 0.11 mol) is added and the mixture stirred for 1 h. The solvent is evaporated and the residue is distilled under vacuum, giving phenyltellurotrimethylsilane (20.5 g (74%); b.p. 77–79°C/2 torr).

o-Chlorobenzaldehyde (0.140 g, 1 mmol) in benzene (5 ml) and MeOH (0.5 ml) is treated dropwise with phenyltellurotrimethylsilane (0.610 g, 2.2 mmol) for 30 min. After evaporation of the solvent the residue is chromatographed on an SiO_2 column (eluting with $CHCl_3$), giving *o*-chlorobenzyl alcohol (0.13 g (92%); m.p. 72.8–73.8°C).

Under the above-described conditions, esters are unaffected and aliphatic ketones are barely reduced. In the presence of zinc iodide, however, aliphatic ketones, as well as aldehydes and aromatic ketones, are converted into methyl ethers in high yields [5]. If alcohols other than methanol are used, the corresponding alkyl ethers are obtained.

R = Me, $PhCh_2$; R^1 = H; R^2 = Ph, n-$C_{11}H_{23}$

R^1 = Ph; R^2 = n-C_3H_7

R^1 = Me; R^2 = n-C_9H_{19}

R^1, R^2 = $(CH_2)_5$

Benzyl dodecyl ether (typical procedure) [5] Phenyltellurotrimethylsilane (0.610 g, 2.2 mmol) is added to a solution of benzaldehyde (0.238 g, 2.2 mmol) in dry benzene (10 ml) under N_2. After stirring for 30 min at room temperature, ZnI_2 (0.016 g, 0.05 mmol) in benzene (3 ml) is added, followed by dodecanal (0.184 g, 1.0 mmol). Stirring is maintained for 3 h and then H_2O is added, the mixture is extracted with benzene, and the benzene extract is dried ($MgSO_4$) and evaporated under vacuum. The residue is chromatographed on SiO_2 (elution with benzene/hexane), giving the product (0.265 g (96%)).

4.1.1.3 With diisobutyl telluride/titanium(IV) chloride

The known property of diorganyl tellurides of reducing metallic salts [6], applied to titanium(IV) chloride, generates a titanium(III) species which is a useful reagent for some selective reductions. By this method benzaldehyde is reduced to dihydrobenzoin, and benzil to benzoin, and successively to desoxybenzoin [7, 8].

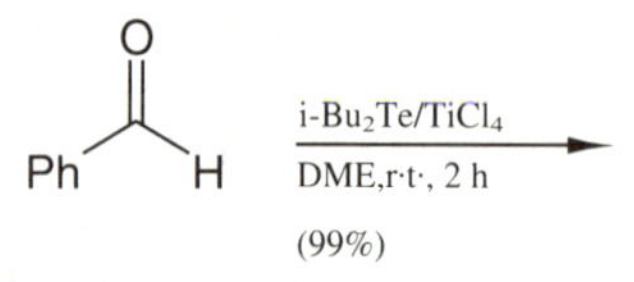

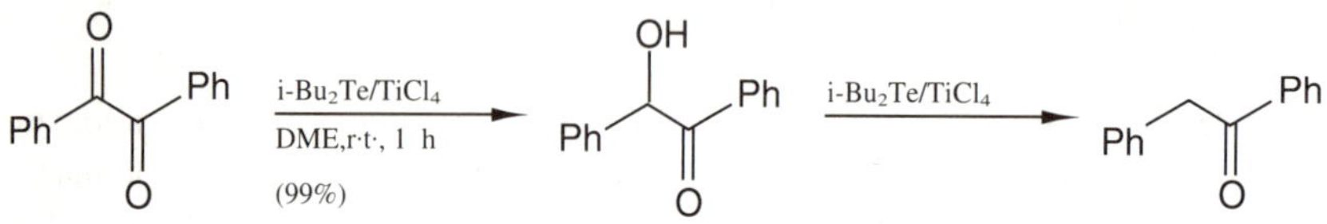

4.1.2 Selective reduction of α,β-unsaturated carbonyl compounds

The regioselective reduction of the C=C bond of α,β-unsaturated carbonyl compounds, one of the most important organic reactions, is achieved by means of hydrogen telluride [1] and phenyl tellurol [3], under appropriate experimental conditions.

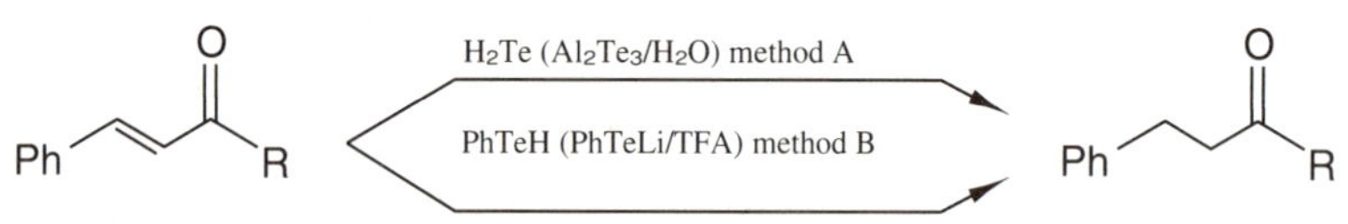

Method A: R = H, Me, Ph (54–84%)

Method B: R = Ph (80%)

When the phenyl tellurol is prepared by methanolysis of phenyltellurotrimethylsilane, both the C=C bond and the carbonyl group can be reduced [4].

Sodium hydrogen telluride was first prepared several years ago [9], and is recognized as a selective and versatile reducing agent. It is inert towards non-conjugated carbonyl, carboxyl, amide, ester, nitrile, haloaryl, non-activated haloalkyl and sulphone groups, but exhibits great selectivity for several functional groups. Sodium hydrogen telluride is a very convenient reagent for the reduction of the C=C bond of α,β-unsaturated carbonyl compounds [3, 10–12].

Representative examples are given in sequence below.

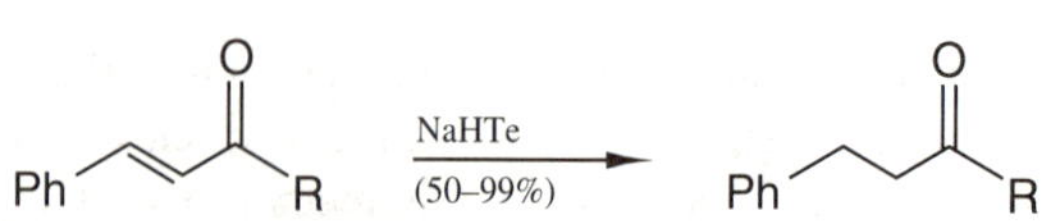

R = H, Me, OMe, OCH_2Ph, NH_2, OH

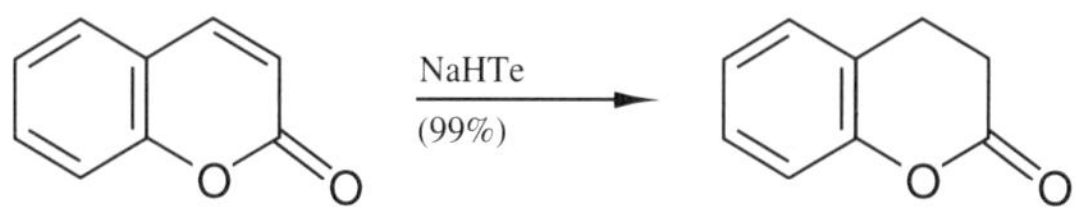

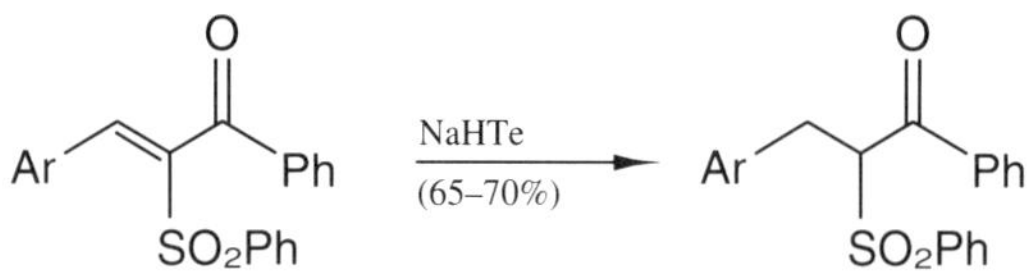

Ar = Ph and halo derivatives

3-Phenylpropanal (typical procedure) [11] *NaHTe* [9]. A stirred mixture of Te powder (0.65 g, 5 mmol) and $NaBH_4$ (0.45 g, 11.8 mmol) in EtOH (20 ml) is heated under argon. After 15 min most of the Te will have dissolved. The mixture is then cooled at −20°C and treated with a deoxygenated solution of HOAc (0.6 ml) in EtOH (2.5 ml).

Reduction. To the above-prepared solution of NaHTe is added 3-phenylpropenal (0.264 g, 2 mmol) in EtOH (2 ml). The mixture is stirred at room temperature for 4–5 h, then filtered from the formed black Te through Celite and the filtrate evaporated. The residue is purified by distillation, giving the pure product (0.265 g (99%)).

The synthetically important isopropylidene malonates (Meldrum acids) can be prepared in a "one-pot" procedure employing this method [13].

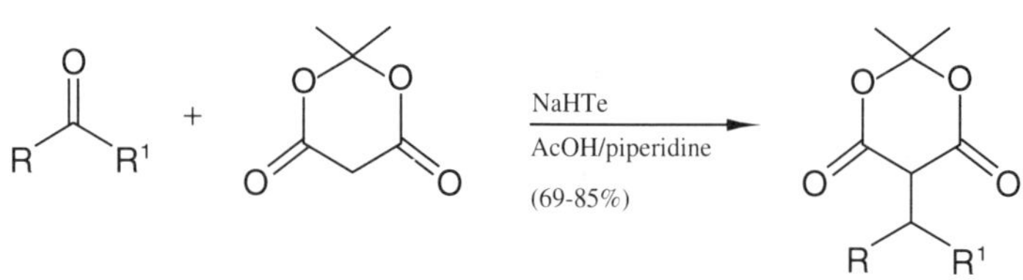

R, R^1 = Me, $(CH_2)_5$

R = H; R^1 = i-Pr, *p*-MeO, *p*-HOC_6H_4

Monosubstituted isopropylidene malonates (typical procedure) [13] To a solution of NaHTe (10 mmol) prepared as above is added a solution of isopropylidene malonate (0.72 g, 5 mmol), benzaldehyde (0.83 g, 7.5 mmol), AcOH (1.5 ml) and piperidine (4 drops) at −30°C under N_2. After 10 min the mixture is warmed to room temperature, stirred for 20 min, filtered and evaporated under vacuum. Aqueous K_2CO_3 (5 ml) is added to the residue. After extraction with ether (3 × 3 ml), the solution of K_2CO_3 is acidified with 6 N HCl (5 ml) and then filtered, giving the pure product (0.99 g (81%); m.p. 80–81°C).

4.1.3 Reduction of conjugated arylalkenes and arylalkynes

In addition to the described reduction of double bonds conjugated to a carbonyl group, sodium hydrogen telluride and phenyltellurol reduce double (and triple) bonds conjugated to aromatic systems [3, 4, 10].

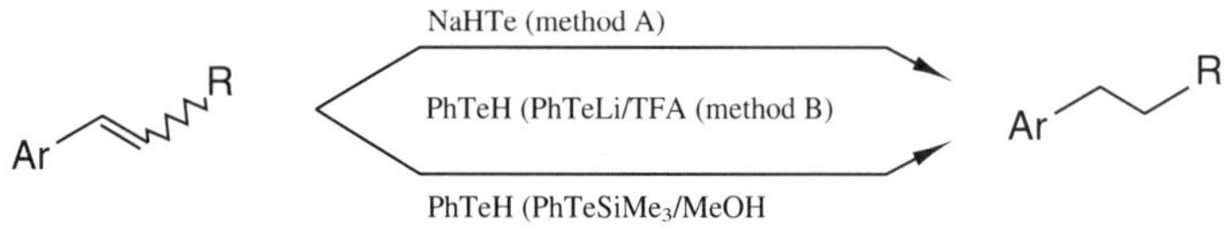

High yields are obtained with terminal alkenes (styrene and α-methylstyrene) and with the strained acenaphthylene, whereas stilbene and β-methylstyrene are barely reduced.

Acenaphthene (typical procedure) [3] *Method A*. To a solution of NaHTe (10 mmol, prepared as previously described) is added acenaphthylene (0.304 g, 2 mmol) in EtOAc (20 ml), and the mixture refluxed for 9 h. During this time a black Te precipitate is formed. After dilution with H_2O (500 ml) the product is extracted with ether, the ethereal solution is dried (Na_2SO_4) and evaporated, and the residue chromatographed on SiO_2, giving acenaphthene (0.303 g (99%)).

Method B. A mixture of diphenyl ditelluride (2.05 g, 5 mmol), $NaBH_4$ (0.397 g, 10.5 mmol) and dry EtOH (20 ml) is heated with stirring and under an N_2 atmosphere for 15 min, when most of the ditelluride will have produced an almost colourless solution. The solution is then cooled to −20°C and TFA (0.81 g, 10. 5 mmol) in EtOH (5 ml) is added, followed by a solution of acenaphthylene (0.305 g, 2 mmol) in EtOAc (20 mol). The mixture is refluxed for 4 h, then diluted with cold H_2O (500 ml) and worked up in the usual manner, including SiO_2 chromatography, to give diphenyl ditelluride (1.93 g (94%)) and acenaphthene (0.300 g (98%)).

Phenylacetylene is reduced by phenyltellurol to ethylbenzene in a yield of 42% [4].

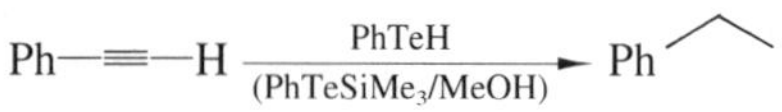

4.1.4 Reduction of imines and enamines

Hydrogen telluride [2, 14], sodium hydrogen telluride [3] as well as phenyltellurol [3] afford the reduction of imines to secondary amines.

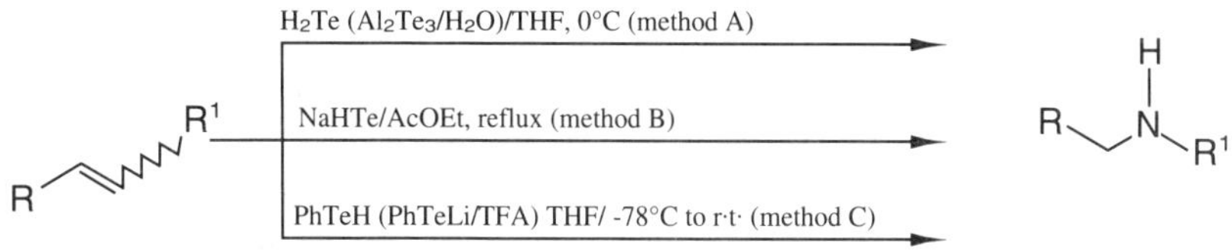

Method A: R = R^1 = Ph (71%)

Method B: R = Ph; R^1 = Et (65%)

Method C: R = R^1 = Ph (93%)

Enamines are susceptible to a similar reduction [14].

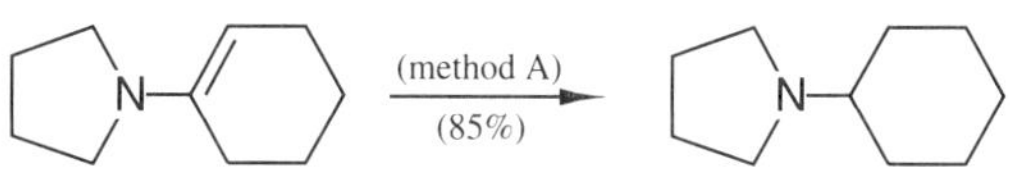

In the reduction of imines and enamines with hydrogen telluride and sodium hydrogen telluride, hydrolysis leading to primary amines and carbonyl compounds is frequently competitive with the reduction. This undesired side-reaction is minimized by the addition of triethylamine, in the case of hydrogen telluride [14].

Control of the pH to within the 6–7 range has been shown to be important in the reduction of imines with sodium hydrogen telluride. At a higher pH (10–11) no reduction occurs [15].

A valuable application of the above-described reduction is the one-step reductive alkylation of amines with carbonyl compounds [14, 16].

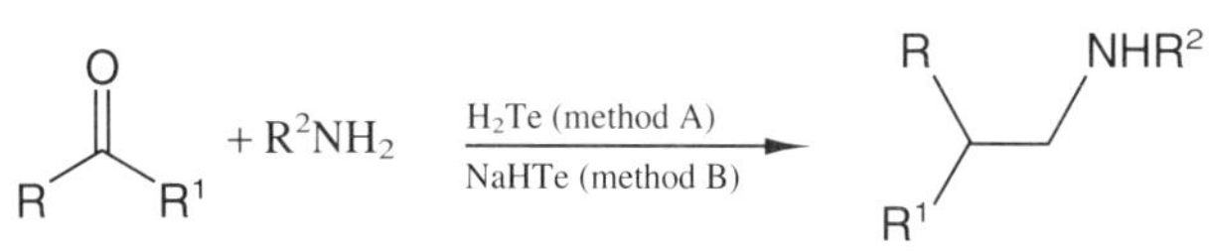

Good yields of secondary amines are achieved using both the methods in the reactions of aromatic and aliphatic aldehydes as well of dialkyl ketones and cycloalkanones with aliphatic and alicylic amines (and ammonia). Anilines give low yields but, when 2 eq is used in the sodium hydrogen telluride method, the yields are improved. In the reaction of ammonia with aldehydes, symmetrical secondary amines are obtained, whereas glutaraldehyde and amines [16] lead to *N*-substituted piperidines.

Reductive amination of carbonyl compounds *Method A—*N*-octylbenzylamine (typical procedure)* [14]. A THF (30 ml) suspension of benzylamine (1.67 g, 15 mmol), octanal (1.55 g, 15 mmol), Et_3N (7.0 ml, 50 mmol) and Al_2Te_3 (6.1 g, 14 mmol) is refluxed. The reaction is monitorized by GLC and H_2O (3 ml, 168 mmol) is added at 15°C after 30 min. The heterogeneous mixture is refluxed for another 30 min. The black solid is removed by filtration and the filtrate dried ($CaCO_3$) and evaporated. The residue is distilled, giving the product (2.7 g (80%); b.p. 108–112°C/1 torr).

Method B (general procedure) [16]. To a solution of NaHTe (5 mmol) prepared as described previously [9] (see Section 4.1.2) the carbonyl compounds (2 mmol) and the amine (2 mmol) are added at –20°C. The mixture is stirred for 20 h at room temperature. The precipitated black Te is filtered and the filtrate evaporated under vacuum. The residue is dissolved in EtOAc (20 ml) and acidified with dilute HCl. The combined aqueous layers are made basic with NaOH and extracted with EtOAc (2 × 30 ml). The extract is dried (Na_2SO_4), concentrated and purified by TLC (elution with hexane/EtOAc).

4.1.5 Reductive desulphuration of aromatic thioketones

Under aprotic conditions, aromatic thioketones are reduced by sodium telluride to the corresponding hydrocarbons [17].

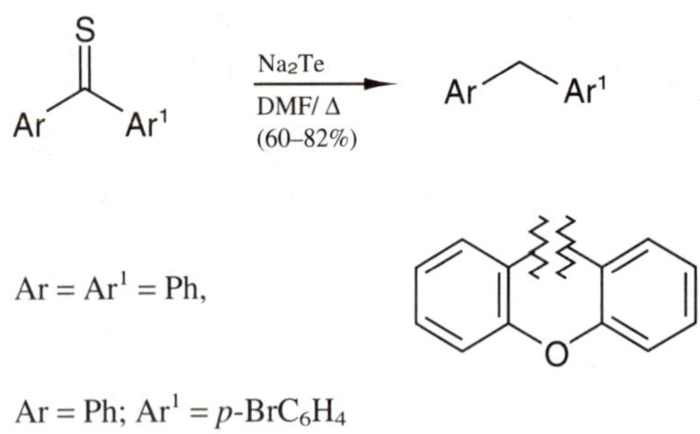

Reductive desulphuration of aromatic thioketones: xanthene (typical procedure) [17] To a suspension of Na_2Te (prepared by heating powdered Te (0.256 g, 2.0 mmol) and NaH (0.10 g, 4.2 mmol) in DMF (3 ml) at 140°C for 1 h under N_2) is added xanthene-9-thione (0.210 g, 1.0 mmol) in THF (5 ml), and the mixture is heated under gentle reflux. The colour changes from green to orange to blue and finally to black. After 2 h, excess Na_2Te is destroyed by exposure to air, and H_2O (20 ml) and CH_2Cl_2 (20 ml) are added. After filtration, the organic layer is washed with 10% HCl and evaporated. The residue is purified by SiO_2 chromatography, giving xanthene (0.161 g (82%); m.p. 96–99°C).

Under aqueous conditions (Na_2Te/Rongalite method) no reduction occurs, the thioketone being converted to the ketone [17].

4.1.6 Reduction of nitro compounds

Hydrogen telluride [2, 18] and sodium telluride in a protic solvent [19] easily reduce nitroarenes to anilines. In contrast, sodium telluride in an aprotic solvent exhibits a milder reducing capacity and reduces the nitro compounds only to azo compounds [17].

The reducing power of sodium hydrogen telluride towards nitro compounds is dependent on their structure. Unhindered nitrobenzenes are reduced to azoxybenzenes whereas sterically hindered nitrobenzenes are reduced to anilines [20, 21] by simple stirring with 5 mol eq of the reagent at room temperature for 1 h.

Under alkaline conditions, sodium hydrogen telluride reduces nitrobenzene to *N*-phenylhydroxylamine [22].

A more practical procedure for the reduction of nitroarenes to *N*-arylhydroxylamines is the use of $NaBH_4$ in the presence of a catalytic amount of tellurium [22]. No over-reduction products are observed.

The reduction of nitroalkanes and dinitroalkanes with sodium hydrogen telluride gives, respectively, diazenes (dimers of nitrosoalkenes) and olefins [20].

Reduction of nitrobenzene with H_2Te: aniline (typical procedure) [18] Powdered Al_2Te_3 (674 mg, 2 mmol), stored under N_2 in the dark, is added to a THF (10 ml) solution of nitrobenzene (123 mg, 1 mmol) under N_2 at room temperature. The temperature is then raised to 66°C. To the vigorously stirred reaction mixture is slowly added an excess of H_2O (0.43 ml, 24 mmol), dropwise by a syringe over a period of 40 min. The reaction mixture is refluxed for another 20 min and worked up, giving the product (0.837 g (90%)).

Reduction of aromatic nitro compounds with Na_2Te (Rongalite method): anilines (general procedure) [19] To Te powder (0.013 g, 0.10 mmol) and Rongalite (0.77 g, 5 mmol) thoroughly flushed with N_2 is added 1 M NaOH (15 ml) followed by the nitro compound (1.0 mmol) in dioxane (3 ml), and the hetero

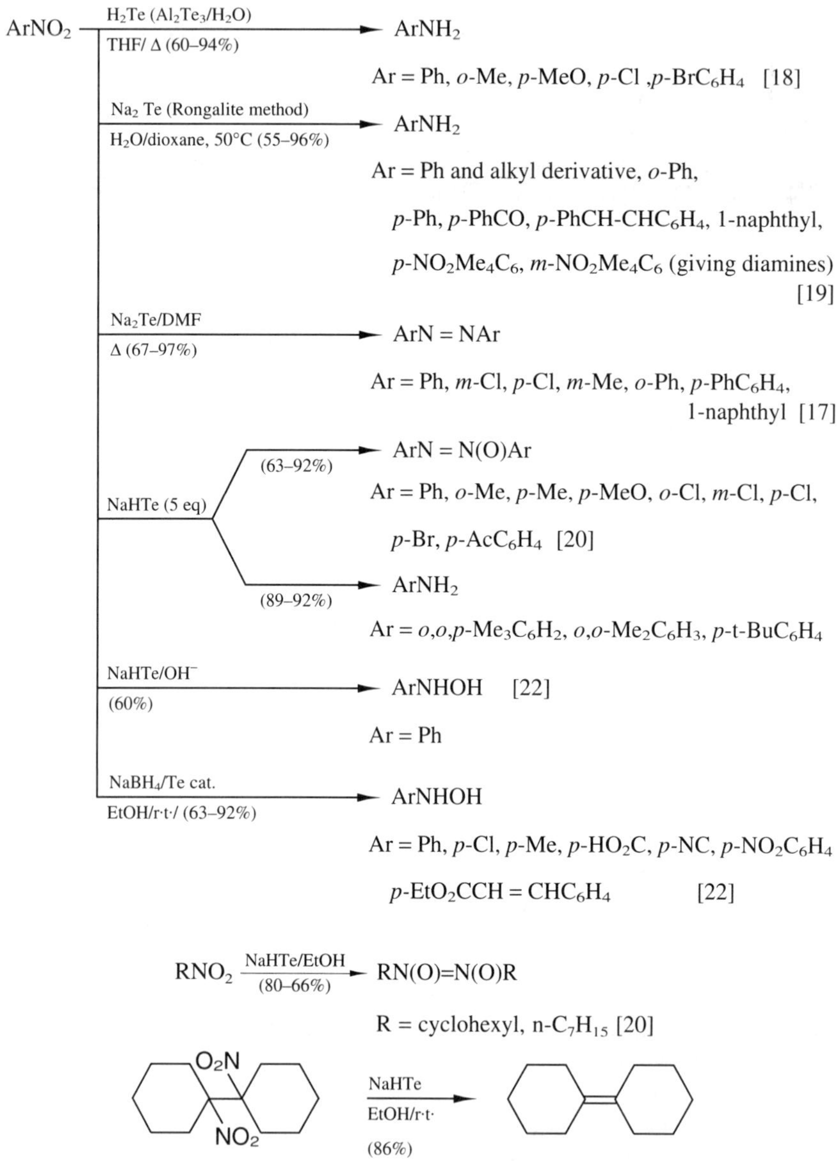

geneous mixture stirred at 50°C for 1–3 h. The progress of the reaction is monitored by TLC. After an appropriate time, when the nitro compound has almost completely disappeared, the reaction is quenched by exposing the mixture to air. NaCl (3–4 g) and $CHCl_3$ (15–20 ml) are added with stirring and the mixture is freed from insoluble inorganic materials by filtration over a thin layer of

Celite. The organic phase is separated and the aqueous solution is extracted with $CHCl_3$. The $CHCl_3$ solution is washed with brine, dried (Na_2SO_4), and evaporated to give the product as an oil or a solid residue.

Dinitro compounds can be similarly reduced to diamines by using twice the amount of reducing agent, increasing the proportion of dioxane in the solvent system, and extracting the product from the aqueous solution with ether. A much longer reaction time (5–7 h) is needed for complete reduction.

Reduction of aromatic nitro compounds with Na_2Te/DMF: azo compounds (general procedure) [17] To a stirred suspension of Na_2Te (2.0 mmol) in DMF (3 ml) is added by a cannula a solution of the nitroarene (1; 1.0 mmol) in THF (2 ml) at room temperature, and under N_2. The mixture is gently heated to 70°C and kept at this temperature for 3–5 h. The progress of the reaction is quenched by bubbling air into the mixture, to which brine (20 ml) followed by ether (20 ml) is then added. The insoluble material is filtered through a thin layer of Celite and the organic phase is separated from the filtrate, washed with 10% HCl to remove any DMF, and evaporated to leave the crude product, which is purified by chromatography on SiO_2 or recrystallization from an appropriate solvent.

Reduction of nitrobenzene with $NaBH_4$/Te catalyst: N*-phenylhydroxylamine (typical procedure)* [22] To a stirred mixture of powdered Te (51 mg, 0.4 mmol) and $NaBH_4$ (1.52 g, 40 mmol) in EtOH (25 ml) under argon is added nitrobenzene (492 mg, 4 mmol) as a single portion, and the mixture is stirred at room temperature (22°C) for 1.5 h. Then the mixture is poured into ice–H_2O, while bubbling argon through it, and acidified to pH 6 with 10% HCl, and immediately extracted with ether. The extract is washed with brine and dried ($MgSO_4$). After evaporation of the solvent under reduced pressure at room temperature, the crude product is purified by short SiO_2 column chromatography (eluting with CH_2Cl_2) and subsequent recrystallization (benezene/petroleum ether) to give the product (0.27 g (63%); m.p. 80–82°C).

Phenyltellurol is another valuable reagent for the reduction of nitroarenes (and nitrocycloalkanes) to the corresponding amines.

The reagent is easily generated *in situ* by the previously described methods (methods A [3] and B [4, 23]), or by the reduction of diphenyl ditelluride with $NaBH_4$ [23] (method C). In the last method, the procedure can be carried out with a catalytic amount of ditelluride, since the ditelluride is continuously reformed during the reaction.

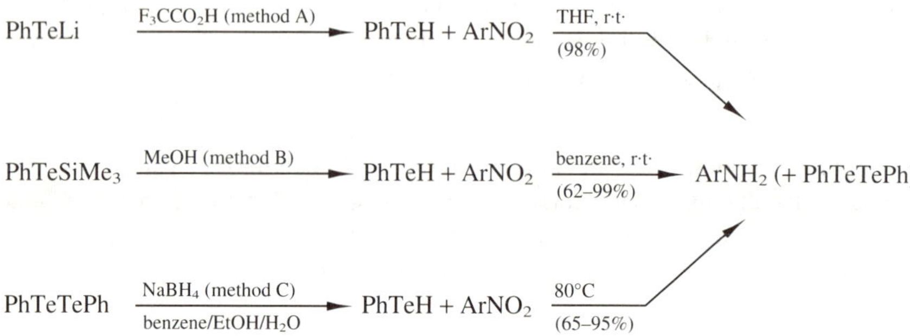

method A: Ar = Ph

methods B and C: Ar = Ph, *o*-Me, *m*-Me, *p*-Me, *o*-MeO, *p*-MeO,
o-Cl, *o*-Br, *o*-I, *m*-$NO_2C_6H_4$
o, *p*-$Me_2C_6H_3$, 1-naphthyl, 1-(5-NO_2)naphthyl
(giving diamine)
o-NC-C_4H_4 (giving anthranilamide)*

Reduction of aromatic nitro compounds with phenyltellurol ($NaBH_4$/$(PhTe)_2$ catalyst): aniline (typical procedure) [23] Nitrobenzene (0.246 g, 2.00 mmol) and PhTeTePh (0.164 g, 0.4 mmol) are dissolved in benzene (10 ml). To this mixture is added $NaBH_4$ (0.378 g, 10.0 mmol) dissolved in EtOH/H_2O (v/v 2:1, 10 ml) dropwise under an N_2 atmosphere. The reaction mixture is heated to 80°C for 15 h with stirring. After cooling, the mixture is washed with saturated brine and dried (Na_2SO_4). The crude product is purified by SiO_2 column chromatography (solvent: benzene/ether) and then gel permeation liquid chromatography. Aniline is isolated (134 mg (72%)).

An additional method for the reduction of nitrobenzene to aniline involves the previously described i-Bu_2Te/$TiCl_4$ system (see Section 4.1.1.3) [7, 8].

$$ArNO_2 \xrightarrow[TiCl_4,\ CH_2,\ Cl_2 \ (42\text{–}98\%)]{i\text{-}Bu_2Te} ArNH_2$$

Ar = Ph, *m*-CN, *p*-Cl, *p*-Br, *p*-Ph,
p-$PhCOC_6H_4$, *m*, *p*-$Me_2C_6H_3$

Reduction of aromatic nitro compounds with i-Bu_2Te/$TiCl_4$: aniline (typical procedure) [8] To a mixture of i-Bu_2Te (0.36 g, 1.5 mmol) and a 1 M solution

*Nitrocyclohexane is also reduced to cyclohexylamine in a yield of 62%.

of $TiCl_4$ in CH_2Cl_2 (4 ml) is added nitrobenzene (0.06 g, 0.5 mmol) in CCH_2Cl_2 (2 ml) at room temperature and under N_2. An instantaneous reaction occurs and the colour of the solution turns dark brown. After stirring for 0.5 h, NaCl (3–4 g) followed by 1 M aqueous NaOH is added until the aqueous phase is alkaline. The organic phase is extracted with Et_2O and the extract is washed with brine, dried (Na_2SO_4) and evaporated. The residue is purified by SiO_2 chromatography (elution with CH_2CL_2/Et_2O 5:1), giving the almost pure amine.

Sodium phenyl tellurolate is a milder reducing agent than phenyltellurol. The reduction of nitroarenes is therefore limited at the intermediate azoxy or azo compound stage, depending on the reaction temperature [24, 25]. The reagent is prepared *in situ* by treatment of diphenyl ditelluride with $NaBH_4$ in ethanol in the presence of sodium hydroxide. Since the diphenyl ditelluride is continuously regenerated from the formed phenyl tellurenate (PhTeONa) by reduction with the excess $NaBH_4$, it is used only in catalytic amounts.

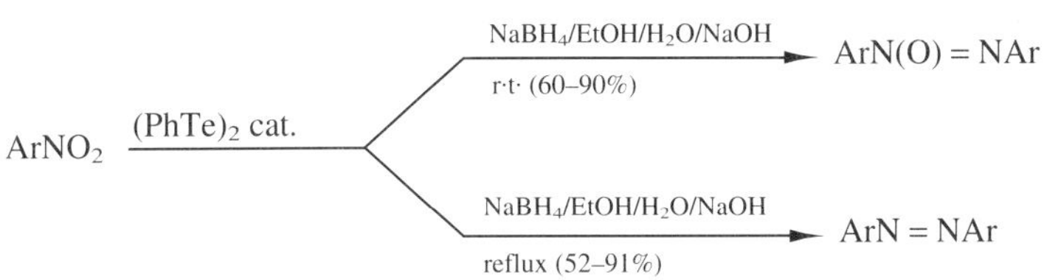

Ar = Ph, *o*-Me, *m*-Me, *p*-Me, *p*-MeO, *p*-Cl, *p*-Br, *p*-AcC_6H_4, *o*, *p*-$Me_2C_6H_3$

In the reflux procedure, minor amounts of the corresponding arylhydrazines are formed as by-products.

Reduction of nitrobenzene with $NaBH_4/(PhTe)_2$ catalyst/EtOH/aqueous NaOH at room temperature: azoxybenzene (typical procedure) [25] To a mixture of diphenyl ditelluride (0.021 g, 0.05 mmol) and $NaBH_4$ (0.378 g, 10 mmol) are added EtOH (10 ml) and 5 M aqueous NaOH (2.5 ml) successively at room temperature under an N_2 atmosphere. An EtOH (5 ml) solution of nitrobenzene (0.246 g, 2.0 mmol) is then added to the colourless, homogeneous solution and the mixture is stirred at room temepratture for 10 h, during which time the yellow colour turns to orange. The mixture is poured into brine and then extracted with $CHCl_3$ (3 × 50 ml), and the extract is dried ($MgSO_4$). Evaporation of the solvent leaves a yellow oil, which is subjected to column chromatography on SiO_2 using hexane/EtOAc (10:1) as the eluent, to give the pure product as a yellow solid (0.181 g (91.3%)).

Reduction of p*-bromonitrobenzene with $NaBH_4/(PhTe)_2$ catalyst/EtOH/aqueous NaOH under reflux: bromoazobenzene (typical procedure)* [25] To a mixture containing diphenyl ditelluride (0.021 g, 0.05 mmol) and $NaBH_4$ (0.378 g, 10 mmol) is added EtOH (10 ml) followed by 5 M aqueous NaOH (2.5 ml) at reflux. A solution of *p*-bromonitrobenzene (0.404 g, 2.0 mmol) in EtOH (5 ml) is then added to the colourless homogeneous solution, and the mixture is stirred under reflux for 15 h, during which time the yellow colour turns to red. After cooling, the mixture is poured into H_2O and extracted with $CHCl_3$ (3 × 50 ml), and the extract is then dried ($MgSO_4$). The products, bromoazobenzene (0.224 g (66%); m.p. 206–207°C) and *p*, *p'*-dibromo-*N*, *N'*-diphenylhydrazine (0.065 g (19%)), are isolated by column SiO_2 chromatography, eluting with hexane/$CHCl_3$ (3:1 and 1:1, respectively).

4.1.7 Reduction of other nitrogenated compounds

Hydrogen telluride reduces hydroxylamines to anilines, and nitroso, azo and azoxy compounds to the corresponding hydrazo compounds [1, 2].

The reduction of azo compounds to hydrazo compounds is also achieved by means of aryltellurols or sodium hydrogen telluride [3]. The last reagent (generated from $NaBH_4$ and Te catalyst) reduces tertiary amine *N*-oxides to the corresponding amines [26]. Azides are reduced to primary amines by sodium hydrogen telluride [27].

$$\mathrm{PhNHOH} \xrightarrow[\text{(98\%)}]{\mathrm{H_2Te\ (Al_2Te_3/H_2O)}} \mathrm{PhNH_2}$$

$$\mathrm{PhNO} \xrightarrow[\text{(78\%)}]{\text{as above}} \mathrm{PhNH\text{ - }NHPh + PhNH_2\ (4{:}1)}$$

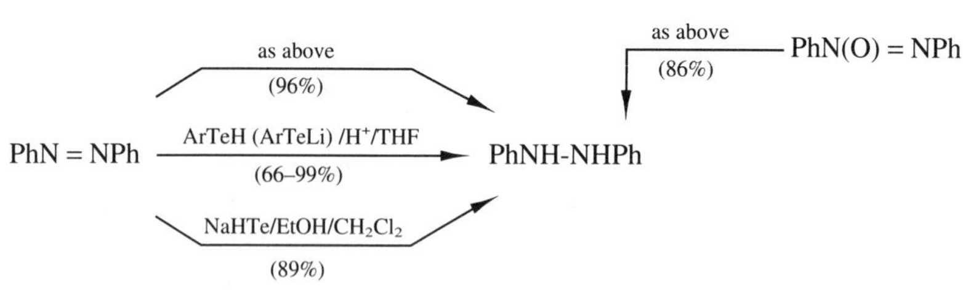

$$\mathrm{RR^1R^2NO} \xrightarrow[\text{EtOH (70–38\%)}]{\mathrm{NaHTe\ (NaBH_4/Te\ cat.)}} \mathrm{RR^1R^2N}$$

$$\mathrm{RN_3} \xrightarrow[\text{EtOH/ether (69–100\%)}]{\text{NaHTe}} \mathrm{RNH_2}$$

4.1.8 Deoxygenation of oxyranes with alkali *O,O*-dialkyl phosphorotellurolates

A useful complement to the known methods for the deoxygenation of epoxides to alkenes [28] is the reaction of epoxides with alkali *O,O*-diethyl phosphorotellurolates [29, 30].

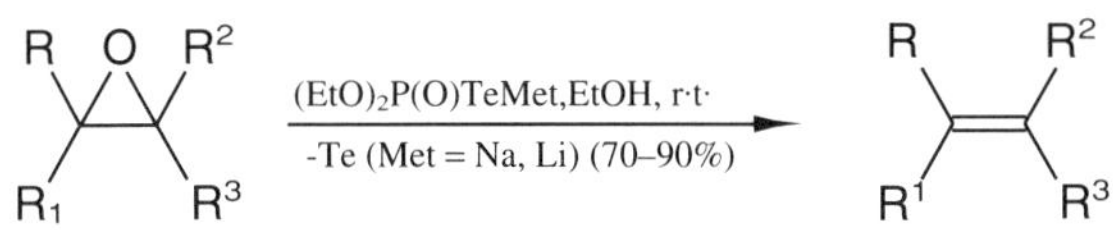

substrates: terminal epoxyalkanes, cyclohexenoxide and derivatives, cyclopentenoxide derivatives, *cis*-4,5-epoxyoctane

The reagent can be used in stoichiometric amounts or generated more conveniently *in situ*, employing only catalytic amounts of tellurium, since the metal is continuously regenerated during the reaction. Important points to note are:

- The tellurium reagents are more effective than their selenium analogues. The lithium salts seem to be the most reactive.
- Terminal oxiranes are more reactive than other types, (*Z*) isomers more than (*E*) isomers and cyclohexene oxides more than cyclopentene oxides.
- The deoxygenation is stereospecific, the (*Z*)-oxiranes giving the (*Z*)-olefins.

An epitelluride has been postulated as the intermediate.

Deoxygenation of 1,2-epoxyoctane with $(EtO)_2P(O)TeNa$ (typical procedure) [30] To Te powder (214.9 mg, 1.684 mmol) and 1,2-epoxyoctane (12.180 g, 95.01 mmol) is added from a syringe, under N_2 with magnetic stirring, a stock EtOH solution of $(EtO)_2P(O)Na$ (107.4 mmol) over 11 h. After stirring overnight, the mixture is filtered through cotton wool, diluted with H_2O (1000 ml), and extracted with isopentane (3 × 300 ml). The organic extract is dried (Na_2SO_4), concentrated, and distilled in a spinning band apparatus to give 1-octene as a colourless liquid (7.76 g (72%); b.p. 112°C); purity 99% by GLC (silver ion-impregnated columns).

4.1.9 Reductive opening of oxiranes with sodium hydrogen telluride and sodium telluride

Sodium hydrogen telluride reacts with epoxides, in accordance with an S_N2 displacement, giving rise to telluro-alcohols. These products are useful intermediates since they are easily converted to the corresponding alcohols and ketones by treatment with nickel boride followed by oxidation (reaction (a)) or to alkenes via the corresponding tosylates (reaction (b)) [31].

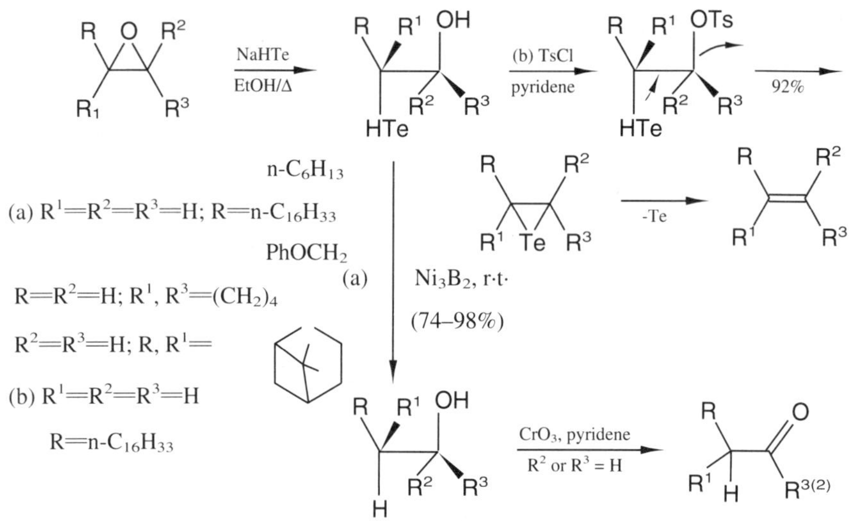

α,β-Epoxyketones are reduced by sodium hydrogen telluride to the corresponding β-hydroxy ketones [32].

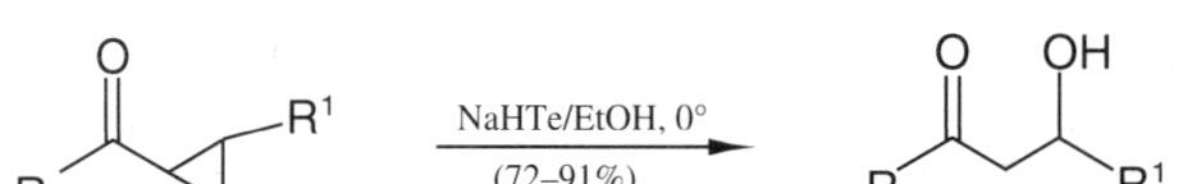

R = Me, Ph; R^1 = Ph

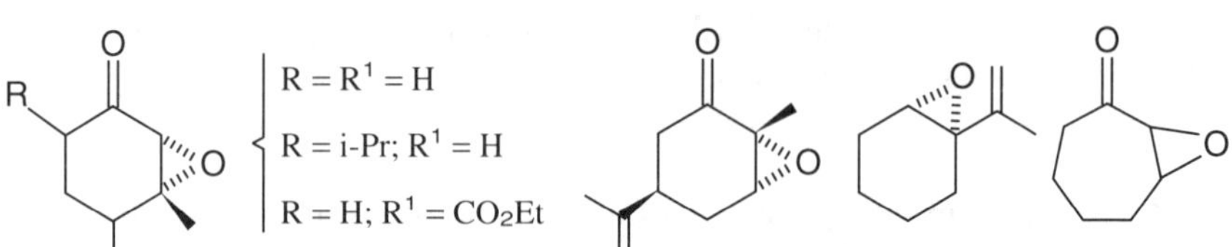

The following mechanism has been proposed for this reaction:

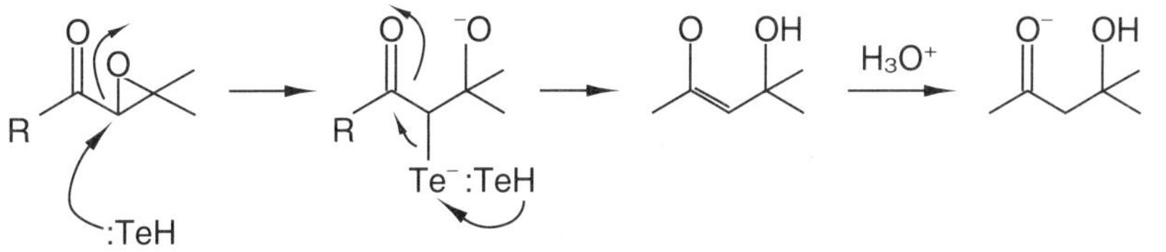

Reduction of α,β-epoxy ketones with NaHTe (typical procedure) [32] *β-Epoxy ketones* [32]. To a solution of NaHTe (prepared from Te (1.30 g, 10 mmol) and $NaBH_4$ (0.90 g 0.24 mmol)) in EtOH is added isophorone oxide (0.616 g, 4 mmol) in EtOH (4 ml). An instantaneous reaction occurs, the colour of the mixture changing to deep black. After 15 min H_2O (50 ml) is added and the mixture exposed to air at 0°C. The solution becomes clear after 1 h with deposition of black Te, and is filtered through Celite. The filtrate is extracted with CH_2Cl_2, and the extracts washed with H_2O, dried (Na_2SO_4) and evaporated, leaving a crystalline solid which is recrystallized from hexane to give 3-hydroxy-3,5,5-trimethylcyclohexanone (0.578 g (83%); m.p. 79–79.5°C).

Epoxides bearing a leaving group in a suitable position, such as chloromethyl epoxides, react with sodium telluride (prepared by the Rongalite method), giving allylic alcohols [33].

An unstable epitelluride has been assumed once more as an intermediate.

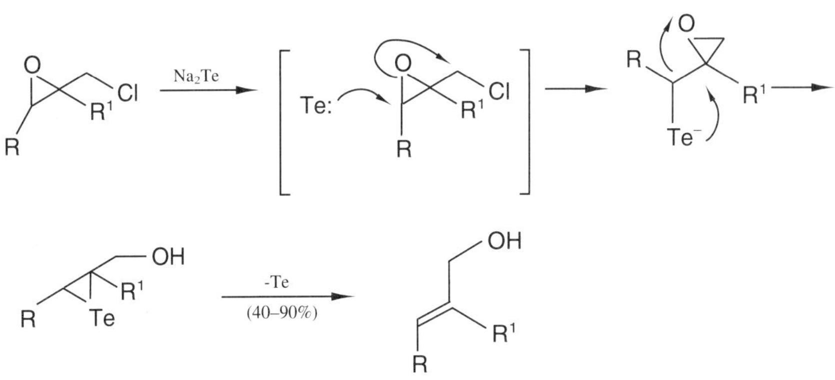

R = H; R[1] = Ph, Et, ⁄═, *p*-MeOC_6H_4

R = Ph; R[1] = H

Bischloromethylcarbinols are converted directly into allylic alcohols [33].

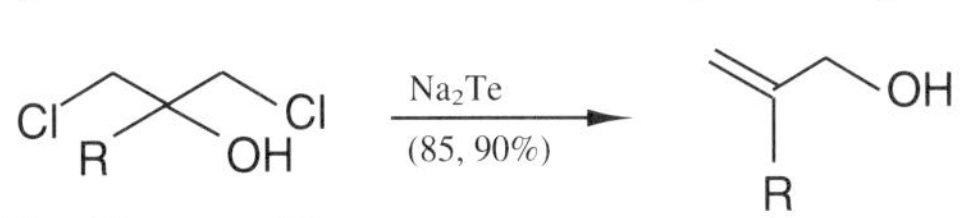

R = Ph, ≡—Ph

Reduction of chloromethyl epoxides with Na_2Te: allylic alcohols (general procedure) [33] To a Na_2Te solution (prepared from Te (0.52 g, 6.2 mmol), Rongalite (0.70 g, 6.2 mmol) and aqueous NaOH (2.5 ml) by heating at 55°C for 30 min) is added the chloromethyl epoxide (or the dichloropropanol) in dioxane (20 ml). After refluxing for 4 h the mixture is filtered through Celite, extracted with $CHCl_3$ and the solvent evaporated. The residue is purified by flash chromatography or TLC on SiO_2 (elution with Et_2O/hexane 3:7).

4.1.10 Correlate reaction—tellurium-mediated resolution of racemic allyl alcohols

The combination of the preceding method of obtaining allyl alcohols with the Sharpless kinetic resolution (SKR) of secondary allyl alcohols allows conversion of the original racemic allyl alcohol into a pure enantiomer with a 100% theoretical yield [34]. By this procedure, the glycidol obtained by the SKR epoxidation of the secondary allyl alcohol is converted into the corresponding mesylate and then treated with the Te^{2-} ion, furnishing the allylic alcohol with the same configuration of the enantiomer in the SKR which remains unreacted. An illustrative scheme is provided.

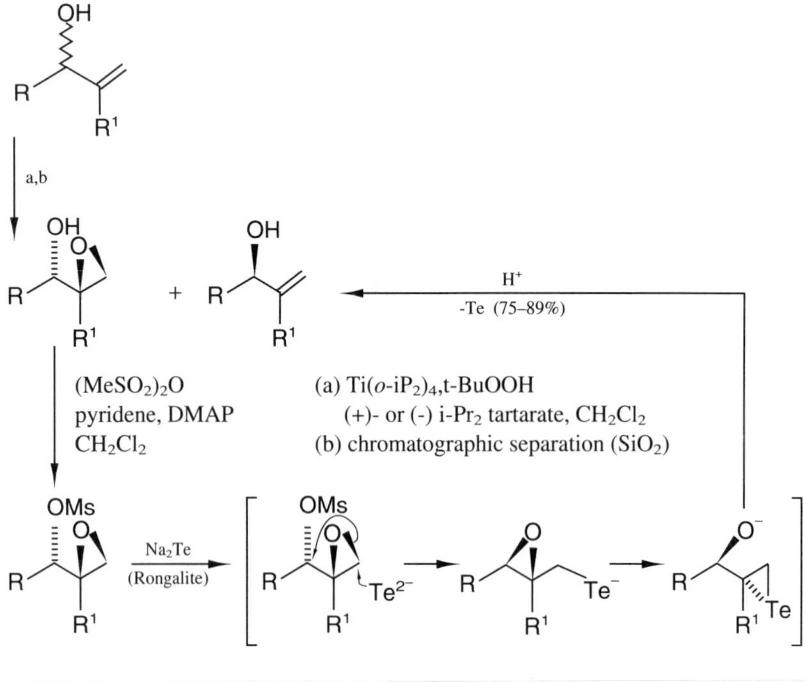

$R = Et$, n-Bu, $Me_2CH = CHCH_2$; $R^1 = Me$

$R = CH_2 = CH(CH_2)_3$; $R^1 = Et$

$R = Ph(CH_2)_2$; $R^1 = H$

The SKR of *cis* secondary allyl alcohols is often unsatisfactory since low enantioselectivity is observed.

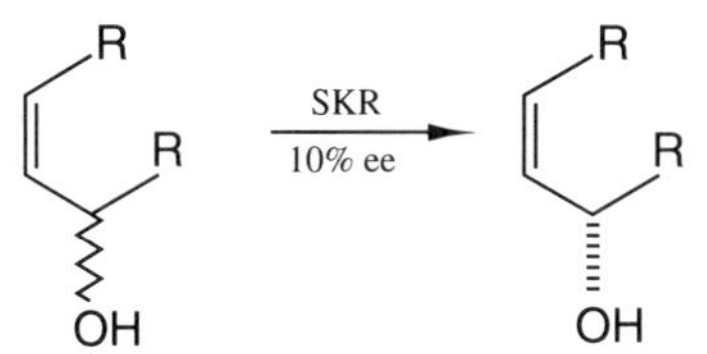

A valuable alternative for enhancing the selectivity is based on telluride-mediated transposition applied to the *trans*-allylic alcohols [35].

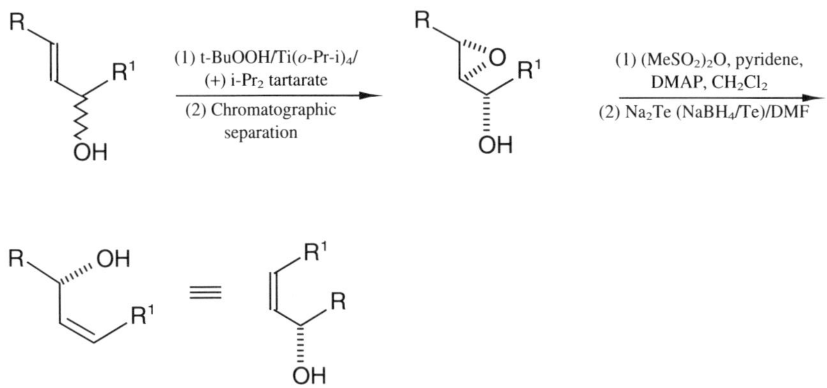

The transposition of the C=C bond and the alcohol functionality is highly stereospecific, *erythro*-glycidol sulphonates giving *cis*-allylic alcohols and the *threo* isomer the *trans*-allylic alcohol.

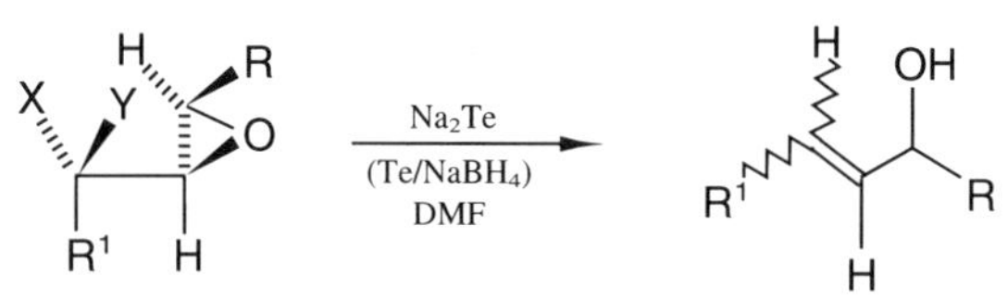

X = OMS; Y = H = *threo* ⟶ *trans*

X = H; Y = OMS = *erythro* ⟶ *cis* (80–91%)

R^1	R
Me	n-C_5H_{11}
cyclohexyl	Me
n-Pr	n-Pr
CH_2	Me

Resolution of racemic allyl alcohols (general procedure) [35] Elemental powdered Te (2.0–2.5 eq) and $NaBH_4$ (4.0–5.0 eq) in DMF (0.5–1.0 M in Te) are mixed under an inert atmosphere and heated to 70–80°C for 0.5 h. The grey suspension turns a deep purple colour. The reaction is cooled to room temperature and the glycidyl sulphonate (1.0 eq based on Te) in DMF is added. The reaction is monitored by ^1H NMR analysis of reaction aliquots and is generally complete in 3–4 h. Heating to 40°C is sometimes required for complete reaction. At completion, H_2O (about 0.5 × solution volume) is slowly added. Gas is evolved, and in some instances the solution becomes clear and colourless. When the reaction mixture is exposed to air, elemental Te precipitates as a greyish-black solid. H_2O_2 (five drops, 30% v/v) is cautiously added to complete the oxidation of the telluride ion to tellurium. The mixture is filtered through a bed of SiO_2 or Celite, and the Te is removed. The mixture is extracted with ether (3 × 20 ml) and the combined organic phases are washed with saturated aqueous $Na_2S_2O_3$, dried ($MgSO_4$), filtered, and the solvents removed by a rotary evaporator. Residual DMF is easily removed by flash chromatography.

4.1.11 Debromination of *vic*-dibromides with tellurium reagents

Several methods achieving the debromination of *vic*-dibromides by means of tellurium reagents are well established. These methods are particularly advantageous compared with the conventional ones [36] in terms of the mildness of the experimental conditions, good yields, lack of important side-reactions, and inertness of several functionalities to the employed reagents. A relevant characteristic of these reactions is the high E_2-type stereospecificity demonstrated by the formation of olefins with (*Z*) and (*E*) geometry from *threo*- and *erythro*-dibromides, respectively.

The following should be noted:

Method A [37]. Diaryltellurium dibromides are formed together with the olefinic product. The starting tellurides can easily be recovered by reduction of the dibromides (see Section 3.1.4).

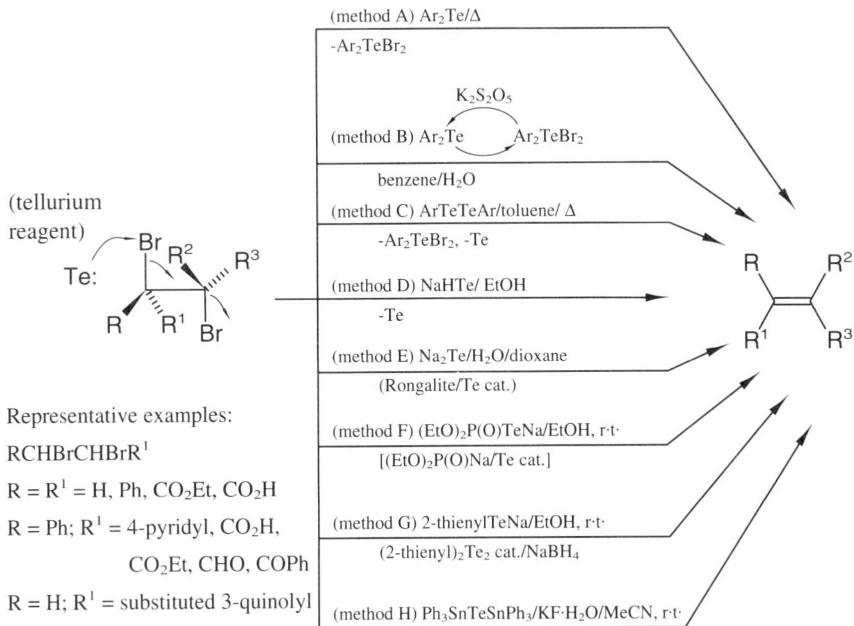

Representative examples:
$RCHBrCHBrR^1$
$R = R^1$ = H, Ph, CO_2Et, CO_2H
R = Ph; R^1 = 4-pyridyl, CO_2H, CO_2Et, CHO, COPh
R = H; R^1 = substituted 3-quinolyl
R, R^1 = $(CH_2)_4$, $(CH_2)_{10}$ and dibromocholesterol

Method B [38]. This is an improved modification of method A, effecting the debromination in a catalytic cycle by means of potassium bisulphite and catalytic amounts of tellurium.

Method C [39]. This method is attractive due to the easy one-step preparation of the starting ditelluride. The reaction can be rationalized via an aryltellurenyl bromide as an intermediate.

$$[ArTeBr] \longrightarrow Ar_2TeBr_2 + Te$$

Method D [10, 40]. Sodium hydrogen telluride furnishes higher yields under milder experimental conditions compared with the previously employed thio and selenium analogues. The tellurium metal can be recovered almost quantitatively and re-employed.

Methods E [41] *and F* [42]. Since elemental tellurium is separated during the reaction, the debromination can be performed using only catalytic amounts of tellurium, the reagent being continuously regenerated at the expense of excess Rongalite (method E) and $(EtO)_2P(O)Na$ (method F).

Method G [43]. A catalytic cycle is also involved in this method. Sodium thienyl tellurolate, generated from the ditelluride and $NaBH_4$, attacks a bromine atom and the formed tellurenyl bromide reacts with another tellurolate molecule, regenerating the starting ditelluride.

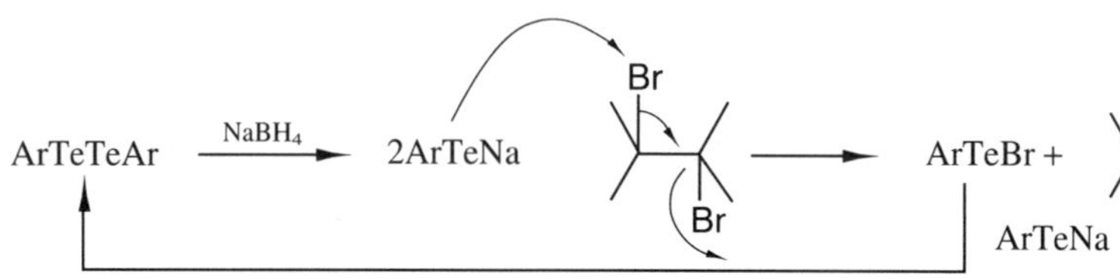

Method H [44]. The reagent is prepared by reacting NaHTe with chlorotriphenylstannane [45]. The reaction can be rationalized as shown.

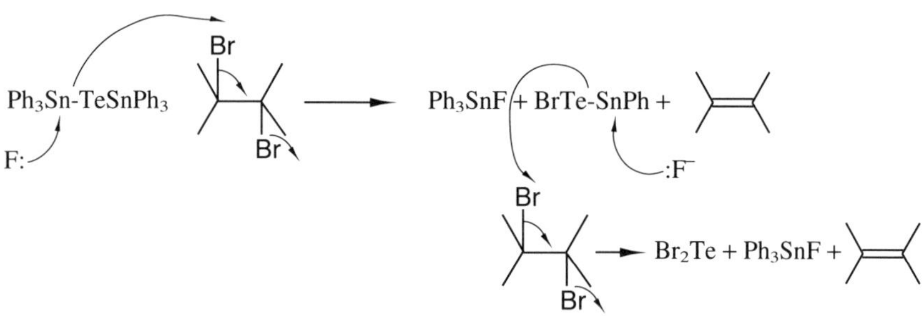

Debromination of vic-*dibromides Method B (general procedure)* [38]. To a solution of vicinal dibromide (1 mmol) and bis(4-methoxyphenyl) telluride (0.017 g, 0.05 mmol) in benzene (30 ml) is added $K_2S_2O_5$ (2.22 g, 10 mmol) in H_2O (20 ml). The mixture is refluxed with vigorous stirring, monitoring the progress of the reaction by GLC. After completion of the reaction, the mixture is cooled to room temperature, and the organic layer is separated, washed with H_2O, dried ($Na_2S_2O_4$) and evaporated. The crude product is purified by chromatography on SiO_2.

Method C (general procedure) [39]. The vicinal dibromide (1–2 mmol) and the diaryl ditelluride (1–2 mmol) are refluxed in toluene or HOAc for 1–4 h (1,2-dibromoethane is used neat). After cooling, the mixture is filtered to remove the Te (quantitative yield) and the diaryltellurium dibromide is precipitated by addition of petroleum ether 30–60°C. The alkene is obtained by evaporation of the filtrate. (Cinnamic and fumaic acids, both insoluble in petroleum ether, are separated from the tellurium dibromide with EtOH.)

Method D (general procedure) [40]. To a solution of NaHTe (10 mmol), prepared *in situ* from Te (1.30 g) and $NaBH_4$ (0.90 g) in EtOH, is added the dibromide (10 mmol). An instantaneous reaction occurs and the colour of the reaction mixture changes to deep black. The mixture is heated under reflux for 2 h and then poured into ice-cold H_2O. The product is extracted with $CHCl_3$, dried and evaporated. The residue is purified by distillation or by recrystallization as appropriate.

Method E (typical procedure) [41]. A solution of 1,2-dibromohexadecane (1.152 g, 0.3 mmol) in EtOH (5 ml) is added to a mixture of Te powder (0.382 g, 3 mmol), Rongalite (0.70 g, 6.0 mmol) and 1 M NaOH (15 ml), under N_2 and at room temperature. After stirring at 70°C until the colour of the solution turns deep red, excess reagent is destroyed by addition of 2 M HCl (10 ml). The resulting black suspension is filtered through a thin bed of Celite 535 and the filtrate extracted with ether. The extracts are washed with H_2O, dried (Na_2SO_4) and evaporated, leaving a colourless oil which is distilled using a Kugelrohr apparatus under vacuum to give the pure product (0.551 g (82%)).

Method F (typical procedure) [42]. $(EtO)_2P(O)Na$ (1.10 g, 6.87 mmol) in EtOH (10 ml) is added to Te powder (0.025 g, 0.2 mmol) under N_2. The mixture is stirred, and when all the black powder disappears *meso*-stilbene dibromide (0.68 g, 2 mmol) is added. The reaction mixture immediately becomes orange, and then black. The reaction is monitored by SiO_2 TLC (eluting with petroleum ether/ether 9:1) until the reaction is complete (about 6 h). H_2O (20 ml) is added and the reaction system is opened to the atmosphere to precipitate Te. Half an hour later, the solid is filtered with the aid of ethyl ether (20 ml). The filtrate is extracted with ether (3×20 ml). The organic layer is washed with brine and dried ($MgSO_4$). The solvent is removed and the residue is purified by SiO_2 column chromatography (eluent: petroleum ether/ethyl ether 9:1) to give a white crystalline product (0.32 g (88.9%); m.p. 123.5–125°C).

Method G (typical procedure) [43]. $NaBH_4$ (5%) in 5% aqueous NaOH is added dropwise under N_2 to a solution of 5α,6β-dibromocholestan-3β-ol (1.50 g, 2.73 mmol) and bis(2-thienyl) ditelluride (0.10 g, 0.24 mmol) in EtOH (20 ml) at room temperature until the red colour of the ditelluride disappears. Air is then introduced to the system to oxidize the catalyst to the ditelluride. The mixture is diluted with ether and H_2O, and the organic phase is washed with H_2O, dried and evaporated. Chromatography of the residue on SiO_2 (eluting with CH_2Cl_2/MeOH 95:5) furnishes cholesterol (0.95 g (90%); m.p. 149–150°C).

The tellurolate-promoted *anti* debromination by method G is extended to the preparation of conjugated dienes, starting from 1,2,3,4-tetrabromoalkanes and cycloalkanes, 1,4-dibromo-2-alkenes and allylic dibromides [46].

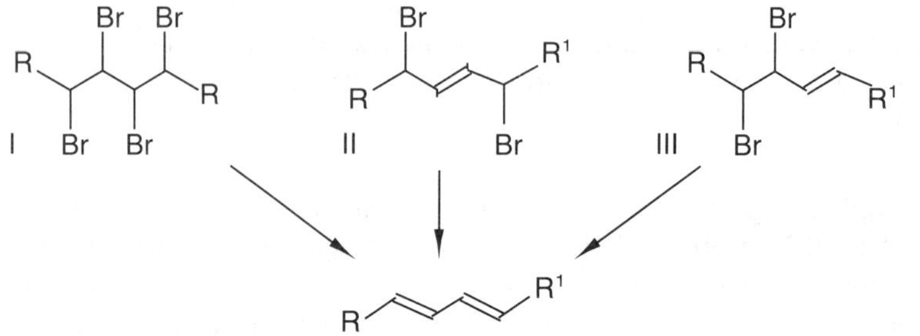

Representative examples:

I 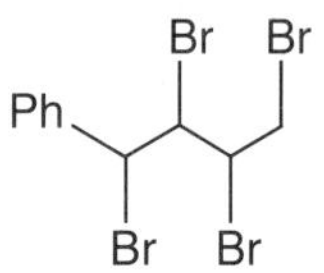

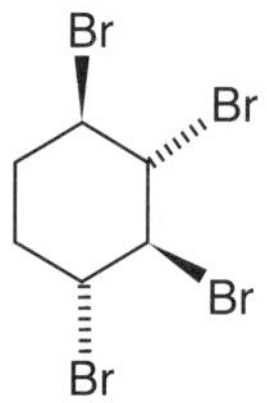

II R = H: R^1 = Me,Ph

III R = H; R^1 = Et, n-Bu, n-hexyl, $(CH_2)_n$OAc (n = 7, 8)

Conjugated dienes (general procedure) [46] $NaBH_4$ (10% in 0.1% aqueous NaOH) is added dropwise at room temperature under N_2 to a solution of the allylic dibromide in MeOH or THF (1.5–2 mmol/15 ml solvent) containing bis(2-thienyl) ditelluride (5–10 mol %) until disappearance of the orange-red colour of the ditelluride. Two and up to three equivalents of $NaBH_4$ are required in the MeOH and THF runs, respectively. The mixture is then poured into H_2O and extracted with pentane. During this process the ditelluride is usually re-formed and extracted into the organic layer. In some cases (synthesis of terminal dienes) the ditelluride is not reformed, but the formation of a white precipitate is observed. The combined organic extracts are washed several times with H_2O, dried and evaporated. The diene products are purified by distillation or column chromatography.

Another related reaction is the debromination of α,α'-dibromo-o-xylenes by sodium aryl tellurolates. The formed o-quinodimethane is allowed to react *in situ* with dienophiles to give the Diels–Alder adducts [47].

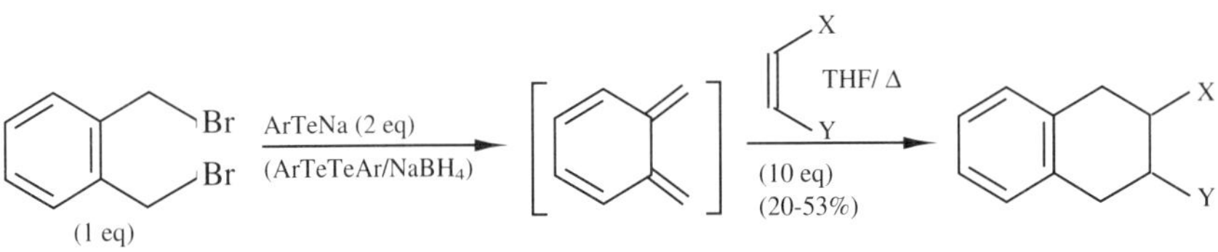

Ar = Ph, *p*-Me, *p*-MeO, *m*-ClC_6H_4; X = H; Y = CO_2Et, COMe, CN
X = Y = CO_2Et

The debromination step has been rationalized (in view of the 1:2 ratio of the reagents) as involving an attack by the tellurolate ion at the benzylic carbon, followed by an attack by a second tellurolate at the tellurium of the intermediate product.

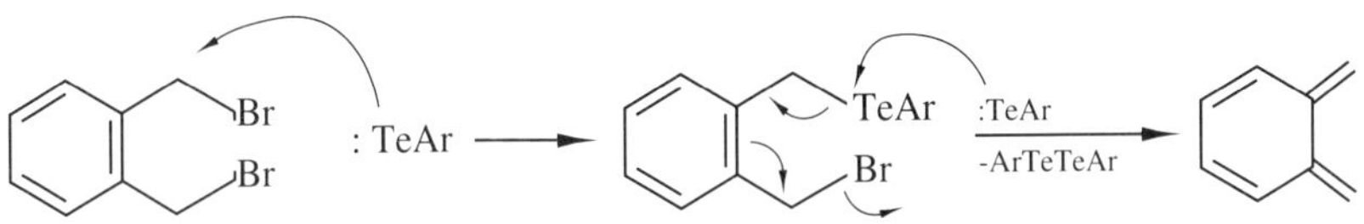

The reactivity of the tellurolate ion is peculiar, since the corresponding aryl selenolate gives only the disubstituted *o*-selenoxylenes, and sodium hydrogen telluride gives only poor yields.

Ethyl-2-tetralin carboxylate (general procedure) [47] To a solution of sodium phenyl tellurolate (from diphenyl ditelluride (2.05 g, 5 mmol) and $NaBH_4$ (0.400 g, 10.5 mmol) in EtOH (30 ml) at 25°C) is added a solution of ethyl acrylate (5.0 g, 50 mmol) and α,α′-dibromo-*o*-xylene (1.32 g, 5 mmol) in THF (15 ml). The mixture is refluxed for 2 h with stirring, and then stirred overnight in contact with air at room temperature. The red colour fades and a white precipitate is formed. The precipitate is filtered, the product extracted with ether, and the extracts dried ($MgSO_4$) and evaporated. The residue is chromatographed on SiO_2 (eluting with benzene), to give the product (0.46 g (45%)).

4.1.12 Reductive fission of carbon–heteroatom bonds

4.1.12.1 Reductive removal of electronegative α substituents from ketones, acids and derivatives

The removal of functional groups from the α carbon of carbonyl compounds is an important transformation in organic synthesis [48]. Anionic tellurium reagents offer additional useful methods to attain this.

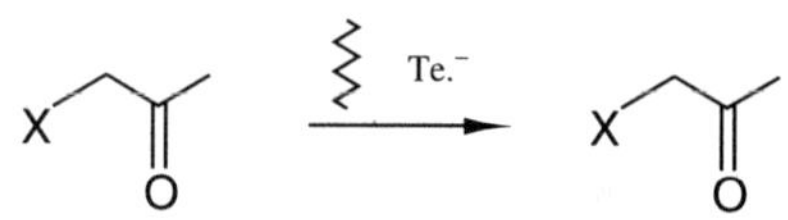

Sodium hydrogen telluride [49], sodium *O,O*-diethyl phosphorotellurolate [50], alkali 2-thienyl tellurolates [51] and bis(triphenylstannyl) telluride/KF [52] reduce α-haloketones to the corresponding ketones.

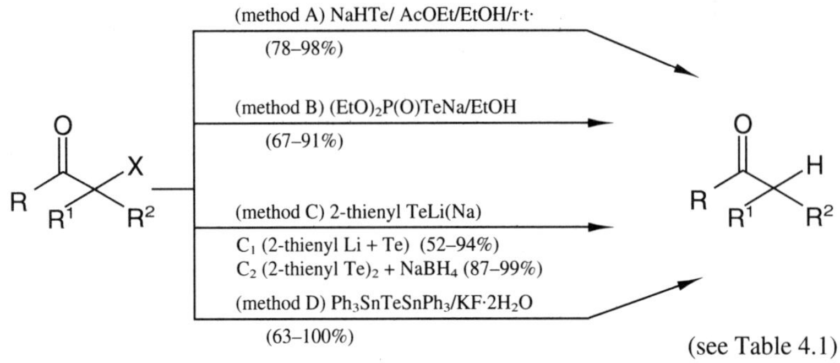

The following mechanisms have been proposed to rationalize the described reactions:

Method A

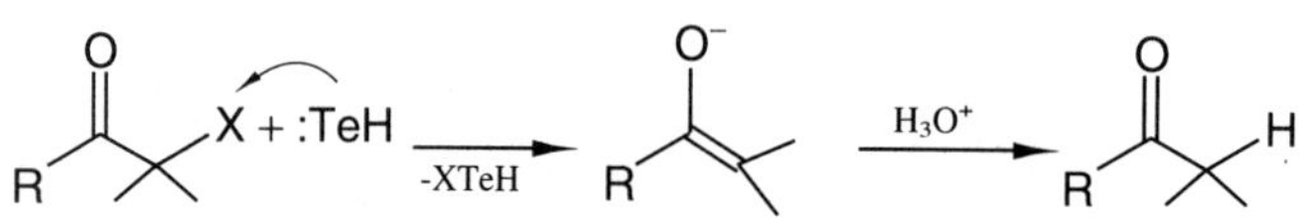

Method B

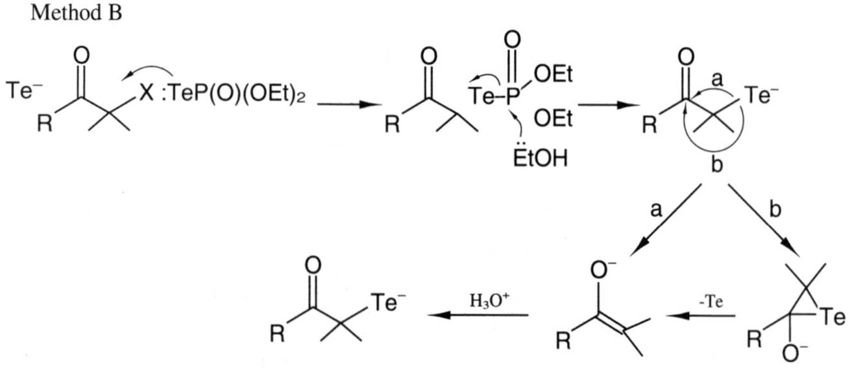

TABLE 4.1

α-Haloketones and the methods used for their reduction

α-Haloketone	Method	α-Haloketone	Method
$PhCOCH_2X$ X = Cl	A, B, C, D	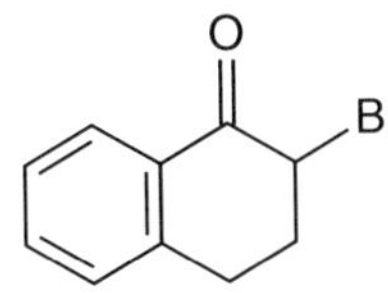	A, B
X = Br	A, C, D		
X = I	C		
p-$YC_6H_4COCH_2Br$ Y = Ph, Br, HO	A	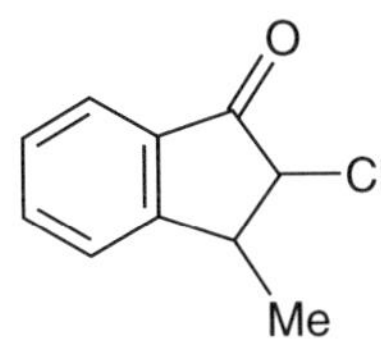	C
Y = Br	B, C		
$PhNHCOCH_2Br$	C	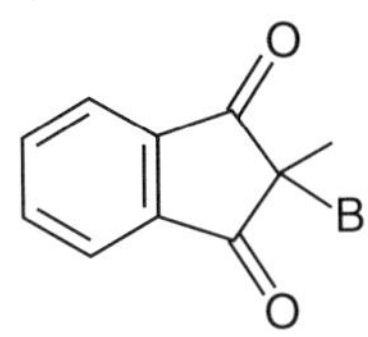	A
$PhCH_2OCOCH_2Cl$	A		
PhCOCHPhCl	C		
$Me_2CCOCMe_2Br$	B[a]		
CH_3COCH_2Cl	D	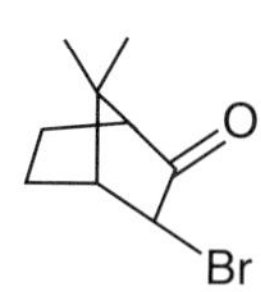	A
$EtCOCH_2Br$	D		
$PhCOCH(Me)CH_2Br$	D		
$PhCOCMe_2CH_2Br$	D		
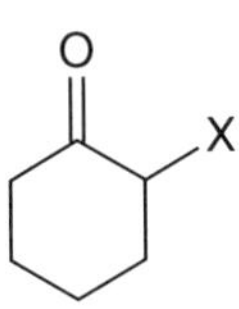 X = Cl	B[b], D	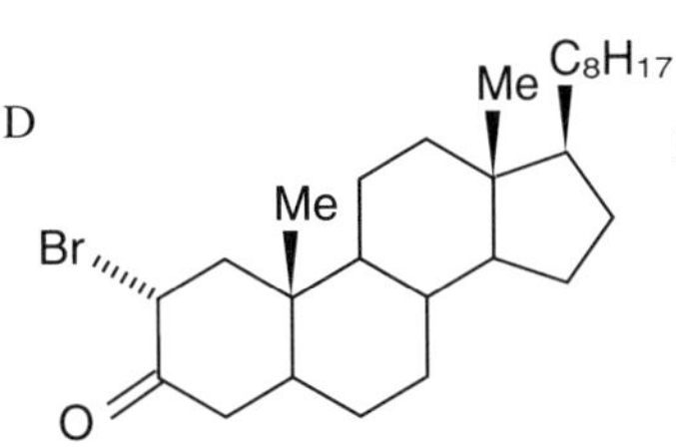	B, C
X = Br	B[c]		
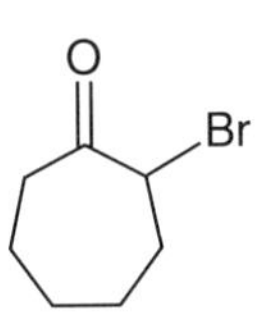	B		
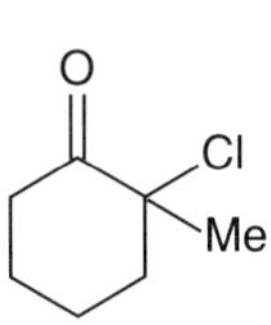	B[d]	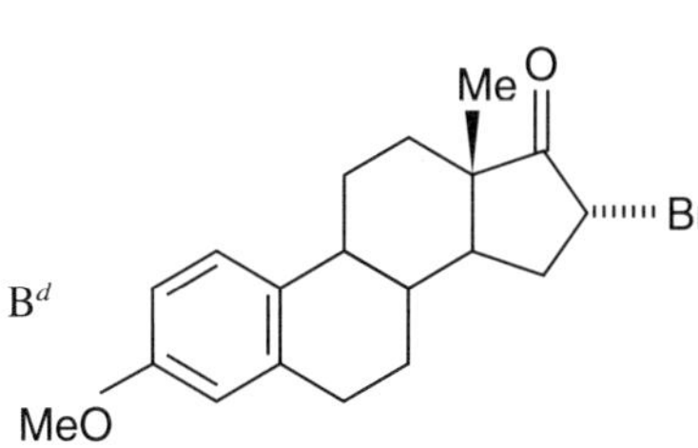	B

[a] Yield 48%. [b] Yield 36%. [c] Yield 56%. [d] Yield 19%.

Method C

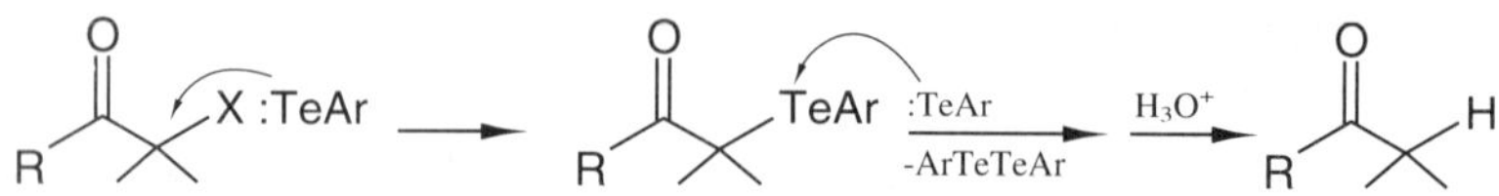

Method D

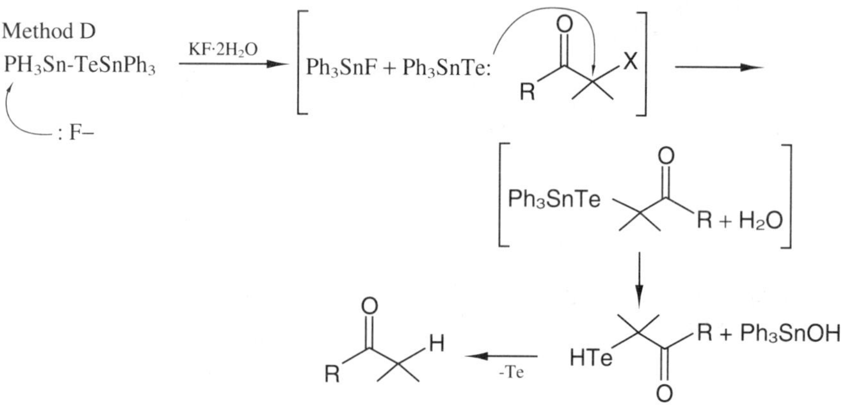

Since in method C di(2-thienyl) ditelluride is formed as a by-product and $NaBH_4$ seems to reduce ditellurides preferentially to the carboxyl group, a catalytic procedure can also be employed in which $NaBH_4$ is added to the halo compound in the presence of a catalytic amount (0.1 eq) of the ditelluride (which is continuously regenerated).

2-Thienyl tellurolates also remove several different substituents such as acetoxy, mesyloxy, phenylthio and 2-thienyltelluro groups, as well as effecting the reductive dehalogenation of bromoacetanilide, and of α-haloacids (such as α-bromophenylacetic acid, α-bromo-1-naphthylacetic acid and α-chlorodiphenylacetic acid) [51].

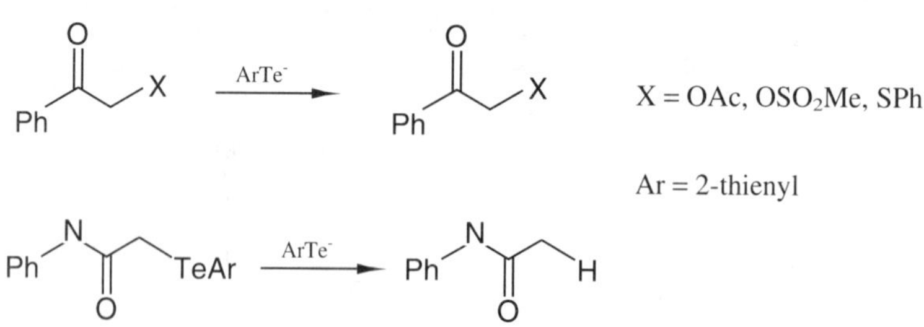

Dehalogenation of α-haloketones Method A (typical procedure) [49]. A solution of *p*-bromophenacyl bromide (1.112 g, 4 mmol) in EtOAc (20 ml) is added to a solution of NaHTe, prepared from Te (1.3, 10 mmol) and $NaBH_4$ (0.9 g, 22.5 mmol) in EtOH (20 ml). An instantaneous reaction occurs and the colour of the reaction mixture changes to black. After stirring for 1 h at room temperature the mixture is filtered through Celite 545, the filtrate diluted with ether, and the organic layer washed with H_2O and dried (Na_2SO_4). Evaporation of the solvent gives *p*-bromoacetophenone, which is recrystallized from EtOH (0.737 g (93%); m.p. 50–52°C).

Method B (general procedure) [50]. Finely divided Te (0.128 g, 1.0 mmol) is treated with an EtOH solution of sodium diethyl phosphite (1.5 mmol) in an apparatus which has been previously evacuated and flushed with argon. After a clear solution has formed, the α-haloketone (1.0 mmol) in dry EtOH or THF (2.0 ml) is injected. The mixture is stirred (25 or 80°C, 1–15 h) and then filtered through Celite with the aid of ether. The orange filtrate is evaporated at room temperature under vacuum (water pump) and the residue applied to an SiO_2 column, proceeding to elution (hexane/EtOAc) only after 30 min, to allow decomposition of the residual tellurium species. If necessary the products are further purified by crystallization or distillation.

Method C1 (typical procedure) [51]. n-BuLi solution (1.6 M, 7.5 ml, 12.0 mmol) is added under N_2 to a stirred solution of thiophene (1.12 g, 13.3 mmol) in dry THF (50 ml) at 0°C. After 50 min at room temperature, finely powdered Te (1.50 g, 11.8 mmol) is rapidly added under N_2. After 30 min all the Te will have been consumed, and a solution of *p*-bromophenacyl bromide (1.6 g, 5.75 mmol) in THF (5 ml) is then added dropwise over 10 min. The solution turns from light yellow to deep red as bis(2-thienyl) ditelluride is formed. After 30 min the solvent is evaporated and the residue partitioned in H_2O/ether. The organic phase is dried ($CaCl_2$) and evaporated to give a red solid. Chromatography (SiO_2, CH_2Cl_2/hexane 1:1) gives *p*-bromoacetophenone (0.82 g (72%); m.p. 49–50°C).

Method C2 (typical procedure) [51]. A 5% solution of $NaBH_4$ in 5% NaOH is added dropwise to a stirred suspension of bis(2-thienyl) ditelluride (0.75 g, 1.78 mmol) in EtOH under N_2, until disappearance of the red colour of the solution. 2-Bromocholestan-3-one (0.80 g, 1.72 mmol) in EtOH (5 ml) is then added dropwise over 10 min, causing an immediate red coloration. After 30 min the mixture is poured into H_2O/diethyl ether, and the organic phase is washed several times with H_2O and dried ($CaCl_2$). Evaporation of the solvent and chromatography (SiO_2/CH_2Cl_2) gives cholestan-3-one (0.62 g (93%); m.p. 127–128°C).

Method C2 (typical catalytic procedure) [51]. A 5% solution of $NaBH_4$ in 5% aqueous NaOH is added dropwise under N_2 to a solution of α-bromophenylacetic acid (1.50 g, 7.0 mmol) and bis(2-thienyl) ditelluride (0.30 g, 0.71 mmol) in EtOH (40 ml) until the red colour of the ditelluride just disappears. At this point air is introduced in the system to oxidize the catalyst back to the ditelluride. The mixture is then partitioned in ether/5% aqueous NaOH. The aqueous phase is separated, acidified with 2 M HCl and extracted with ether. The extract is dried ($CaCl_2$) and evaporated to give phenylacetic acid (0.93 g (98%); m.p. 77°C).

Method D (typical procedure) [52]. A mixture of phenacyl bromide (0.796 g, 0.4 mmol), bis(triphenylstannyl) telluride (3.30 g, 0.4 mmol) and $KF{\cdot}2H_2O$ (0.103 g, 1.2 mmol) in MeCN (10 ml) is stirred at room temperature, under N_2. Very soon a black solid is deposited. After stirring for 5 h, the reaction mixture is filtered through Celite and the filtrate evaporated. The product (acetophenone) is isolated by flash chromatography (hexane/EtOAc 20:1).

4.1.12.2 Dehalogenation of polyhalogenated organic compounds

Some investigations directed to the dehalogenation of polyhalogenated organic derivatives by means of tellurium reagents have been reported.

(a) 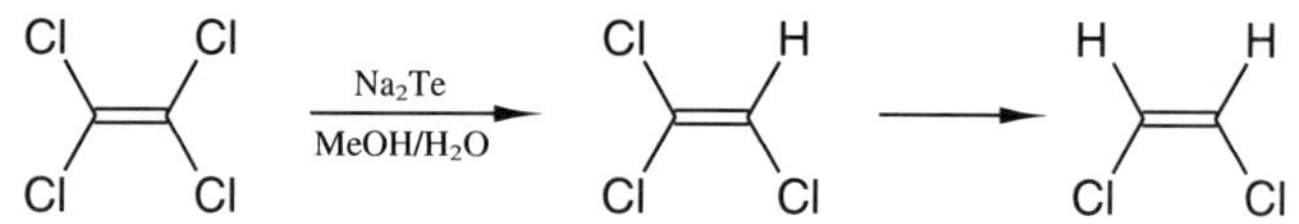 [53]

(b) hexachlorobenzene $\xrightarrow[\text{procedure}]{\text{same}}$ 1,2,4-$Cl_3C_6H_3$ [53]

(c) Cl_nthiophenes $\xrightarrow[\text{procedure}]{\text{same}}$ Cl_{n-1} thiophenes [53]

(d) 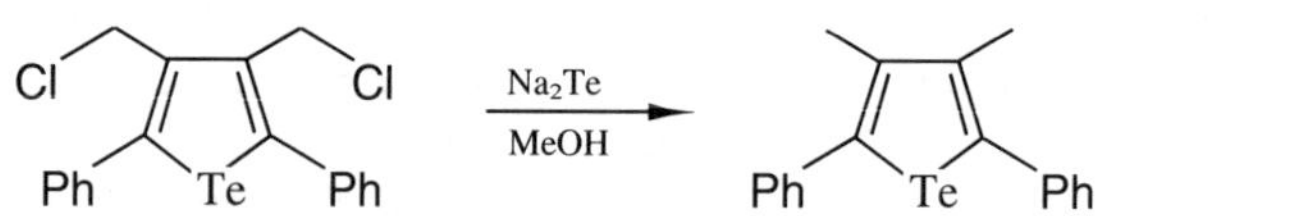 [54]*

*Similar reactions performed with Na_2S and Na_2Se lead to the heterocyclic compounds

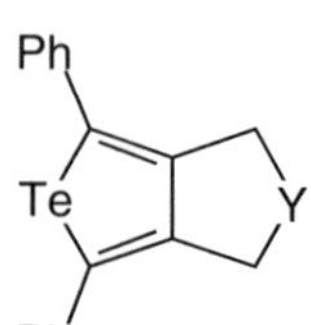 Y = S, Se

(e) 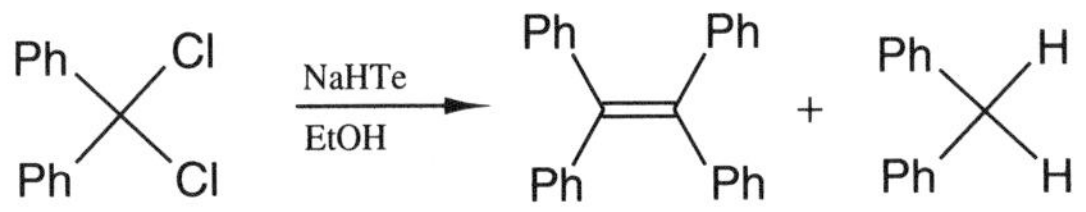 [55]

(f) *gem*-dibromocyclopropanes $\xrightarrow[\text{EtOH}]{\text{NaHTe}}$ monobromocyclopropanes [55]

4.1.12.3 Reductive removal of tertiary nitro groups

Sodium hydrogen telluride replaces tertiary nitro groups by hydrogen in a very simple procedure which seems to be free from the drawbacks of previously described methods [56]. The viability of the reaction is dependent on the presence of a carbonyl or an ester group at the α or β position to the nitro group. α-Nitrocumene derivatives undergo a similar reductive fission. A reaction is not observed if the activating group is situated at the γ position.

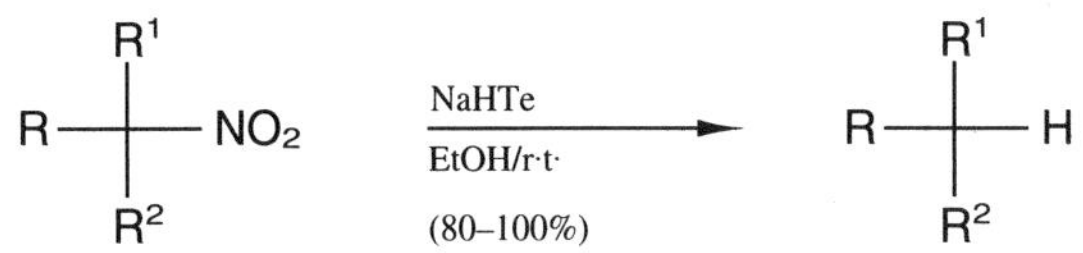

R = Me; R^1 = CO_2Et; R^2 = COMe

R^1 = R^2 = Me; R = $MeC(CO_2Et)_2$, $MeC(CO_2Et)(COMe)$, $MeC(CO_2Et)CN$, *p*-MeCO, *p*-NC, *p*-$PhSO_2$, *p*-$EtO_2CC_6H_4$

The reaction has been suggested as involving a one-electron transfer from sodium hydrogen telluride to the nitro compound followed by the detachment of a nitrite anion from the resulting radical anion.

Reductive removal of tertiary nitro groups (general procedure) [56] A solution of the tertiary nitro compound (4 mmol) in EtOH (4 ml) is added at room temperature under N_2 to a solution of NaHTe, prepared *in situ* by reacting Te powder (1.30 g, 10 mmol) and $NaBH_4$ (40 ml). An instantaneous reaction occurs and the colour of the reaction mixture changes to deep black. After 5 min, H_2O (50 ml) is added, and the resulting mixture is left exposed to air, with stirring. The solution becomes clear after 1 h, with the deposition of Te powder. It is then filtered through Celite 545 and the filtrate is extracted with three portions of CH_2Cl_2 (30 ml). The combined extracts are washed with H_2O, dried (Na_2SO_4) and evaporated to give the crude product, which is purified by Kugelrohr distillation or recrystallization.

4.1.12.4 Reductive dealkylation of quaternary ammonium salts

Sodium hydrogen telluride efficiently dealkylates quaternary ammonium salts to the corresponding tertiary amines in high yields [31].

$$[\mathrm{Ph\overset{+}{N}Me_3}]\mathrm{I}^- \xrightarrow[\text{-MeTeH}]{\text{NaHTe/EtOH}} \mathrm{PhNMe_2}$$

$$[\mathrm{R_2\overset{+}{N}R^1R^2}]\mathrm{I}^- \xrightarrow{\text{NaHTe/EtOH}} \mathrm{R_2NR^1}$$

R^1 = Me; R^2 = Et; R = 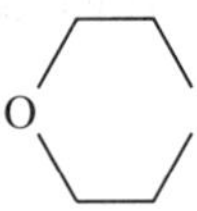

$R^1 = R^2$ = Me; R = cyclohexyl 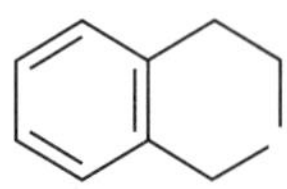 and derivatives

4.1.12.5 Reductive desulphonylation of β-ketosulphones

β-Ketosulphones, inert towards sodium hydrogen telluride in EtOH [27], are easily desulphonylated by the same reagent in the aprotic solvent DMF [12, 57], or under phase transfer catalysis [57].

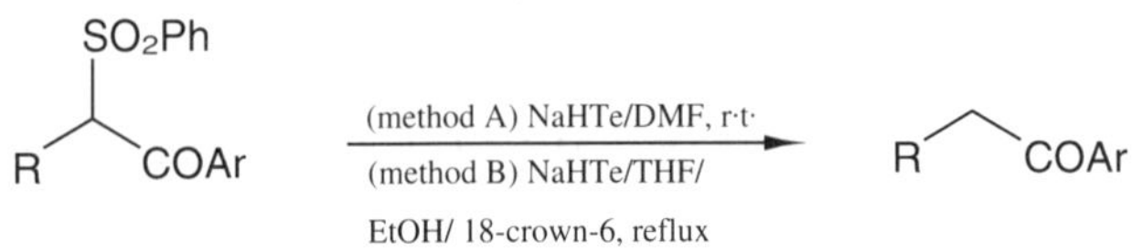

Ar = Ph; R = Me, $PhCH_2$, *p*-Me, *p*-MeO, *p*-Cl, *p*-$BrC_6H_4CH_2$

Ar = *p*-Me, *p*-$MeOC_6H_4$; R = $PhCH_2$

Desulphonylation of β-ketosulphones (typical procedure) [57] *Method A*. To a violet solution of NaHTe, prepared by heating a mixture of Te powder (0.77 g, 6 mmol) and $NaBH_4$ (0.28 g, 8 mmol) in DMF (10 ml) at 70–80°C for 30 min under N_2, is added a solution of the ketosulphone (Ar = *p*-MeC_6H_4, R = $PhCH_2$) (2 g, 5 mmol) in THF at room temperature. The end of the reaction is detected by TLC (SiO_2, benzene). The mixture is then quenched with H_2O (20 ml) and filtered. The filtrate is extracted with $CHCl_3$ (3 × 20 ml). The organic phase is washed with H_2O (3 × 20 ml) and brine (3 × 20 ml), and dried ($MgSO_4$). After evaporating the solvent, the crude product is obtained and recrystallized from EtOH to give (2-phenyl)ethyl *p*-tolyl ketone (0.83 g (72.1%); m.p. 66–67°C).

Method B. To a solution of NaHTe (5 mmol), prepared by heating a mixture of powdered Te (0.77 g, 5 mmol) and $NaBH_4$ (0.56 g, 12 mmol) in EtOH (15 ml), is added a solution of the same ketosulphone (2 g, 5 mmol) and catalytic amounts of 18-crown-6 (0.02 g) in THF (20 ml) at room temperature under N_2. The final work-up is similar to that of method A to give the ketone (1.05 g (90.1%); m.p. 66–67°C).

The above-described desulphonylation, combined with the previously described selective reduction of α,β-unsaturated carbonyl compounds, can be applied to α-alkylidene β-ketosulphones (easily prepared by a conventional Knovenagel condensation of β-ketosulphones with aldehydes), achieving the direct one-pot synthesis of unsymmetrical ketones [12].

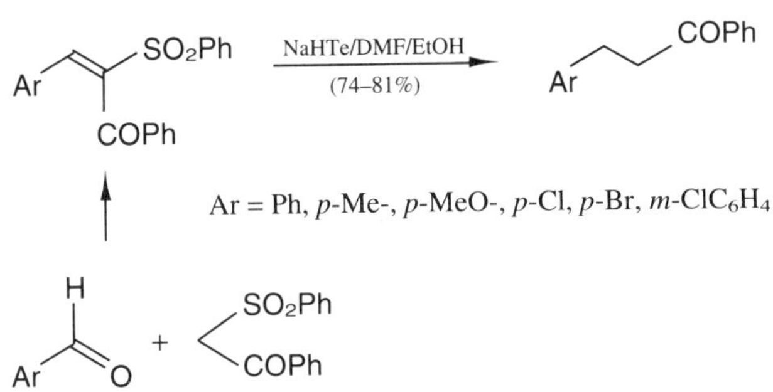

Reductive desulphonylation of α-alkylidene β-oxosulphones (typical procedure) [12] To a solution of NaTeH, prepared from Te (0.65 g, 5 mmol) and $NaBH_4$ (0.45 g, 12 mmol) in EtOH (20 ml) under N_2, is added, while stirring, a solution of α-(*p*-bromobenzylidene) β-ketosulphone (0.73 g, 2 mmol) in DMF (15 ml). The solution, which immediately turns red-black, is stirred at room temperature for 3 h, and after addition of H_2O (30 ml) is exposed to air for 30 min, to precipitate Te. The mixture is filtered, the filtrate extracted with Et_2O (3 × 40 ml), and the combined ether extracts dried ($MgSO_4$) and concentrated *in vacuo* to give the crude product, which is purified by SiO_2 chromatography (elution with benzene/EtOAc 10:1), giving pure 2-(*p*-bromophenyl)ethyl phenyl ketone (0.46 g (80%), m.p. 65–66°C).

4.1.12.6 Desulphonylative condensation of α-cyanosulphones with aldehydes

α-Alkilidene β-cyanosulphones are desulphonylated by sodium hydrogen telluride, giving α,β-unsaturated nitriles with maintenance of the C=C bond and retention of its configuration [58].

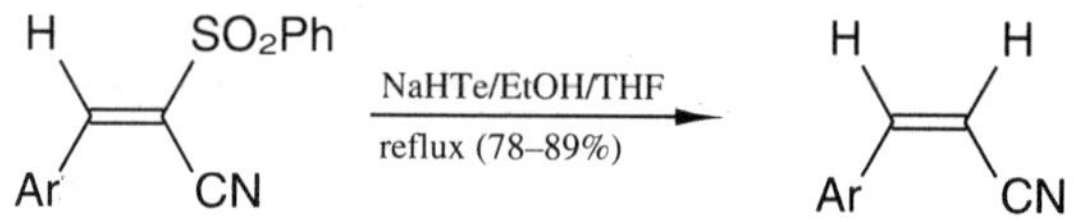

(method A)

Ar = Ph, *p*-MeO, *p*-Cl, *o*-Cl, *m*-Cl, *p*-Br, *m*-Br, *p*-$Me_2NC_6H_4$

Otherwise, the same products can be obtained by a Knovenagel-type reaction between β-cyanosulphones and aromatic aldehydes in the presence of sodium telluride (behaving as a base). This result is consistent with the above-formulated desulphonylation of an intermediate α-alkylidene β-cyanosulphone at the expense of the sodium hydrogen telluride formed during the reaction.

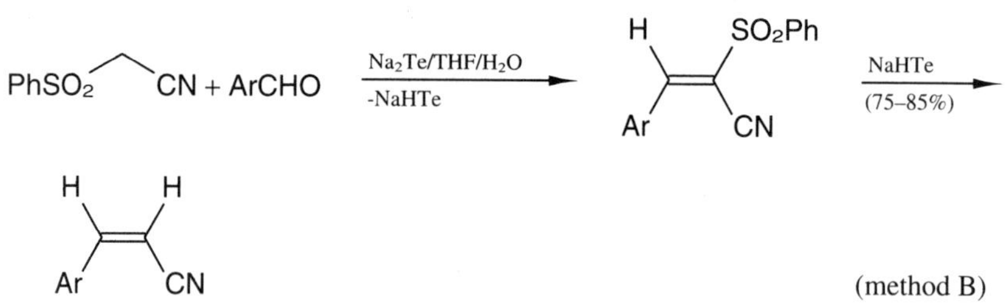

(method B)

α,β-Unsaturated nitriles (typical procedure) [58] *Method A*. To a solution of NaHTe, prepared from Te (1.3 g, 10 mmol) and $NaBH_4$ (0.9 g, 24 mmol) in EtOH (20 ml), under an N_2 atmosphere, is added a solution of (2*E*)-α-cyano-β-(*p*-chlorophenyl) ethenyl phenyl sulphone (3.1 g, 10 mmol) in THF (20 ml), and the mixture is stirred under reflux for 30 min. The reaction is quenched with H_2O (20 ml), the mixture is filtered and the filtrate is extracted with $CHCl_3$ (3 × 20 ml). The combined organic extracts are dried ($MgSO_4$) and concentrated to give the crude product, which is purified by SiO_2 column chromatography (eluent: benzene), giving pure (*Z*)-*p*-chlorocinnamonitrile (1.40 g (85%); m.p. 63–65°C).

Method B. To a solution of Na_2Te, prepared by heating a mixture of Te (1.30 g, 10 mmol), $NaBH_4$ (0.9 g, 24 mmol) and H_2O (10 ml) at 60–70°C under an N_2 atmosphere, is added a solution of α-cyanomethyl phenyl sulphone (1.81 g, 10 mmol) and *p*-chlorophenyl aldehyde (1.8 g, 12 mmol) in THF (20 ml). An instantaneous reaction occurs, and the colourless reaction mixture becomes violet. The end of the reaction is detected by TLC (SiO_2, benzene as the eluent). The remaining work-up is analogous to method A and gives the pure product (1.3 g (79%); m.p. 63–65°C).

4.1.12.7 Correlate reaction—desulphonylation of α-nitrosulphones

By treating a quaternary α-nitrosulphone with sodium hydrogen telluride, the sulphonyl group is displaced preferentially to the nitro group, as illustrated in the example [56].

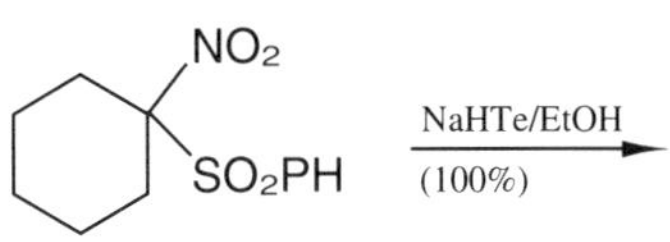

4.1.12.8 Monodesulphuration of diaryl thioketals and bis-sulphenylated β-dicarboxyl compounds

The title reactions have been performed using sodium telluride in aprotic solvents [59].

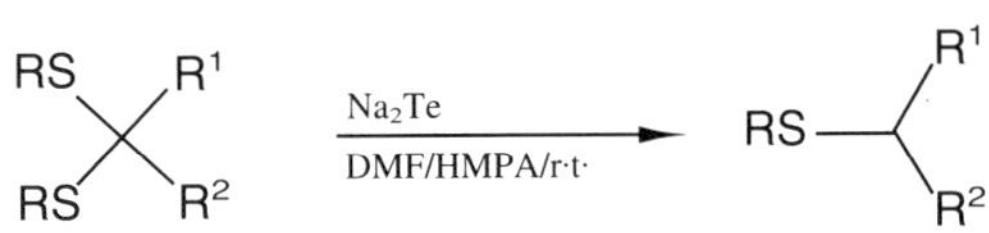

(a) R = Et; R^1 = Ph; R^2 = Ph, *p*-MeO, *p*-Cl, *p*-BrC_6H_4;

R^1, R^2 = 9-fluorenyl (41–75%)

(b) R = Ph; R^1, R^2 = MeCO, CO_2ET, COPh; R^1 = MeCO; R^2 = PhCO (22–67%)

The method employs mild experimental conditions and is therefore competitive with the existing methods for the preparation of benzhydryl sulphides and sulphenylated β-dicarbonyl compounds [59].

Benzhydryl ethyl sulphides (general procedure) [59] The diaryldiethylthioacetal derivative (2.0 mmol) in dry DMF (3 ml) is added to a pale yellow suspension of freshly prepared Na_2Te (1 mmol) in a mixture of DMF (3 ml) and HMPA (3 ml) under N_2 at room temperature. The colour gradually changes to deep violet and the mixture is stirred at room temperature for 12 h. The reaction is quenched by the addition of saturated aqueous NH_4Cl (10 ml) and the reaction mixture is extracted with benzene (10 ml). The extract is freed from inorganic insolubles by filtration through a thin bed of Celite and the aqueous layer is extracted with benzene. The combined benzene extracts are washed with H_2O, dried (Na_2SO_4) and evaporated to give the crude product. This is purified by chromatography over SiO_2 using hexane as the eluent.

Monosulphenylated β-dicarbonyl compounds (general procedure) [59]. To a suspension of Na_2Te (1 mmol) in dry DMF (3 ml) under N_2 at room temperature is added a solution of bis-sulphenylated active methylene compound (1 mmol) in dry DMF (3 ml). Immediate separation of Te occurs and the reaction goes to completion in 1 h. The reaction is quenched by the addition of 0.5 M H_2SO_4 saturated with NaCl followed by benzene (10 ml). The reaction mixture is stirred for 30 min and then filtered through a thin bed of Celite to remove insolubles. The aqueous layer is extracted with benzene and the combined organic extracts are washed with H_2O, dried (Na_2SO_4), and the solvent removed under reduced pressure to leave the crude product. This is purified by chromatography over SiO_2 using hexane as the eluent to give diphenyl disulphide. Further elution, with a mixture of hexane and CH_2Cl_2, gives the monosulphenylated compound.

REFERENCES

1. N. Kambe, K. Kondo, S. Morita, S. Murai and N. Sonoda, *Angew. Chem.* **19**, 1009 (1980) (in English).
2. N. Kambe, T. Inagaki, N. Miyoshi, A. Ogawa and N. Sonoda, *Nippon Kagaku Kaishi*, 1152 (1987).
3. M. Akiba and M. P. Cava, *Synth. Commun.* **14**, 1119 (1984).
4. Y. Aso, T. Nishioka, M. Osuka, K. Nagakawa, K. Sasaki, T. Otsubo and F. Ogura, *Nippon Kagaku Kashi*, 1490 (1987).
5. K. Nagakawa, M. Osuka, K. Saski, Y. Aso, T. Otsubo and F. Ogura, *Chem. Lett.*, 1331 (1987).
6. M. Moura Campos, E. L. Surany, H. de Andrade and N. Petragnani, *Tetrahedron* **20**, 2797 (1964).
7. H. Suzuki, H. Manabe, R. Enokiya and Y. Hanazaki, *Chem. Lett.*, 1339 (1986).
8. H. Suzuki and Y. Hanazaki, *Chem. Lett.*, 549 (1986).
9. D. H. R. Barton and S. W. McCombie, *J. Chem. Soc. Perkin Trans I*, 1574 (1975). For mechanistic rationales of reductions with NaHTe see D. H. Barton, L. Bohe and X. Lusinchi, *Tetrahedron Lett.* **28**, 6609 (1987).
10. K. Ramasamy, S. K. Kalyanasundaram and P. Shanmugam, *Synthesis*, 545 (1978).
11. M. Yamashita, Y. Kato and R. Suemitsu, *Chem. Lett.*, 847 (1980).
12. X. Huang and H. Zhang, *Synth. Commun.* **19**, 97 (1989).
13. X. Huang and C. Xie, *Synth. Commun.* **16**, 1701 (1986).
14. N. Kambe, T. Inagaki, N. Miyoshi, A. Ogawa and N. Sonoda, *Chem. Lett.*, 1275 (1987).
15. D. H. R. Barton, A. Fekih and X. Lusinchi, *Tetrahedron Lett.* **26**, 3693 (1985).
16. M. Yamashita, M. Kadokura and R. Suemitsu, *Bull. Chem. Soc. Jpn* **57**, 3359 (1984).
17. H. Suzuki, H. Manabe, T. Kawaguchi and M. Inouye, *Bull. Chem. Soc. Jpn* **60**, 771 (1987).
18. N. Kambe, K. Kondo and N. Sonoda, *Angew. Chem.* **19**, 1009 (1980) (in English).
19. H. Suzuki, H. Manabe and M. Inouye, *Chem. Lett.*, 1671 (1985).
20. A. Osuka, H. Shimizu and H. Suzuki, *Chem. Lett.*, 1373 (1983).
21. The reduction of nitrobenzene to aniline in 23% yield, by means of NaHTe has also been reported by M. Akiba and M. P. Cava, *Synth. Commun.* **14**, 1119 (1984).
22. S. Uchida, K. Yanada, H. Yamaguchi and H. Meguri, *Chem. Lett.*, 1069 (1983).
23. N. Ohira, Y. Aso, T. Otsubo and F. Ogura, *Chem. Lett.*, 853 (1964), and personal communication.

24. K. Ohe, H. Takayashi, S. Uemura and S. Sugita, *J. Chem. Soc. Chem. Commun.*, 591 (1988).
25. K. Ohe, S. Uemura, N. Sugita, H. Masuda and T. Taga, *J. Org. Chem.* **54**, 4169 (1989).
26. D. H. Barton, A. Fekih and X. Lusinchi, *Tetrahedron Lett.* **26**, 4603 (1985).
27. H. Suzuki and K. Takaoka, *Chem. Lett.*, 1733 (1984).
28. I. T. Harrison and S. Harrison, *Compendium of Organic Synthetic Methods*, Vol. I, Section 204, p. 501, Vol. II, Section 204, p. 200. Wiley, Chichester, 1971.
29. L. J. Clive and S. M. Menchen, *J. Chem. Soc. Chem. Commun.*, 658 (1977).
30. C. J. Clive and S. M. Menchen, *J. Org. Chem.* **45**, 2347 (1980).
31. D. H. R. Barton, A. Fekih and X. Lusinchi, *Tetrahedron Lett.* **26**, 6197 (1985).
32. A. Osuka, K. Taka-Oka and H. Suzuki, *Chem. Lett.*, 271 (1984).
33. G. Polson and D. C. Dittmer, *Tetrahedron Lett.* **27**, 5579 (1986).
34. R. P. Discordia and D. C. Dittmer, *J. Org. Chem.* **55**, 1414 (1990).
35. R. P. Discordia, C. K. Murphy and D. C. Dittmer, *Tetrahedron Lett.* **31**, 5603 (1990).
36. *Houben–Weyl—Methoden der Organischen Chemie*, 4th edn, Vol. V/1b, p. 180. Georg Thieme, Stuttgart, 1972; D. W. Young, *Protective Groups in Organic Chemistry*, p. 309. Plenum Press, London, 1973; I. M. Mathai, K. Schug and S. I. Miller, *J. Org. Chem.* **35**, 1733 (1970); J. E. Gordon and V. S. K. Chang, *J. Org. Chem.* **38**, 3062 (1973), and references therein.
37. M. Moura Campos and N. Petragnani, *Tetrahedron Lett.*, 5 (1960).
38. H. Suzuki, A. Kondo and A. Osuka, *Bull. Chem. Soc. Jpn* **58**, 1335 (1985).
39. N. Petragnani and M. Moura Campos, *Chem. Ber.* **94**, 1759 (1961).
40. K. Ramasamy, S. K. Kalyanasundaram and P. Shanmugam, *Synthesis*, 311 (1978).
41. S. Suzuki and M. Inouye, *Chem. Lett.*, 225 (1985).
42. K. Huang and Y. Q. Hou, *Synth. Commun.* **18**, 2201 (1988).
43. L. Engman, *Tetrahedron Lett.* **23**, 3601 (1982).
44. C. J. Li and D. N. Harp, *Tetrahedron Lett.* **31**, 6291 (1990).
45. F. B. W. Einstein, C. H. W. Jones, T. Jones and R. D. Sharma, *Can. J. Chem.* **61**, 2611 (1983).
46. L. Engman and S. E. Bystrom, *J. Org. Chem.* **50**, 3170 (1985).
47. N. Kambe, T. Tsukamoto, N. Miyoshi, S. Murai and N. Sonoda, *Bull. Chem. Soc. Jpn* **59**, 3013 (1986).
48. See methods reported in refs 49–52.
49. A. Osuka and H. Suzuki, *Chem. Lett.*, 119 (1983).
50. D. L. Clive and P. L. Beaulieu, *J. Org. Chem.* **47**, 1124 (1982).
51. L. Engman and M. P. Cava, *J. Org. Chem.* **47**, 3946 (1982).
52. C. J. Li and D. N. Harpp, *Tetrahedron Lett.* **32**, 1545 (1991).
53. W. Mack, *Angew. Chem.* **6**, 1083 (1967) (in English).
54. E. Luppold and W. Winter, *Chem. Ztg.* **101**, 303 (1977).
55. A. Osuka, K. Takechi and H. Suzuki, *Bull. Chem. Soc. Jpn* **57**, 303 (1984).
56. H. Suzuki, K. Takaoka and A. Ogura, *Bull. Chem. Soc. Jpn* **58**, 1067 (1985).
57. X. Huang and J. H. Pi, *Synth. Commun.* **20**, 2297 (1990).
58. X. Huang, J. H. Pi and Z. Z. Huang, *Phosphorus Sulfur Silicon* **67**, 177 (1992).
59. S. Padmanabhan, T. Ogawa and H. Suzuki, *J. Chem. Research (S)*, 266 (1989).

4.2 TELLURIUM-MEDIATED FORMATION OF ANIONIC SPECIES AND THEIR REACTIONS WITH ELECTROPHILES

4.2.1 Reformatsky-type reactions

By treatment with sodium telluride, bromoacetic esters react with aromatic as well as non-enolizable aldehydes in aprotic solvents, giving α-unsaturated esters [1, 2].

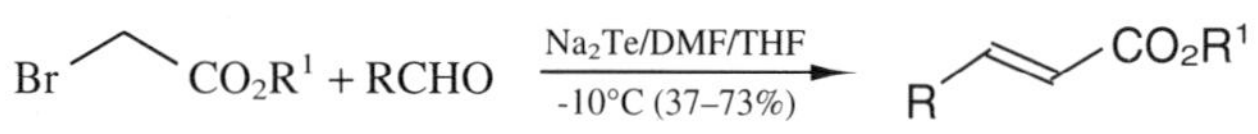

R^1 = Me. Et; R = Ph, *o*-Me, *p*-Me_2N, *p*-Cl, *p*-FC_6H_4

PhCH = CH, Me_2CPh, Me_2C = CH $(CH_2)_2CMe$ = CH

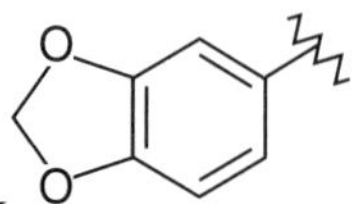

In accordance with a Reformatsky reaction mechanism, the telluride ion attacks the bromine atom to generate an ester enolate which reacts in sequence with the aldehyde to afford the α-unsaturated ester.

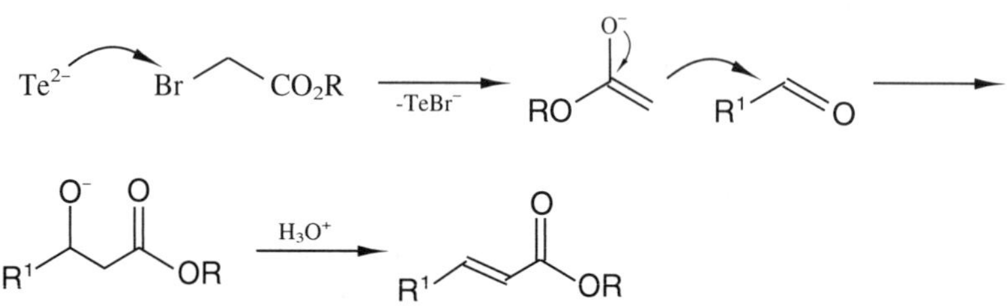

$TeBr^- \longrightarrow Te + Br^-$

The reaction exhibits the following features: chloroacetates give lower yields than bromoacetates; the (*E*) configuration is observed for the formed C=C bond; α-unsaturated carbonyl compounds lead only to 1,2 addition products; ketones are unreactive under the described conditions; a large excess of the bromoacetate and sodium telluride is required to assure high conversions.

Reformatsky-type reaction (typical procedure) [1] To a solution of Na_2Te (prepared from Te (0.7 g, 4 mmol), NaH (0.20 g, 8.6 mmol) and DMF (7 ml) by heating at 140°C for 30 min under N_2) is added dropwise at −20°C a solution of benzaldehyde (0.10 g, 1 mmol) in THF (1.5 ml) followed by ethyl bromoacetate (0.67 g, 4 mmol) in THF (1.5 ml). Te precipitates instantaneously. After stirring for 1 h, 0.5 M H_2SO_4 (5 ml) saturated with NaCl is added followed by Et_2O (10 ml). Excess Na_2Te is destroyed by exposing the mixture to air and stirring for 1 h. The mixture is filtered through Celite, the organic phase is separated, and the aqueous layer extracted twice with Et_2O. The combined extracts are washed with brine, dried (Na_2SO_4) and evaporated, leaving a yellow oil, which is purified by Kugelrohr distillation under reduced pressure to give ethyl cinnamate (0.10 g (60%); b.p. 95–110°C/1.5 torr).

Haloacetonitrile and *p*-cyanobenzyl bromide react similarly, leading to α-unsaturated nitriles [2] accompanied by minor amounts of the alcohols.

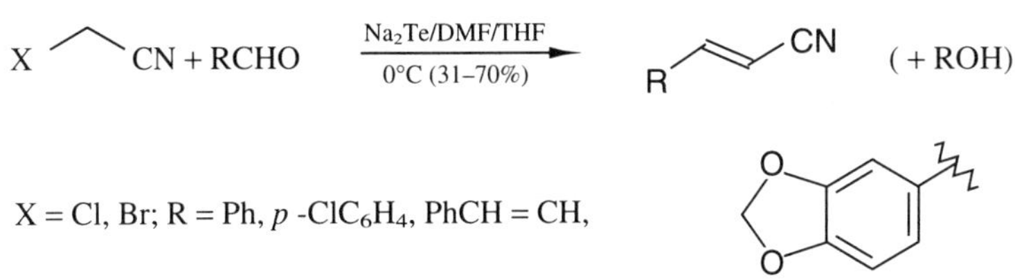

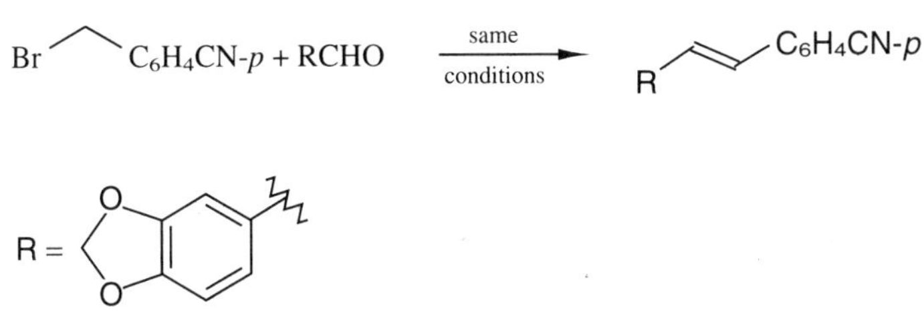

The previously described diisobutyl telluride/titanium(IV) system (see Section 4.1.1.3) also promotes similar Reformatsky-type reactions [3]. In this case, however, the products are β-chloroesters.

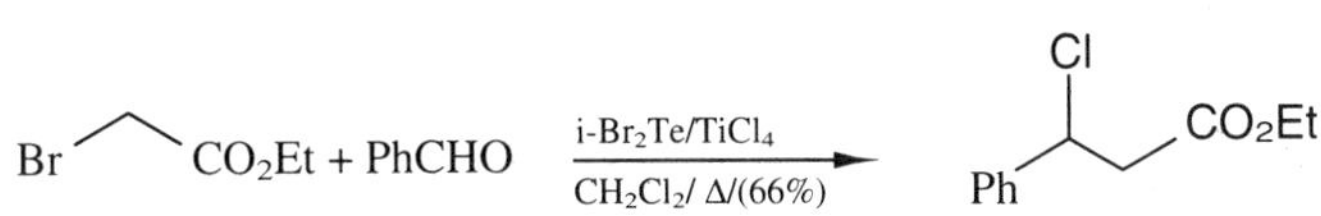

4.2.2 Alkylidenation of aldehydes and cyclopropanation of α,β-unsaturated carbonyl compounds with dibromomalonic esters

Dibromomalonates react with aldehydes under the assistance of dibutyl telluride (2 eq) to afford alkylidene malonates [4].

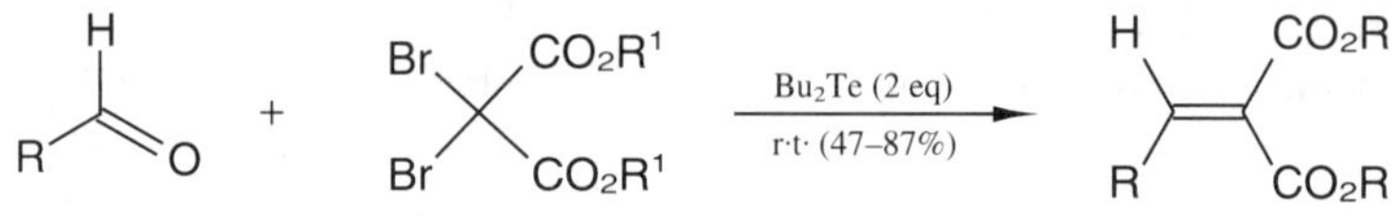

R = Ph, n-hexyl, $CH_2 = CH$, MeCH = CH, $CH_2 = CMe$, *trans*-PhCH = CH

R^1=Me, Et

Otherwise, dibromalonates (and nitriles) react with α-unsaturated carbonyl compounds in the presence of only 1 eq of dibutyl telluride, giving cyclopropane derivates.

R = CO_2Me, CO_2Et, CO_2Bu, COMe, CN; Y = CO_2Me, CO_2Et

R = COMe; Y = CN

These reactions have been rationalized as involving the addition of an intermediate tellurium enolate to the carbonyl group or to the β carbon, respectively, followed by a telluride-assisted elimination step or by an intramolecular cyclopropanation.

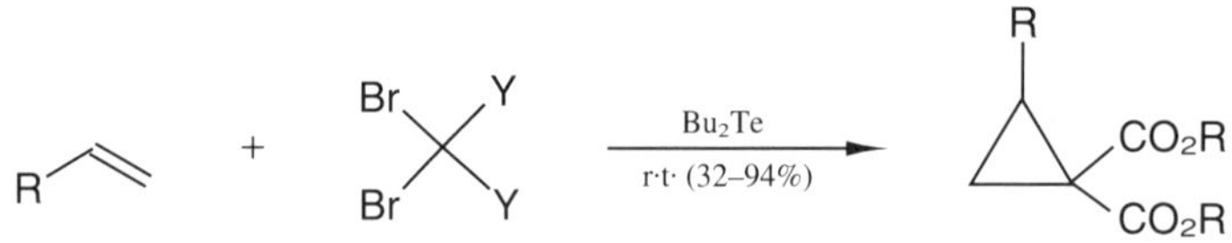

Alkylidenation of aldehydes with dibromomalonates (typical procedure) [4] A mixture of diethyl dibromomalonate (0.48 g, 1.5 mmol) and acrylaldehyde (0.087 g, 1.5 mmol) is treated with dibutyl telluride (0.73 g, 3 mmol) under argon and with stirring. After 30 min the mixture is extracted with $CHCl_3$ (30 ml), washed with H_2O, dried ($MgSO_4$) and concentrated *in vacuo*. The residue is diluted in hexane (20 ml), insoluble material is filtered off and the filtrate chromatographed on an SiO_2 column (88–125 mesh, elution with $CHCl_3$), giving dibutyltellurium dibromide (0.03 g (5%)) and then diethyl 2-propenylidenemalonate (0.26 g (87%)), which is purified by Kugelrohr distillation (b.p. 65°C/0.12 torr).

Cyclopropanation of α,β-unsaturated carbonyl compounds with dibromomalonates (typical procedure) [4] To a mixture of diethyl dibromomalonate (0.95 g, 3 mmol) and methyl vinyl ketone (0.27 g, 3.1 mmol) is added dibutyl telluride (0.73 g, 3 mmol) under argon and with stirring. The exothermic reaction is completed within 1 h. The mixture is chromatographed on an Al_2O_3 column (70–

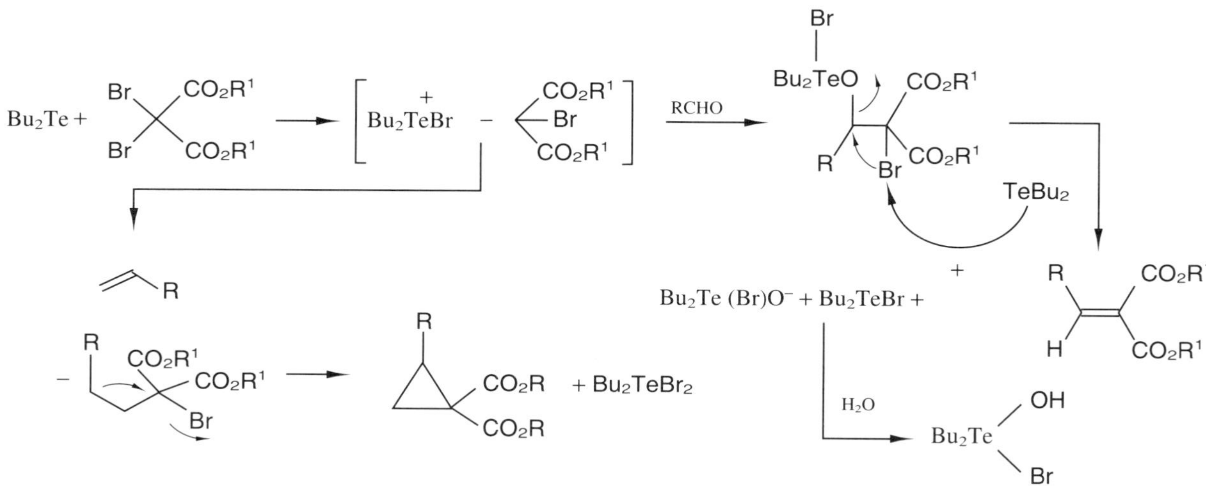

Bu₂Te +
Br
Br
CO₂R¹
CO₂R¹
+
Bu₂TeBr
–
Br
RCHO
Bu₂TeO
R
TeBu₂
+
Bu₂Te (Br)O⁻ + Bu₂TeBr +
H
H₂O
Bu₂Te
OH
R
+ Bu₂TeBr₂
CO₂R

230 mesh, elution with EtOAc), giving dibutyltellurium dibromide (1.01 g, 84%) and then 1-acetyl-2,2-bis(ethoxycarbonyl)cyclopropane, which is purified by Kugelrohr distillation (0.59 g (86%); b.p. 88–90°C/0.08 torr).

4.2.3 Telluride-assisted sulphenylation and sulphonylation reactions

α-Phenylthiocarbonyl compounds and alkyl aryl sulphones are of great synthetic importance [5–7]. New methods for their preparation are therefore the object of continuous interest.

These two classes of compounds can be prepared by treating diphenyl disulphide [5] and aryl sulphonyl chlorides [6] with sodium telluride followed, respectively, by α-halocarbonyl compounds and alkyl halides.

These methods involve the polarity reversal of the formerly electrophilic thio reagents, giving the thiolate and sulphinate anions through an electron transfer from the telluride ion.

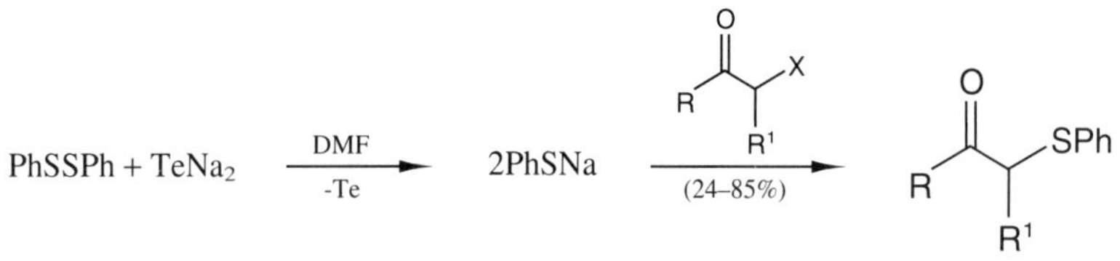

R = aryl, alkyl; R^1 = H, SO_2Ph, COPh, SPh, CO_2Et

R, R^1 = -O-CH_2CH_2CH

$$2ArSO_2Cl + TeNa_2 \xrightarrow[-Te]{H_2O/THF/PTC} 2ArSO_2Na \xrightarrow[(41–71\%)]{RX} ArSO_2R \text{ (method A)}$$

Ar = *p*-Me, *p*-BrC_6H_4, 1-naphthyl, 2-naphthyl

RX = alkyl bromides and iodides, $PhCH_2Cl$

O-,*O*-diethyl phosphorotellurolates can be employed instead of sodium telluride, the alkylation occurring even with α-withdrawing group-substituted halides [7].

$$ArSO_2Cl + RX \xrightarrow[(77–94\%)]{(EtO)_2P(O)TeNa/THF/\ EtOH/TEBA[Et_3PhCH_2N^+]Cl^-} ArSO_2R \text{ (method B)}$$

Ar = Ph, *p*-MeC_6H_4

RX = MeI, $ClCH_2Ph$, $BrCH_2Y$ (Y = CN, CO_2Et, CONHPh, COPh)

α-Bis-sulphenylated carboxyl compounds can be prepared by the inverse addition of the disulphide to the preformed enolate.

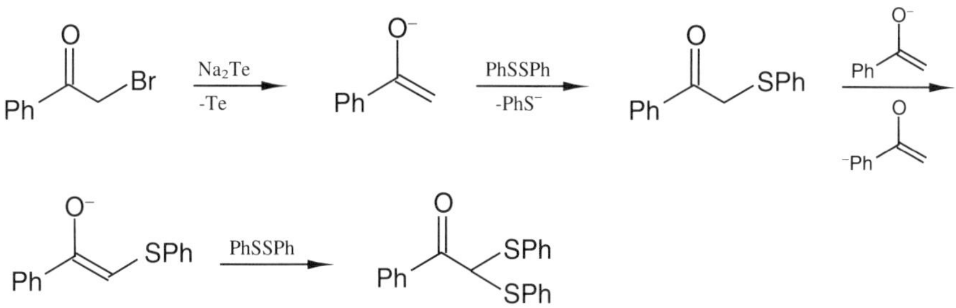

α-Sulphenyl carbonyl compounds (general procedure) [5] To a suspension of Na_2Te (0.346 g, 2.0 mmol), prepared from Te and NaH in DME, is added diphenyl disulphide (0.876 g, 4.0 mmol) in DMF (5 ml) followed by the α-halocarbonyl compound (2.0 mmol) is the same solvent (3 ml). The dark reduction mixture is kept at –40°C for 2 h and then at room temperature for 1 h, with stirring. The reaction is quenched with saturated aqueous NH_4Cl (10 ml) and the mixture is extracted with benzene. The extract is filtered through Celite, washed with 0.5 M H_2SO_4 and H_2O, and dried (Na_2SO_4). The solvent is evaporated and the crude product purified by column chromatography on SiO_2 (eluent: hexane), giving the unchanged diphenyl disulphide. Further elution with hexane and CH_2Cl_2 gives the sulphenylated compound.

Alkyl aryl sulphones (typical procedure) *Method A* [6]. A solution of Na_2Te (prepared by heating a mixture of Te (0.13 g, 1 mmol), Rongalite (0.79 g, 5 mmol) and aqueous 1 M NaOH (15 ml)) is added dropwise to a stirred solution of TsCl (1 mmol) and $BzEt_3NCl$ (0.022 g, 0.1 mmol) in THF (10 ml) at room temperature under N_2. An instantaneous reaction occurs and the colour of the mixture changes to deep black. After stirring for 5 min, benzyl chloride (0.632 g, 5 mmol) in THF (3 ml) is added and the mixture kept at 90°C for 5 h. After cooling the solvent is evaporated under reduced pressure and the residue treated with aqueous NH_4Cl and benzene. The organic phase is evaporated after filtration through Celite and worked up as usual to give benzyl *p*-tolyl sulphone (0.179 g (75%); m.p. 143–145°C).

Method B [7]. To a solution of $(EtO)_2P(O)TeNa$ (5 mmol) in EtOH (10 ml) is added a solution of benzene sulphonyl chloride (0.88 g, 5 mmol) in THF (10 ml) under N_2. An instantaneous reaction occurs and the colour of the mixture changes from colourless to deep black. After stirring for 20 min, a solution of benzyl chloride (0.64 g, 5 mmol) and catalytic amounts of $BzEt_3NCl$ (0.01 g) in THF (10 ml) is added. The mixture is refluxed for 4 h. After evaporating the solvent under reduced pressure, the residue is treated with $CHCl_3$ (50 ml). After filtration, the organic phase is evaporated to dryness. The crude product is recrystallized from EtOH, giving the pure product (1.93 g (82.4%); m.p. 145–147°C).

4.2.4 Telluride-mediated aldehyde methylenation

In the presence of dibutyl telluride, iodomethyl triphenylphosphonium iodide reacts with aldehydes, in accordance with a Wittig-type olefination, giving methylenation products [8].

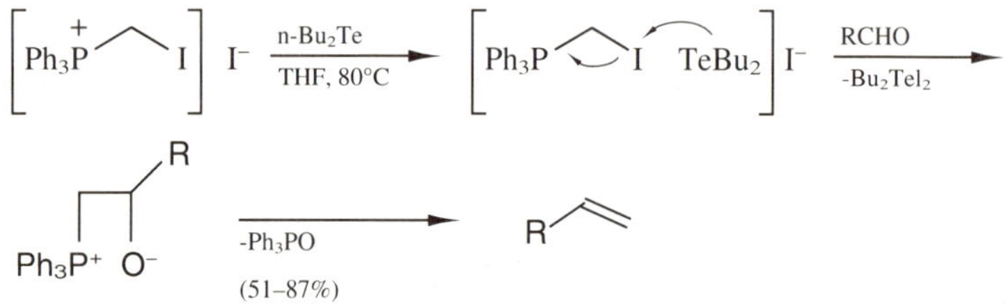

R = Ph, *p*-Cl, *p*-Br, *p*-FC_6H_4, PhCH=CH, 2-naphthyl, n-C_9H_{19}

Methylenation reaction (typical procedure) [8] A mixture of iodomethyl triphenylphosphonium iodide (0.530 g, 1.0 mmol), dibutyl telluride (0.242 g, 1.0 mmol) and *p*-bromobenzaldehyde (0.096 g, 0.5 mmol) in THF (5 ml) is heated under reflux for 30 h, then cooled and filtered. The residue is purified by TLC, giving *p*-bromostyrene (0.080 g (87%)).

REFERENCES

1. S. Suzuki and M. Inouye, *Chem. Lett.*, 403 (1986).
2. H. Suzuki, H. Manabe and M. Inouye, *Nippon Kagaku Kaishi*, 1485 (1987).
3. H. Suzuki, H. Manabe, R. Enokiya and Y. Hanazaki, *Chem. Lett.*, 1339 (1986).
4. T. Matsuki, N. X. Hu, Y. Aso, T. Otsubo and F. Ogura, *Bull Chem. Soc. Jpn* **62**, 2105 (1989).
5. S. Padmanabhan, T. Ogawa and H. Suzuki, *Bull. Chem. Soc. Jpn* **62**, 1358 (1989).
6. H. Suzuki, Y. Nishioka, S. Padmanabhan and T. Ogawa, *Chem. Lett.*, 727 (1988).
7. X. Huang and J. H. Pi, *Synth. Commun.* **20**, 2291 (1990).
8. S. W. Li, Y. Z. Huang and L. L. Shi, *Chem. Ber.* **123**, 1441 (1990).

4.3 TELLURIDE ION-PROMOTED COUPLING OF ALLYLIC HALIDES

Allylic halides are converted into the coupled 1,5-dienes by treatment with telluride anions [1]. The reaction, which gives the best results when employing the reagent prepared *in situ* from elemental tellurium and lithium triethylborohydride, proceeds through the intermediacy of the thermally unstable bis-allylic telluride followed by extrusion of tellurium and coupling of the formed allylic radicals.

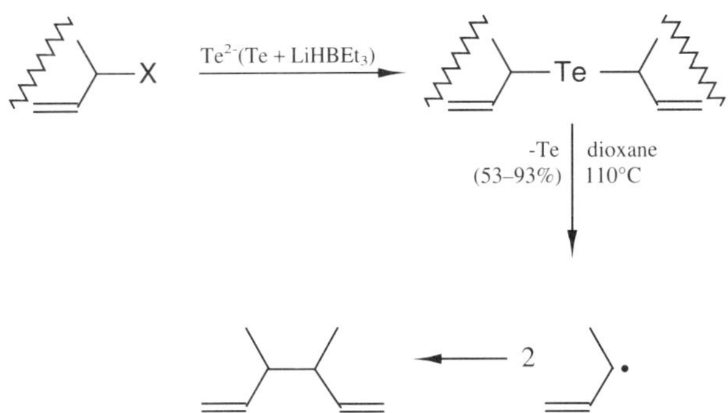

allylic halide:

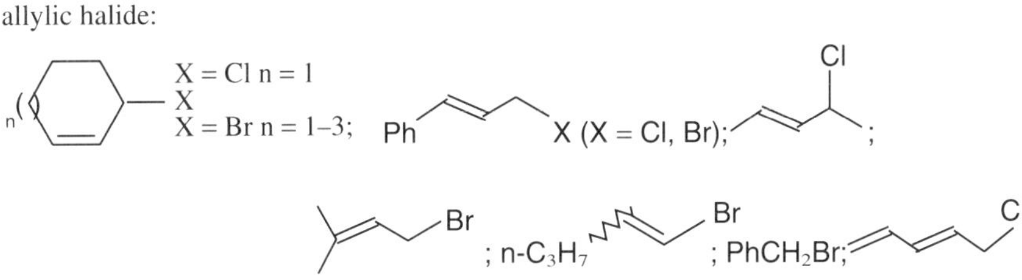

The substituted allylic bromides furnish mixtures of the expected and rearranged products. The same mixture of products is formed independently from the geometry of the double bond (e.g. 1-bromo-2-hexenes).

Coupling of 3-bromocyclohexene (typical procedure) [1] 3-Bromocyclohexene (322 mg, 2.00 mmol) in dry dioxane (4 ml) is added by syringe to the Te reagent (1.00 mmol, prepared adding super-hydride (2.6 eq) to Te powder under argon and stirring at room temperature for 5–7 h). The mixture is immersed in a bath preheated to 110°C and refluxed for 1 h. After cooling, the mixture is filtered through Celite with the aid of CH_2Cl_2. The filtrate is concentrated to an oil and partitioned between CH_2Cl_2 and H_2O (100 ml, 1:1). The aqueous phase is extracted with CH_2Cl_2 (2 × 25 ml) and the combined organic extracts are dried and concentrated to an oil. Chromatography over SiO_2 with hexane, followed by Kugelrohr distillation at 110°C/15 torr gives 2,2′-bicyclohexenyl (140 mg (86%)); purity 96% (by GLC).

REFERENCE

1. D. L. Y. Clive, P. C. Anderson, N. Moss and A. Singh, *J. Org. Chem.* **47**, 1641 (1982).

4.4 DEPROTECTION OF ORGANIC FUNCTIONALITIES BY TELLURIUM REAGENTS

The regeneration of substrates with organic functionalities from certain specific derivatives, which frequently execute a protective function, is an important operation in synthetic methodology [1].

Some tellurium anionic reagents offer a great contribution to these deblocking reactions due to the utilization of mild experimental conditions and a non-hydrolytic medium.

4.4.1 Regeneration of carboxylic acids: cleavage of carboxylic esters

The methods described in this section are of great significance in view of the essentially neutral non-hydrolytic conditions employed.

4.4.1.1 Alkyl carboxylates

Common esters such as alkyl and benzyl carboxylates are easily dealkylated by sodium hydrogen telluride, sodium telluride and sodium ditelluride in DMF [2]. In accordance with a typical S_N2 displacement at the alkoxy group carbon, methyl, ethyl and benzyl esters react smoothly. The nucleophilicity of the reagents is enhanced by the polar aprotic solvent, and the reactivity decreases with higher alkoxy chains due to steric hindrance (e.g. n-Pr and i-Pr esters).

Dealkylation of carboxylic esters (typical procedure) [2] *Method A—with NaHTE.* A mixture of Te (1.17 g, 10 mmol), $NaBH_4$ (0.57 g, 15 mmol) evacuated and purged with N_2 three times, DMF (20 ml) and t-BuOH (0.2 ml, used as a H^+ donor) is heated at 80–90°C under N_2 for 30 min until nearly all the Te has disappeared, giving a deep violet solution. Methyl benzoate (2.47 g,

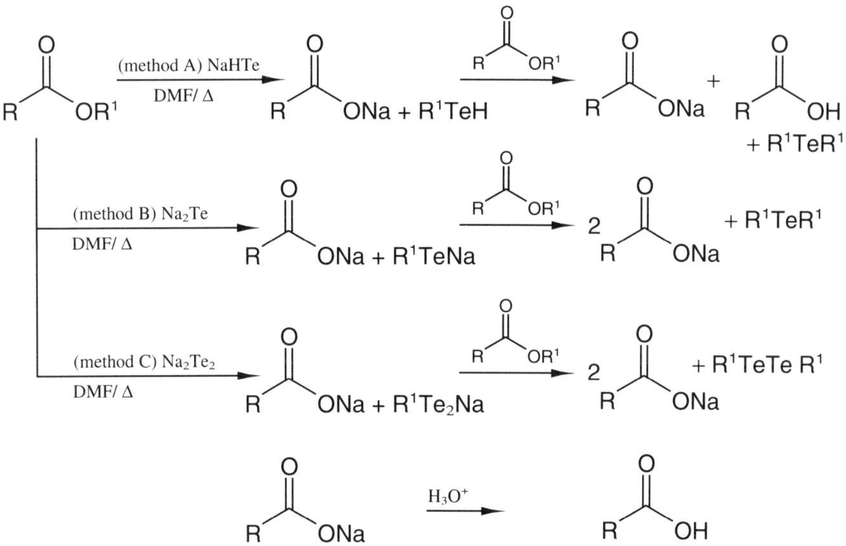

RH = Ph, m-ClC_6H_4, 2-naphthyl, $PhCH_2$, 1-naphthyl-CH_2, n-$C_{11}H_{23}$

R^1 = Me, Et, $PhCH_2$ (yields 88–98%)

18 mmol) is added and the mixture heated at 80–90°C for 15 min (a copious precipitate often appears). H_2O (30 ml) is added and the mixture extracted with Et_2O (3 × 50 ml). The aqueous layer is acidified with HCl, extracted with Et_2O (3 × 20 ml) and dried ($MgSO_4$). The solvent is evaporated, giving pure benzoic acid (2.12 g (95%); m.p. 121–122°C).

Method B—with Na_2Te. A mixture of Te (0.64 g, 5 mmol) and $NaBH_4$ (0.45 g 12 mmol) in DMF (20 ml) is heated at 80–90°C under N_2 for 30 min, giving an almost colourless suspension (which has been proved to be Na_2Te by reaction with 1-bromobutane to give dibutyl telluride). Methyl benzoate is added, and after heating at 80–90°C for 20 min the mixture is worked up as for method A, giving pure benzoic acid (0.99 g (90%); m.p. 121–122°C).

Method C—with Na_2Te_2. A mixture of Te (1.00 g, 7.5 mmol) and $NaBH_4$ (0.19 g, 5.0 mmol) in DMF (20 ml) is heated at 80–90°C under N_2 for 20 min, giving a homogeneous deep purple solution (which has been proved to be Na_2Te_2 by reaction with 1-bromobutane to give dibutyl ditelluride). Methyl benzoate is added, and after 20 min heating at 80–90°C the mixture is handled as in method A, giving pure benzoic acid (0.94 g (85%); m.p. 121–122°C).

4.4.1.2 Phenacyl carboxylates

Phenacyl esters, easily prepared from carboxylic salts and phenacyl bromide under phase transfer catalysis [3], regenerate the original carboxylic acid by treatment with sodium hydrogen telluride in DMF [3].

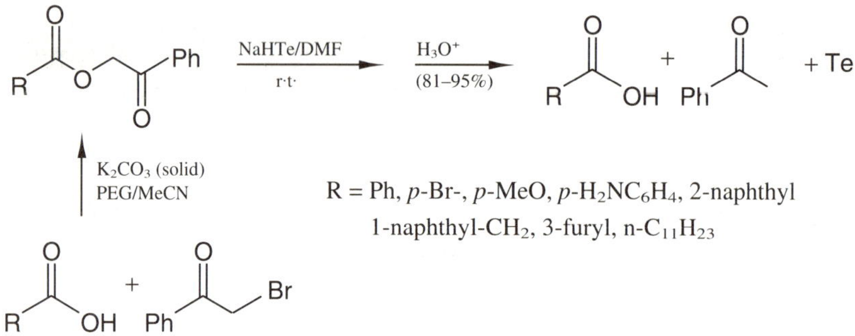

R = Ph, *p*-Br-, *p*-MeO, *p*-$H_2NC_6H_4$, 2-naphthyl 1-naphthyl-CH_2, 3-furyl, n-$C_{11}H_{23}$

The reaction involves the attack by anionic tellurium at the phenacyl methylene group.

Cleavage of phenacyl esters (typical procedure) [3] To a solution of NaHTe (prepared by heating powdered Te (1.28 g, 10 mmol) and $NaBH_4$ (0.75 g, 15 mmol) under N_2 in DMF at 80–90°C for 30 min, and then cooling at room temperature) is added a solution of phenacyl benzoate (1.92 g, 8 mmol) in DMF (30 ml). An instantaneous reaction takes place and the mixture is stirred at room temperature for 20 min. The solvent is evaporated and the mixture poured into H_2O (100 ml) and filtered to remove the Te. The basic filtrate is washed with ether and then the aqueous solution is acidified with 6 N HCl and extracted with ether (3 × 50 ml). The ether extract is dried ($MgSO_4$) and evaporated to give benzoic acid (1.11 g (91%); m.p. 121–122°C).

In general the acids obtained by the above procedure are essentially pure.

4.4.1.3 Allyl carboxylates

The allyl group, a less familiar protective group for carboxylic acids, can be easily removed from esters by treatment with sodium hydrogen telluride [4]. In contrast to the preceding methods, ethanol has been employed as the solvent.

R = Ph, *p*-$MeOC_6H_4$, 1-propenyl

Cleavage of allyl carboxylates (general procedure) [4] To a solution of NaHTe (prepared *in situ* from Te powder (1.30 g, 1.0 mmol) and $NaBH_4$ (0.90 g 2.4 mmol) in EtOH (20 ml) buffered with deoxygenated HOAc (1.2 ml in 5 ml EtOH)) is added the allyl carboxylate (0.005 mol). The mixture is refluxed under N_2 for 2 h, then filtered, the filtrate is evaporated, and the residue is dissolved in H_2O. The aqueous layer is made slightly acidic with a few drops dilute HCl. The aqueous layer is then extracted with $CHCl_3$, and the extract dried and evaporated. The residue is purified by recrystallization or distillation.

4.4.1.4 2-Haloethyl carboxylates

The frequently employed 2-haloethyl protecting group can be removed from carboxylic esters by means of sodium telluride in DMF [5] or sodium hydrogen telluride in ethanol [6].

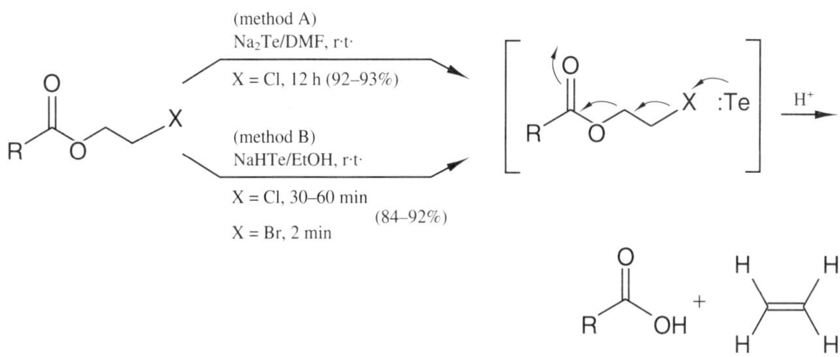

(method A) R = Ph, p-MeC_6H_4, $PhCH_2$; X = Cl

(method B) R = Ph, o-MeO_6H_4, $PhCH_2$; X = Cl

R = Ph, n-$C_{11}H_{23}$; X = Br

The reaction can be rationalized as involving an E_2 fragmentative attack by the telluride anion at the halogen atom. Sodium hydrogen telluride (method B) requires a shorter reaction time than sodium telluride (method A), the 2-bromoalkyl esters reacting very rapidly. Sodium hydrogen selenide reacts similarly but more slowly [5].

Since in the described reactions tellurium is regenerated and sodium hydrogen telluride is prepared from the element and sodium borohydride, a catalytic procedure is also effective, in which only catalytic amounts of preformed reagent are employed [7].

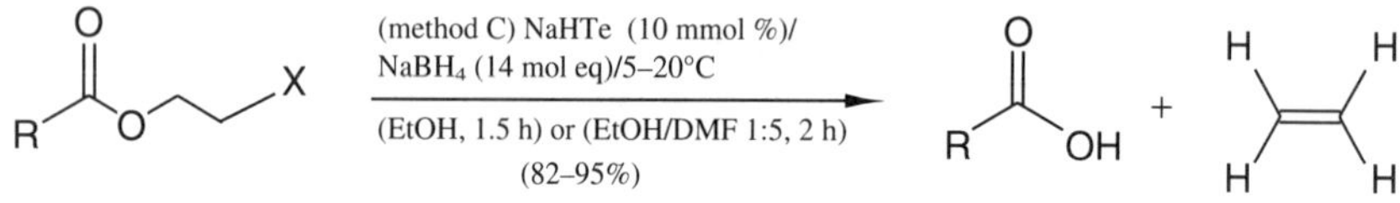

X = Br; R = Ph, *p*-MeC_6H_4, $PhCH_2$, n-C_5H_{11}

X = Cl; R = Ph, *p*-ClC_6H_4, n-$C_{11}H_{23}$

Cleavage of 2-haloethyl carboxylates *Method A (general procedure)* [5]. To Na_2Te (0.346 g, 2.0 mmol) in DMF (3 ml) is added a solution of 2-chloroethyl ester (2.0 mmol) in DMF (2 ml) at room temperature, and the resulting thick mixture is stirred for 12 h, also at room temperature. The reaction is quenched by adding 0.5 M H_2SO_4 (5 ml) and the organic phase is extracted with ether (10 ml). The mixture is filtered through Celite to remove the inorganic material and the ether phase is separated, washed with 0.5 M H_2SO_4 (2 × 5 ml) followed by brine, and dried (Na_2SO_4). The solvent is evaporated under vacuum to give the pure carboxylic acid, which is further purified by crystallization from H_2O containing a little EtOH.

Method B (general procedure) [6]. To an NaHTe solution, prepared from Te powder (1.27 g, 10 mmol) and $NaBH_4$ (0.9 g, 24 mmol) in absolute EtOH (25 ml), is added the 2-haloethyl ester (10 mmol) in EtOH (5 ml). (The reaction takes place instantaneously for 2-bromoethyl esters at room temperature.) The mixture is stirred for about 1 h, then acidified with dilute HCl, extracted with ether, and the organic phase dried and evaporated. The residue is recrystallized from appropriate solvents to give pure carboxylic acids.

Method C (typical procedure) [7]. (i) A violet ethanolic solution of NaHTe is prepared by heating a mixture of powdered Te (0.06 g, 0.5 mmol) and $NaBH_4$ (0.08 g, 2 mmol) in EtOH (5 ml) at 70–80°C for 0.5 h under N_2. After cooling to 5°C, a solution of $NaBH_4$ (0.57 g, 15 mmol) in EtOH (15 ml) and 2-bromoethyl benzoate (1.15 g, 5 mmol) are injected and the mixture is stirred for 0.5 h. A copious precipitate ensues and the mixture is stirred at room temperature for 1 h to ensure completion. The remaining work-up is analogous to that given for method A and affords pure benzoic acid (0.580 g (95%); m.p. 120–122°C).

(ii) A violet solution of NaHTe is prepared by heating a mixture of powdered Te (0.06 g, 0.5 mmol) and $NaBH_4$ (0.57 g, 15 mmol) in DMF (15 ml) at 70–80°C for 0.5 h. After the mixture is cooled to room temperature a solution of 2-chloroethyl benzoate (0.92 g, 5 mmol) in EtOH (3 ml) is injected and the mixture is stirred for 2 h at room temperature. The work-up is analogous to the above method and gives pure benzoic acid (0.567 g (63%); m.p. 119–121°C).

4.4.2 Regeneration of phenols

4.4.2.1 Cleavage of aryl carboxylates and carbonates

Phenols can be regenerated from the corresponding carboxylates [8] or carbonates [9] by reaction with sodium hydrogen telluride.

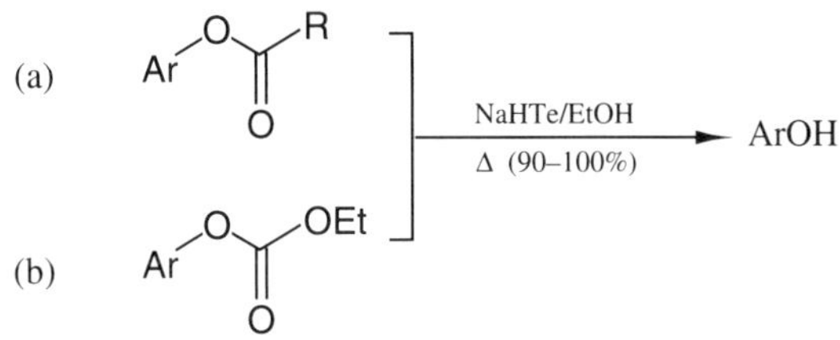

(a) R = Me; Ar = Ph, 2-naphthyl, 3-, 6-, 7-flavonyl, 7-coumaryl
R = Ph; Ar = 2-naphthyl, *o*-$NO_2C_6H_4$, 2-NO_2, 4-5$(MeO)_2C_6H_2$
R = Ar = Ph

(b) Ar = Ph, *m*-Me, *o*-AcC_6H_4, 2-napthyl, 3-flavonyl,
4-coumaryl, 4-(2-hydroxy)quinolyl

Cleavage of aryl carboxylates (general procedure) [8] To a solution of NaHTe (prepared from Te (1.30 g, 10 mmol) and $NaBH_4$ (0.90 g, 24 mmol) in EtOH (20 ml) buffered with AcOH (1.2 ml) in EtOH (5 ml) is added the aryl carboxylate (5 mmol). The mixture is refluxed for 30 min, poured into ice–H_2O (200 ml), extracted with $CHCl_3$, and the combined extracts dried and evaporated. The residue is purified by recrystallization or distillation.

Cleavage of aryl carbonates (general procedure) [9] To a solution of NaHTe prepared *in situ* from Te powder (1.3 g) and $NaBH_4$ (0.9 g) in EtOH (20 ml) buffered with deoxygenated HOAc (1.2 ml in 5 ml EtOH) is added the aryl ethyl carbonate (0.005 mol) and the mixture refluxed under N_2 for 30 min. The reaction mixture is filtered, the filtrate evaporated and the residue dissolved in H_2O. The aqueous layer is slightly acidified with a few drops of HOAc. The respective phenols are obtained in quantitative yields.

4.4.2.2 Cleavage of aryl haloacetates

The haloacetate group has been recently introduced as a protecting group for the phenolic function [5].

By treatment with sodium telluride in DMF, the parent phenols are easily regenerated [5].

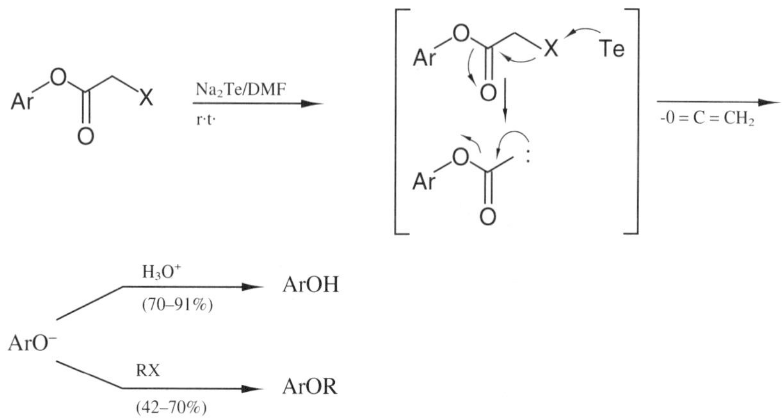

By quenching the reaction with an alkyl halide, alkyl aryl ethers are formed:

X = Br; Ar = 1-naphthyl
X = Cl; Ar = 1-naphthyl, 2-naphthyl, *p*-Ph, *p*-NO_2, *p*-$C_3H_7CO_2C_6H_4$
RX = MeI, EtBr, EtI

The reaction involves an attack by the tellurium anion at the halogen atom followed by a rapid elimination of ketene from the resulting enolate.

Functional groups such as nitro and ester moieties are unaffected.

Cleavage of aryl haloacetates (general procedure) [5] To a suspension of Na_2Te (prepared from Te and NaH in DMF (5 ml) by heating at 140°C for 30 min under N_2) is added a solution of the aryl haloacetate (2.0 mmol) in DMF (3 ml). Immediate separation of Te is observed. After stirring the dark mixture at room temperature for 1 h, saturated aqueous NH_4Cl (5 ml) is added and the organic phase diluted with benzene (10 ml). After stirring the mixture exposed to the atmosphere for some time, it is filtered through Celite, and the organic phase separated, washed with brine, dried and evaporated *in vacuo*. The crude product is purified by SiO_2 chromatography (elution with CH_2Cl_2/hexane).

4.4.2.3 Cleavage of phenyl allyl ethers

Allyl ethers are cleaved with sodium hydrogen telluride with regeneration of the parent phenols [4].

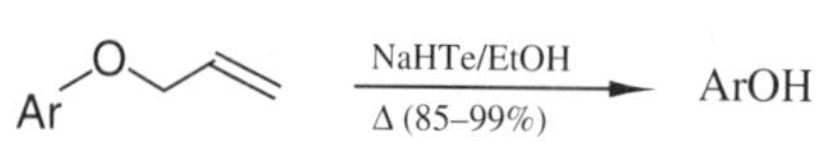

Ar = Ph, 4-coumaryl, 4-(2-hydroxy)quinolyl

The experimental procedure is identical to that described for the cleavage of allylcarboxylates (see Section 4.1.3).

4.4.3 Regeneration of amines by cleavage of trichloro-t-butylcarbamates

Amines are regenerated from the protecting derivatives trichloro-t-butylcarbamates by treatment with sodium 2-thienyl tellurolate [10]. The reagent is generated in a catalytic cycle by reduction of the corresponding ditelluride with sodium borohydride.

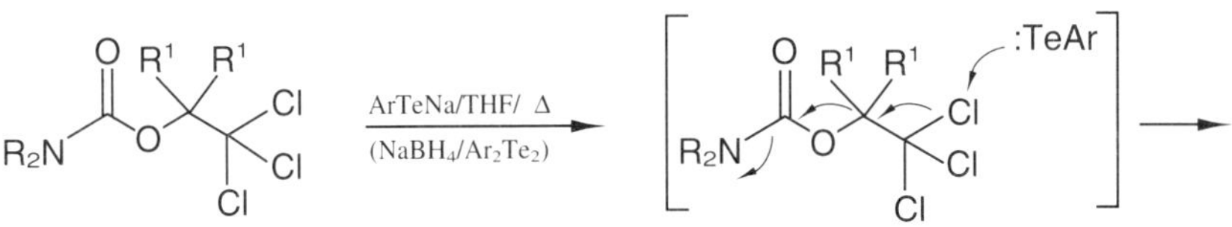

$R_2NH + CO_2 + R_2{}^1C{=}CCl_2$

Ar = 2-thienyl; R^1 = Me

R_2N (representative examples): 1-indolyl, 2,3-dihydro-1-indolyl, *o,p*-$(MeO)_2C_6H_3NH$, *p*-AcC_6H_4NH

Cleavage of trichloro-t-butylcarbamates (general procedure) [10] A 1% solution of the trichloro-t-butylcarbamate derivative with 24 mol % of bis(2-thienyl) ditelluride in THF is heated to 60°C under N_2. A 1% aqueous solution of $NaBH_4$ (stabilized with three drops of a 10% NaOH solution) is added dropwise until the red colour of the catalyst disappears. This usually requires 4 or 5 eq of $NaBH_4$ and is carried out over a period of 30 min. The volatile components are removed *in vacuo*. The mixture is then poured into H_2O and extracted with ether. The amines are isolated by flash chromatography or by acid extraction.

REFERENCES

1. T. W. Greene, *Protective Groups in Organic Synthesis*. Wiley, New York, 1981.
2. J. Chen and X. Y. Zhou, *Synthesis*, 586 (1987).
3. Z. Huang, L. Xie and X. Huang, *Synth. Commun.* **18**, 1167 (1988).
4. N. Shobana and P. Shanmugam, *Indian J. Chem.* **25B**, 658 (1986).
5. H. Suzuki, S. Padmanabhan and T. Ogawa, *Chem. Lett.*, 1017 (1989).
6. J. Chen and X. J. Zhou, *Synth. Commun.* **17**, 161 (1987).

7. Z. Huang and X. J. Zhou, *Synthesis*, 633 (1990).
8. N. Shobana and P. Shanmugam, *Indian J. Chem.* **24B**, 690 (1985).
9. N. Shobana, M. Amirthavalli, V. Deepa and P. Shanmugam, *Indian J. Chem.* **27B**, 965 (1988).
10. M. V. Lakshimikantham, Y. A. Jackson, R. Y. Jones, C. J. O'Malley, K. Ravichandran and M. P. Cava, *Tetrahedron Lett.* **27**, 4687 (1986).

4.5 OXIDATION OF ORGANIC SUBSTRATES BY MEANS OF TELLURIUM REAGENTS

4.5.1 Bis(*p*-methoxyphenyl) telluroxide as a mild and selective oxidizing reagent

Bis(*p*-methoxyphenyl) telluroxide (An_2TeO), an easily accessible (see Section 3.10), stable crystalline compound, soluble in most organic solvents, finds wide use as a mild and selective oxidizing reagent for several functionalities. The reagent exhibits the advantage of being easily regenerated from the parent telluride produced during oxidation reactions.

The oxidation reactions discussed below are well established.

4.5.1.1 Conversion of thio- and selenocarbonyl compounds to their oxo analogues

Y = S, Se [1, 2]

thioesters 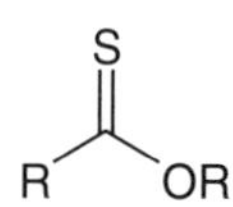 (I a, b, c)

dithioesters 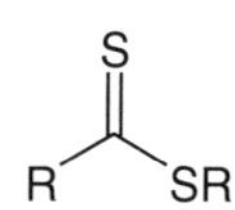 (II)

selenoesters 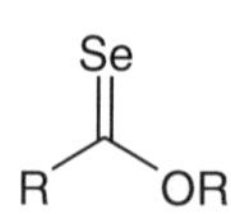 (III)

xanthates 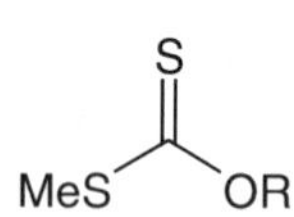 (IV a, b, c)

thionocarbonates 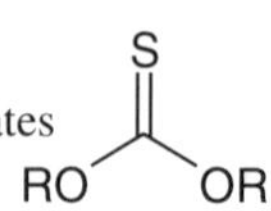 (V)

trithiocarbonates 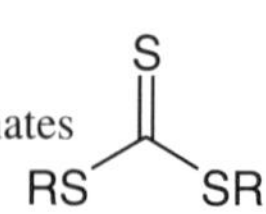 (VI)

thioketones 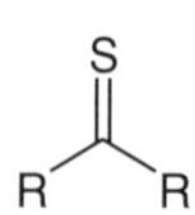 (VII a, b)

thioamides 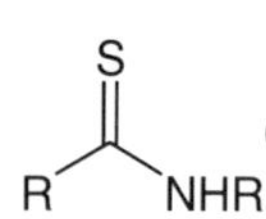 (VIII) thioureas 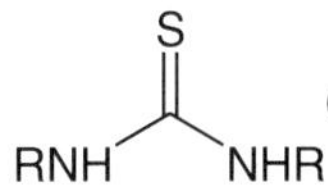 (IX)

Conversion of thio- and selenocarbonyl compounds into their corresponding oxo derivatives (reaction time, percentage yield)

 (4 h, 91)

Ia

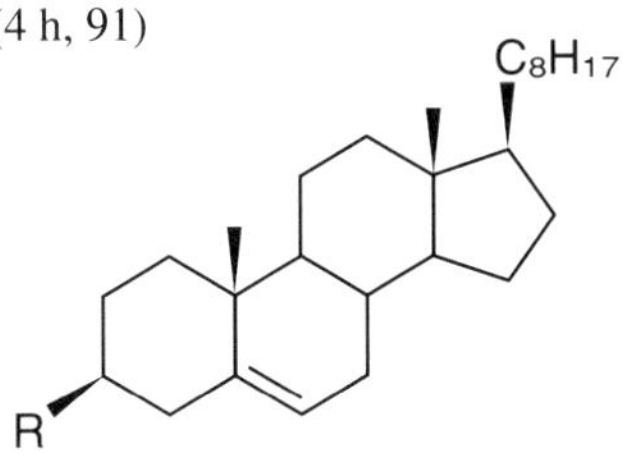

Ib R = 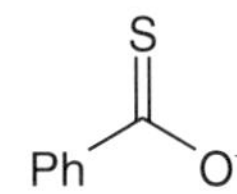 (1·5 h, 53)

Ic R = 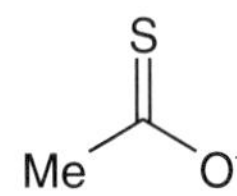 (0·25 h, 84)

II R = 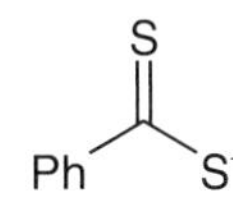 (27 h, 52)

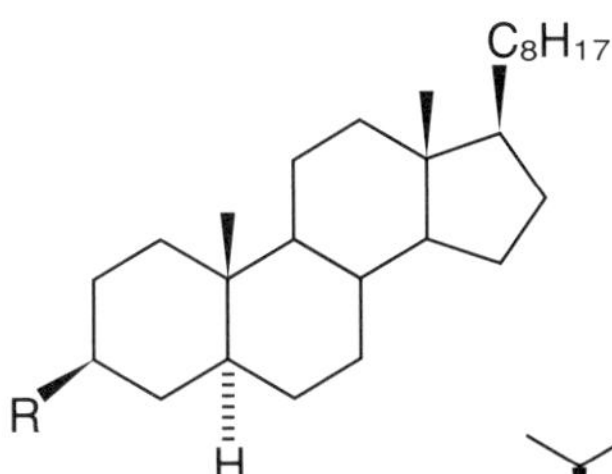

III R = 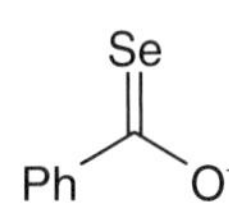 (0·3 h, 80)

IVa R = 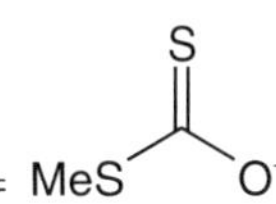 (24 h, 72)

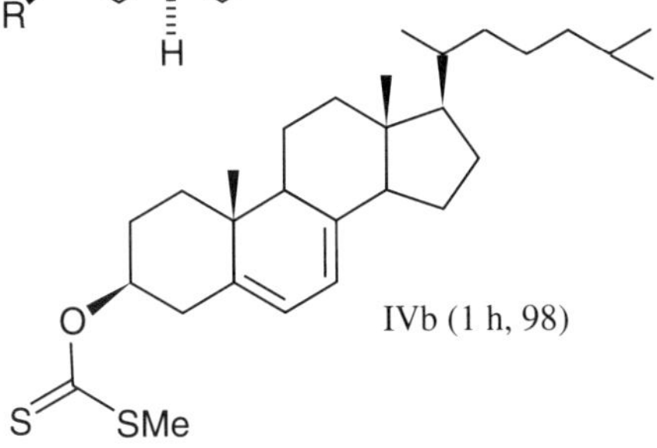

IVb (1 h, 98)

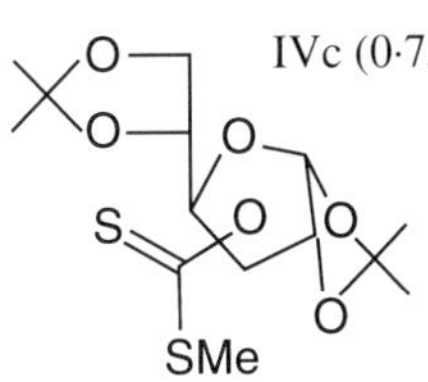

IVc (0·75 h, 87)

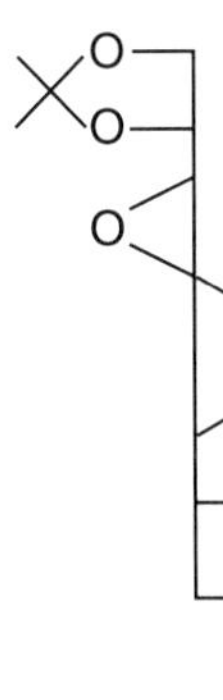

V (0·5 h, 96)

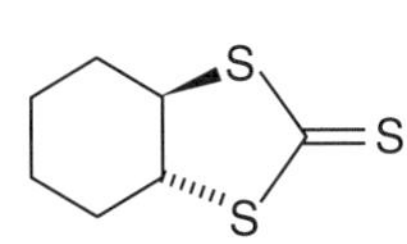

VI (20 h)

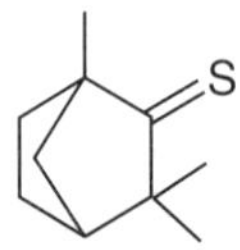

VIIa (42 h, 23)

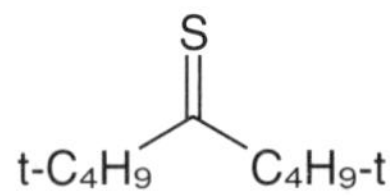

VIIb (0·3 h, 100)

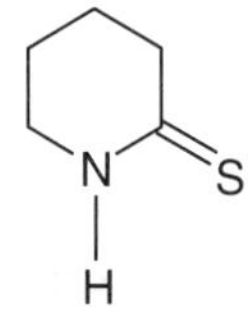

VIII (20 h, 87)

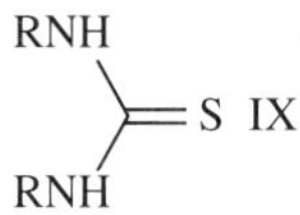

IX R = H (63 h, 100)
R = Ph (16 h, 68)

Thiocamphor, under similar conditions, furnishes a vinylic disulphide as the major product (70%), which slowly undergoes a hetero-Cope rearrangement [2].

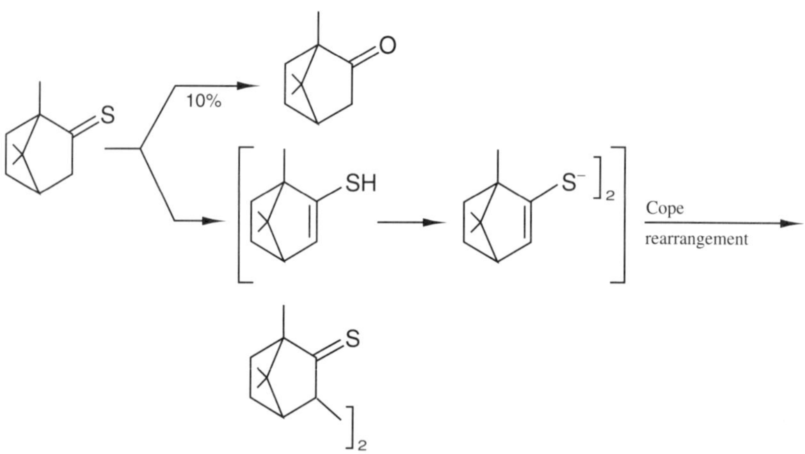

Considering that diaryl tellurides debrominate *vic*-dibromides, forming diaryltellurium dibromides (see Section 4.1.11), which in turn can be easily hydrolysed to the corresponding telluroxides by aqueous bases, a catalytic procedure has been provided for the oxidation of thiocarbonyl compounds.

By this procedure excess 1,2-dibromotetrachloroethane acts as a halogen source, converting a small amount of telluride to the corresponding dibromide, which is then hydrolysed by the basic aqueous medium to the telluroxide [3].

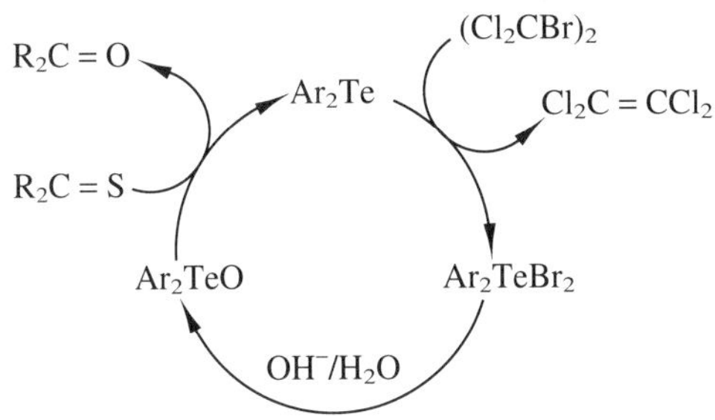

4.5.1.2 Conversion of tertiary phosphines to tertiary phosphine oxides

$$R_3P \xrightarrow{An_2TeO} R_3PO \qquad [2]$$

R = Ph (48 h, 72%); Me (0·5 h, 93%)

4.5.1.3 Conversion of phenyl isothiocyanate to diphenylurea

$2\ PhN=C=S \xrightarrow[0{\cdot}25\ h\ (52\%)]{An_2TeO}$ 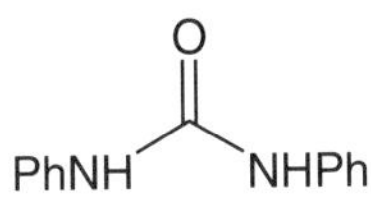 [2]

4.5.1.4 Conversion of thiols to disulphides

$$2\ RSH \xrightarrow{An_2TeO} RSSR$$

[2]

R = PhCH$_2$ (0·5 h; 96%), p-MeC$_6$H$_4$ (0·1 h, 98%), p-H$_2$NC$_6$H$_4$ (1·0 h, 58%), 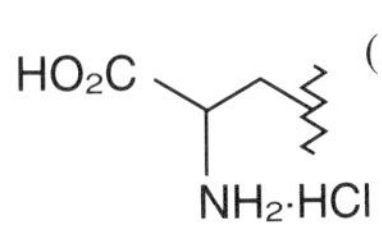 (1·5 h, 79%)

$\xrightarrow[0.25\ h\ (52\%)]{An_2TeO}$

The inertness of functional groups (e.g. NH$_2$) to the reagent is noteworthy.

4.5.1.5 Conversion of *o*- and *p*-diphenols to quinones

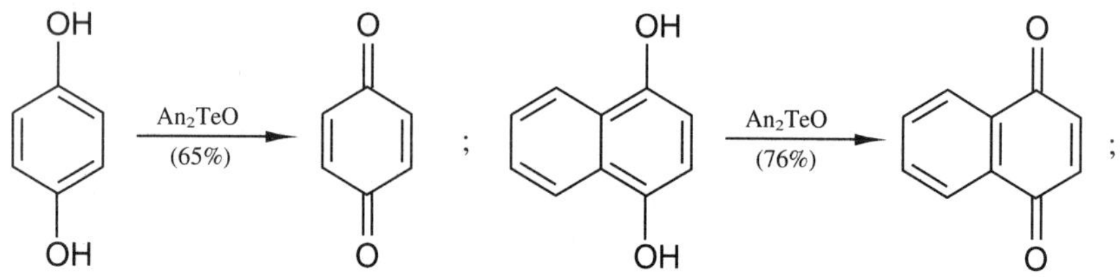

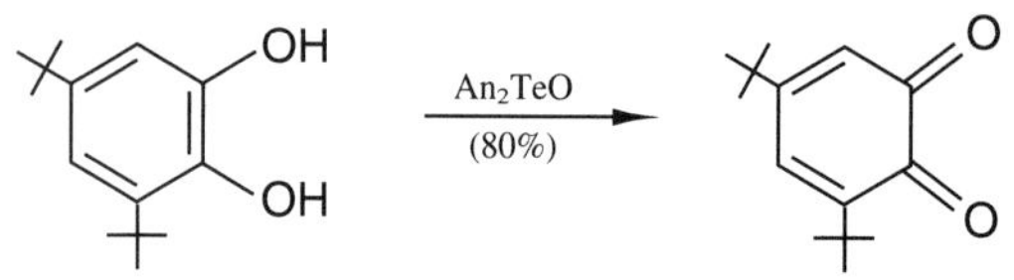

[2]

Simple phenols are unreactive.

4.5.1.6 Conversion of acylhydrazines to acylhydrazides

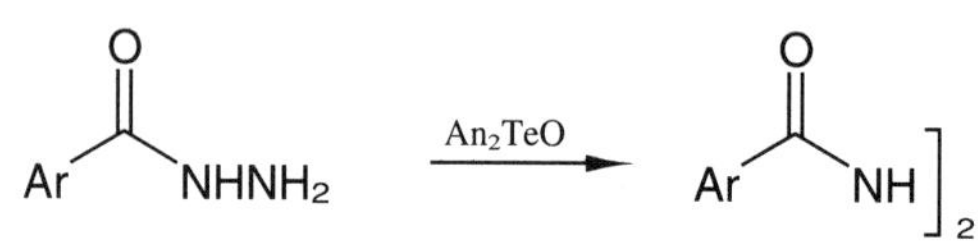

[2]

A = Ph (1 h, 58%), *p*-$MeOC_6H_4$ (24 h, 86%)

p-$NO_2C_6H_4$ (-67%), *p*-$PhOCH_2$ (24 h, 67%)

The oxidation of arylhydrazines with An_2TeO is unsuitable for synthetic purposes, since it furnishes a complex mixture of products.

4.5.1.7 Conversion of *N*-Phenylhydroxylamine to nitrosobenzene

$$\text{PhNHOH} \xrightarrow[\text{0·1 h (90\%)}]{\text{An}_2\text{TeO}} \text{PhNO}$$ [2]

4.5.1.8 Conversion of benzophenone hydrazone to diphenyldiazomethane

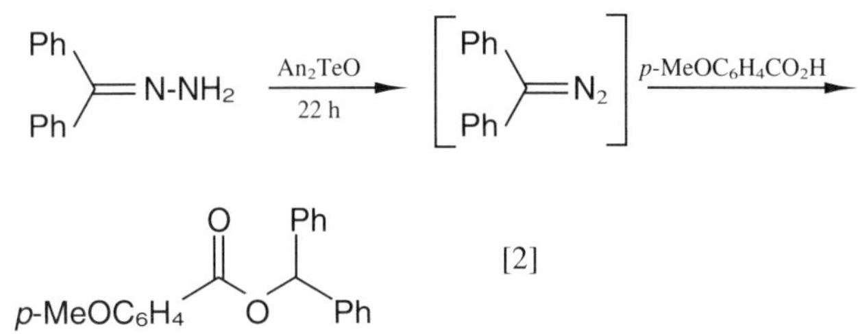

[2]

The primary reaction product, diphenyldiazomethane, is trapped as diphenyl methyl ester.

4.5.1.9 Conversion of glycols to aldehydes

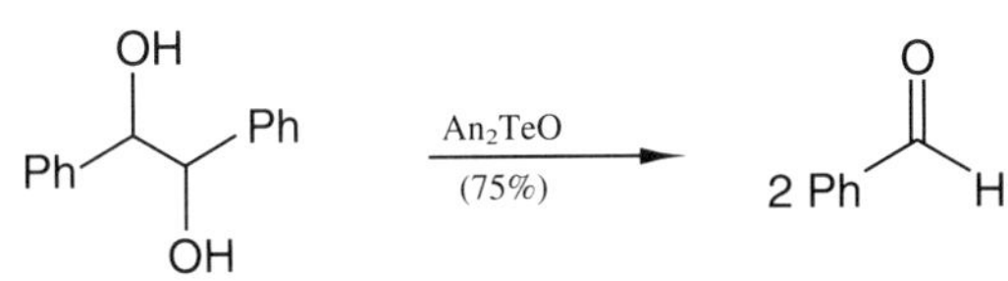

[4]

Simple alcohols are unaffected.

*Oxidation with bis(*p*-methoxyphenyl) telluroxide (general procedure)* [2] All the reactions are performed at room temperature under N_2 in $CHCl_3$ or CH_2Cl_2. Approximately 10 ml of solvent is used for every 100 mg of substrate. For thiocarbonyl derivatives, 1.1 eq of the reagent is used, for thiols 0.55 eq. The mixtures are concentrated by evaporation and submitted to thin layer or column chromatography to isolate the products. An_2Te is always recovered in a yield of 64–96%, while sulphur or selenium is always recovered in near quantitative yields from reaction with thio- or selenocarbonyl derivatives.

The following compounds are unaffected by bis(*p*-methoxyphenyl) telluroxide: dithiolanes, enamines, aldehydes, ketones, alcohols, pyrroles, indoles, amino acids, aromatic amines, monohydroxyarenes, esters, hindered thiocarbonates, isonitriles, oximes, arylhydrazones, sulphides, selenides [2].

4.5.2 Polymer-supported bis(*p*-methoxyphenyl) telluroxide

A useful modification of the bis(*p*-methoxyphenyl) telluroxide reagent is its immobilization on a polymeric resin [5]. This polymer-supported reagent, prepared from *p*-methoxyphenyl tellurocyanate and poly(*p*-lithiostyrene), exhibits several advantages over the monomeric reagent, such as easier product work-up and multiple recycle from the spent reagent.

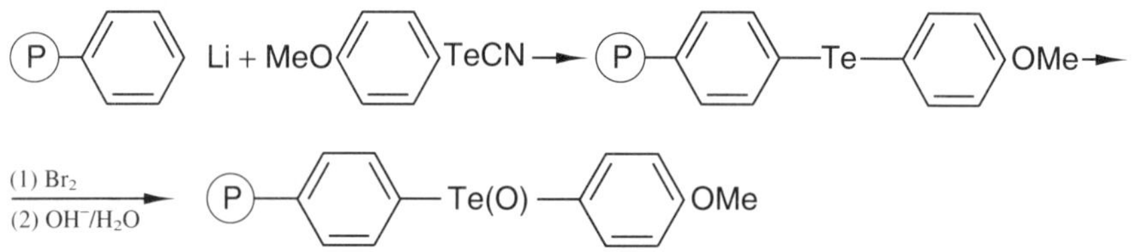

In comparison to the analogous supported selenoxide, the tellurium reagent requires shorter reaction times and milder experimental conditions.

Preparation of the polymeric telluride [5] One per cent divinylbenzene styrene copolymer Bio-Beads S-X_1 (Bio-Rad laboratories)) is converted in poly(*p*-lithiostyrene) according to the described procedure [6].

Poly(*p*-lithiostyrene) prepared *in situ* from 8.0 g of brominated polystyrene (containing 3.0 mol eq Br_2 g^{-1}) is swollen in dry THF (50 ml) in an N_2 atmosphere. A solution of *p*-methoxyphenyl tellurocyanate [7] (8.0 g, 31 mmol; see Section 3.8) in dry THF (10 ml) is added, and the slurry is stirred at room temperature for 30 min and then overnight at 60°C. The resin is collected by filtration and washed in sequence with THF, THF/H_2O, H_2O, THF and finally with MeOH. After evaporating to dryness under reduced pressure at 60°C, a reddish-yellow resin of telluride (8.05 g) is obtained. A solution of Br_2 (1.92 g, 12 mmol) in CCl_4 (10 ml) is added dropwise into a suspension of the telluride resin (8.0 g) swollen in CCl_4 (30 ml). The slurry is stirred at room temperature for 4.5 h. The resulting tellurium dibromide is removed by filtration and washed successively with CCl_4, and THF. It is then mixed with THF (50 ml) and aqueous 15% NaOH (10 ml), and refluxed overnight. The resin is collected by filtration, and washed once with THF, repeatedly with THF/H_2O (1:10) until the washings turn neutral, and finally with MeOH. After evaporating to dryness under vacuum at 60°C a brown resin of telluroxide is obtained (8.1 g).

The spent reagent can be recovered as the reduced species by filtration, and reused after oxidation. The activity is not decreased even after 10 recycles.

Like its monomeric counterpart, the polymeric reagent is inert to simple amines, amides, alcohols and phenols, but easily oxidizes thiols to disulphides, phosphines to phosphine oxides, hydroquinone and catechol to quinones, and thioketones, thioesters and trithiocarbonates to the corresponding oxo derivatives, in dichloromethane, chloroform or acetic acid at room temperature in high yields. The reaction times are 0.5–3.5 h for thiols, thioketones and phosphines, and higher for diphenols (4.5–12 h), thioesters (5 h) and trithiocarbonates (6–36 h).

Thioamides are dehydrosulphurated to nitriles, or converted to 1,2,4-thiadiazoles, depending on if the reactions are performed in non-acidic solvents such as dichloromethane, chloroform and methanol or in acetic acid.

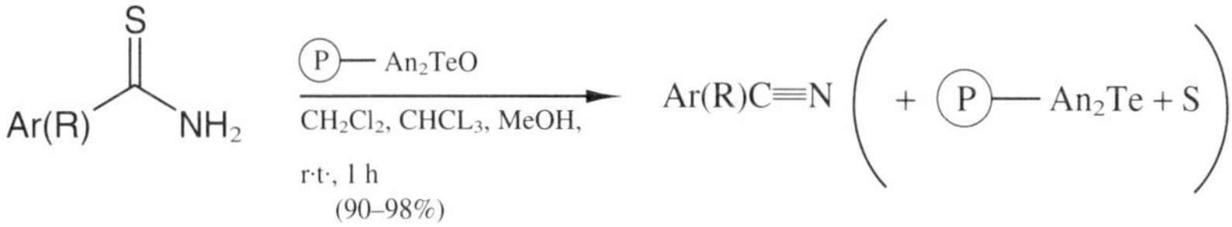

Ar(R) = Ph, *p*-MeO, *p*-Cl, *o*-Cl, *p*-$NO_2C_6H_4$, $PhCH_2$, n-$C_{17}H_{35}$, Ph_2N*

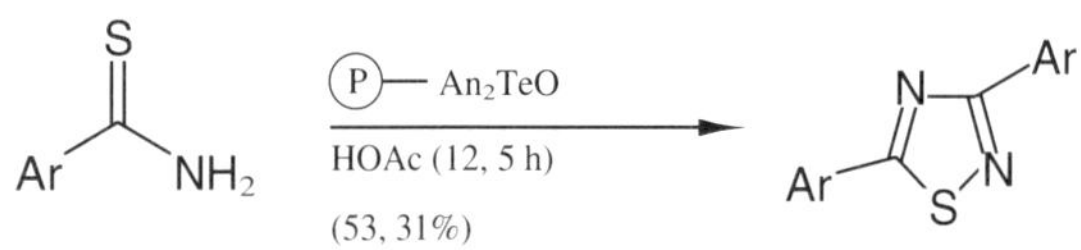

Ar = Ph, *p*-ClC_6H_4

N, N′-Diphenylthiourea, structurally unable to form a nitrile, is converted to the corresponding urea by the first procedure; *N*-phenylthiourea, in acetic acid, gives 5-imino-4,5-dihydro-1,2,4-thiadiazole.

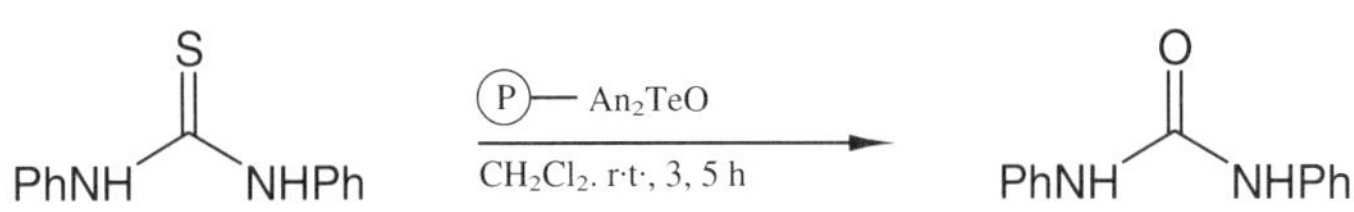

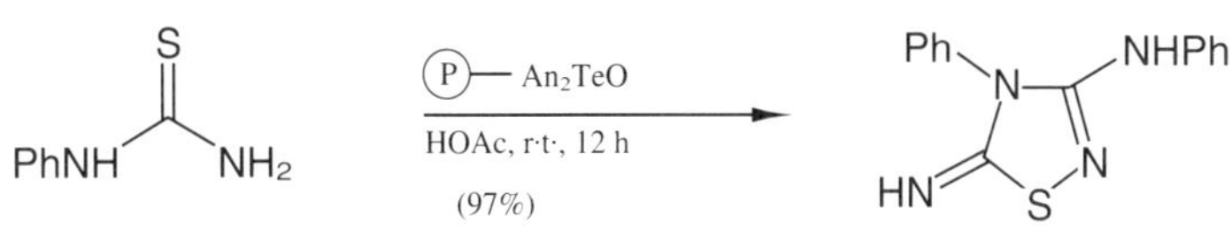

Oxidation with polymeric telluroxide (general procedure) [5] The substrate (0.5 mmol) is mixed with the polymeric telluroxide reagent (20% excess) swollen in 10 ml of the solvent (see text). After completion of the reaction, the

*Reaction time 24 h.

spent reagent is removed by filtration and washed thoroughly with the same solvent as used in the reaction. The combined filtrate and washings are evaporated under vacuum. The residue is purified by short SiO_2 column chromatography to give the pure product.

4.5.3 Bis(*p*-methoxyphenyl) telluride as a mediator in an electrolytic process

Bis(*p*-methoxyphenyl) tellurium diacetate or ditosylate is generated by the electrolysis of the parent telluride in the presence of Bu_4N^+ acetate or tosylate. In analogy to the telluroxide, they accomplish the conversion of thioamides to nitriles and 1,2,4-thiadiazoles. Taking into consideration these facts, the above conversion can be performed in an electrochemical catalytic cycle by submitting to electrolysis a mixture of a thioamide, Bu_4N^+ acetate or tosylate and 5% of the telluride [8].

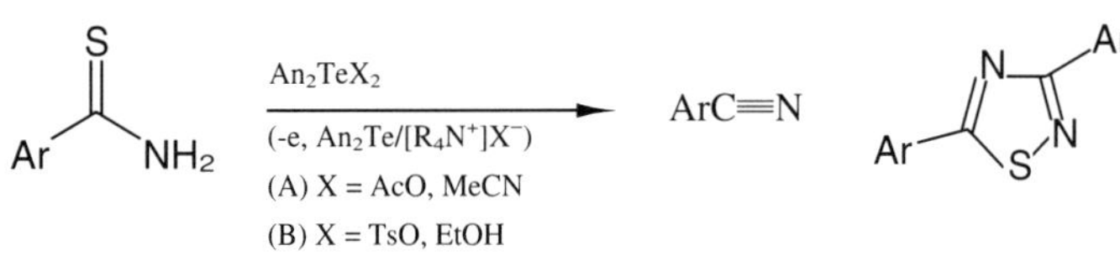

		Percentage yield	
		Nitrile	1,2,4-Thiadiazole
Ar = *p*-$MeOC_6H_4$	(A)	71	0
	(B)	0	73
p-ClC_6H_4	(A)	63	18
	(B)	0	71
PhNH	(A)	80	6
	(B)	4	77

Electrooxidation of thioamides to nitriles and thiadiazoles (typical procedure) [8] A mixture of thiobenzamide (0.0686 g, 0.5 mmol) and An_2Te (0.0086 g, 0.25 mmol) in dry MeCN (16 ml) containing Bu_4N^+OAc (0.602 g) is electrolysed under a constant current density of 4 mA cm^{-2}. The reaction is monitored by HPLC (Microsorb C18 standard column, MeOH as the eluent). After passage of 3.0 F mol^{-1} of electricity (10 h), the starting material is consumed, forming benzonitrile 0.0462 g (89%) and a trace of 3,5-diphenyl-1,2,4-thiadiazole. Both products are isolated as colourless oils and as colourless crystals from column chromatography on SiO_2 (eluent: benzene/hexane 1:1) (m.p. 89.5–90.5 and 89–90°C, respectively).

4.5.4 Bis(*p*-methoxyphenyl) tellurone

The title compound, prepared by the oxidation of the corresponding telluroxide with sodium periodate, exhibits some peculiarities compared with the described reactivity of the telluroxide [4].

It oxidizes thiophenol to diphenyl disulphide, hydroquinone to *p*-benzoquinone, and cleaves not only glycols (in analogy to the telluroxide) but also oxidizes benzylic alcohols.

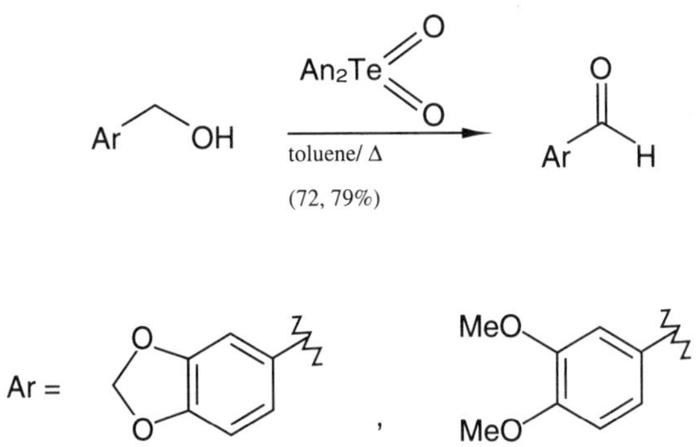

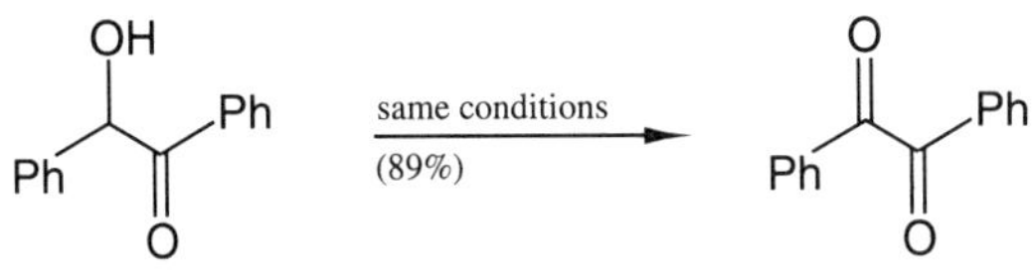

4.5.5 Sodium tellurite as an oxidizing agent for thiols

Sodium tellurite exhibits selective and mild oxidizing properties for thiols under phase transfer conditions, at room temperature [9]. Aromatic and benzylic thiols are oxidized to the corresponding disulphides very quickly, the ease of oxidation decreasing progressively for primary, secondary and tertiary thiols ($PhSH \sim PhCH_2SH > RCH_2SH > RR^1CHSH > RR^1R^2CSH$). Short-chain thiols are more reactive than long-chain ones, probably due to the difference in the solubility in water. Tetrabutylammonium hydroxide is generally used as the phase transfer catalyst, but in the case of less reactive thiols better results are obtained by employing cetyltrimethylammonium bromide.

$$\mathrm{RSH} \xrightarrow[-\mathrm{Te}]{\mathrm{Na_2TeO_3},\ [\mathrm{R_4N^+}]\mathrm{X^-}/\ \mathrm{H_2O,\ benzene,\ r{\cdot}t{\cdot}}} \mathrm{RSSR}$$

	R (reaction time, h/% yield)
$[R_4N^+]X$=n-$Bu_4N^+]OH^-$	Ph (1/87), *o*-$H_2NC_6H_4$ (4/88), *p*-HOC_6H_4 (2/75), $PhCH_2$ (1/88), n-Bu (4/56), n-C_8H_{17} (48/77), i-Pr (48/48), $Me_3CCH_2CMe_2$ (24/0)
$R_4N^+[X$=$CetMe_3N^+]Br^-$	n-C_8H_{17} (4/86), t-Bu (24/0)

The reagent is highly selective, as demonstrated by the inertness of several sensitive functionalities such as amino, hydroxyl, azo, hydrazo, phenol, sulphide, disulphide, sulphoxide, aldehyde moieties, and olefinic and acetylenic carbon–carbon bonds. The ability of the reagent to differentiate thiols may be used to prepare unsymmetrical disulphides.

$$\mathrm{RSH} + \mathrm{R^1SH} \xrightarrow{\text{same conditions}} \mathrm{RSSR^1}$$

	$RSSR^1$ (reaction time, h/% yield)
[n-$Bu_4N^+]OH^-$	PhSSBu-t (24/52), $PhCH_2SSBu$-t (36/74), PhSSPr-i (24/21), $PhCH_2SSPr$-i (24/56)
[cetyl $Me_3N^+]Br^-$	n-$C_8H_{17}SSBu$-t (24/81)

The cross-coupling reaction proceeds through the initial oxidation of the more reactive thiol, followed by the gradual thiolysis of the formed disulphide by the less reactive thiol. The method is therefore useful for the cross-coupling of the reactive aryl and n-alkyl thiols with the less reactive t-alkylthiols.

Oxidative coupling of thiols to disulphides (typical procedure) [9] *Symmetrical disulphide.* To a stirred solution of phenylmethanethiol (0.629 g, 5.06 mmol) in benzene (20 ml) is added dropwise a solution of $Na_2TeO_3 \cdot 5H_2O$ (0.471 g, 1.51 mmol) and 10% aqueous $[Bu_4N^+]OH^-$ (0.28 ml, 0.10 mmol) in H_2O (20 ml) with a syringe, at room temperature (25–30°C). The reaction mixture quickly turns black, and free Te begins to precipitate. Monitoring by GLC indicates that the reaction is almost complete within 1 h. The mixture is filtered through a thin Celite bed and the organic layer is separated. The aqueous layer is extracted with benzene (2 × 20 ml) and the combined organic phase is washed with H_2O (20 ml), dried (Na_2SO_4) and evaporated. An oily residue is purified by chromatography on SiO_2 (using hexane as the eluent) to give dibenzyl disulphide (0.55 g (88%); m.p. 71–72°C).

Unsymmetrical disulphide. A solution of $Na_2TeO_3 \cdot 5H_2O$ (0.465 g, 1.49 mmol) and 10% aqueous $[Bu_4N^+]OH^-$ (0.28 mol, 0.10 mmol) in H_2O (20 ml) is added dropwise to a vigorously stirred solution of phenylmethanethiol (0.307 g, 2.47 mmol) and 2-methyl-2-propanethiol (0.251 g, 2.79 mmol) in benzene (20 ml). The resulting mixture is stirred for 36 h at room temperature and then worked up as described above to give pure benzyl t-butyl disulphide (0.39 g (74% based on phenylmethanethiol)) as an oil (b.p. 97°C/0.9 torr).

4.5.6 Arenetellurinic anhydrides

The readily accessible arenetellurinic anhydrides (see Section 3.6) exhibit oxidizing properties similar, in many aspects, to those of the diaryl telluroxides [10–12].

From competitive experiments performed to establish the relative oxidizing capacity of some anhydrides and telluroxides the following order was determined: $(2\text{-naphthylTeO})_2O > (p\text{-MeOC}_6H_4)_2TeO > (p\text{-MeOC}_6H_4TeO)_2O > (p\text{-PhOC}_6H_4TeO)_2O$. As the reactions proceed, the tellurinic anhydrides are reduced to the corresponding ditellurides.

Pyrrolidine, indoline, phthaloylhydrazide and 4-phenylurazole are unaffected by tellurinic anhydrides.

In analogy to *p*-methoxyphenyl telluroxide, tellurinic anhydrides do not affect phenols but oxidize hydroquinones to the corresponding quinones, and oxidize thiols, thioesters, thioamides and phosphines to disulphides, esters, nitriles and phosphine oxides (see below). In contrast to the telluroxide, which oxidizes *N,N'*-diphenylthiourea to *N,N'*-diphenylurea, tellurinic anhydrides give *N,N'*-diphenylcarbodiimide.

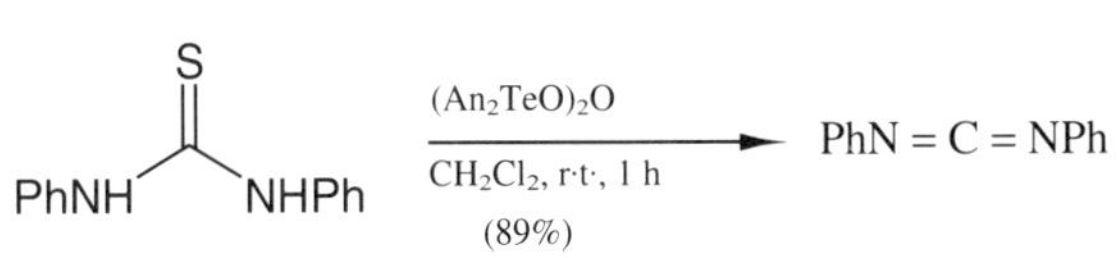

Oxidations with aryltellurinic anhydrides

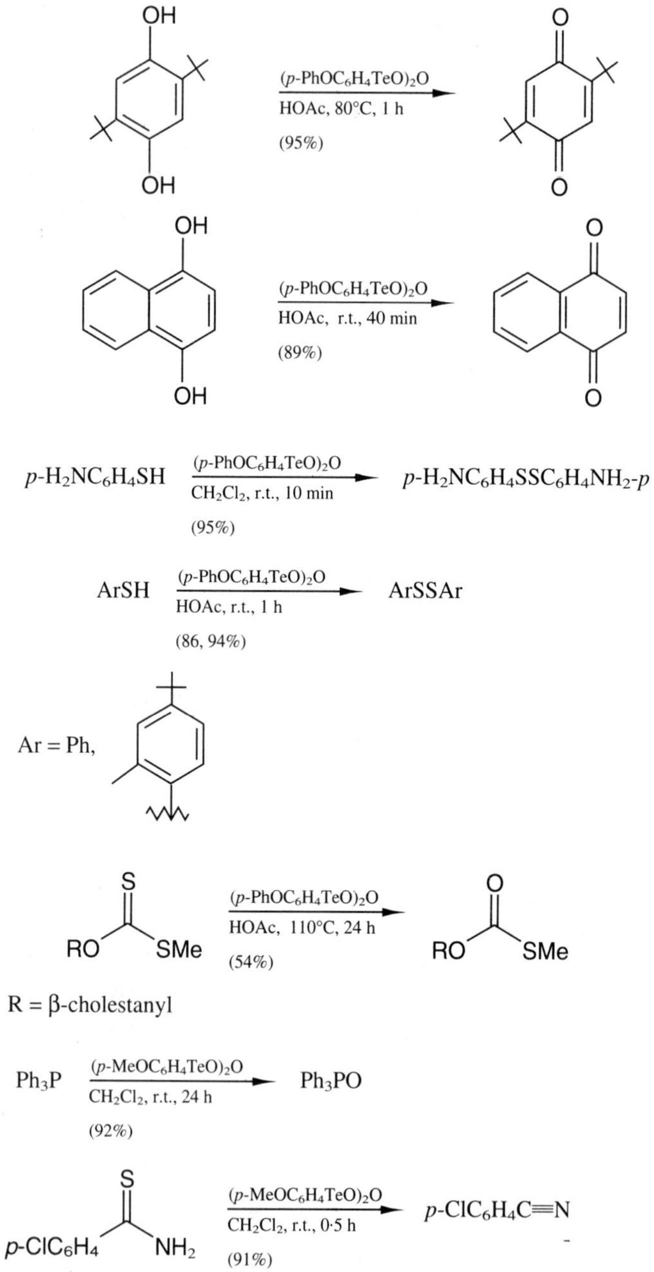

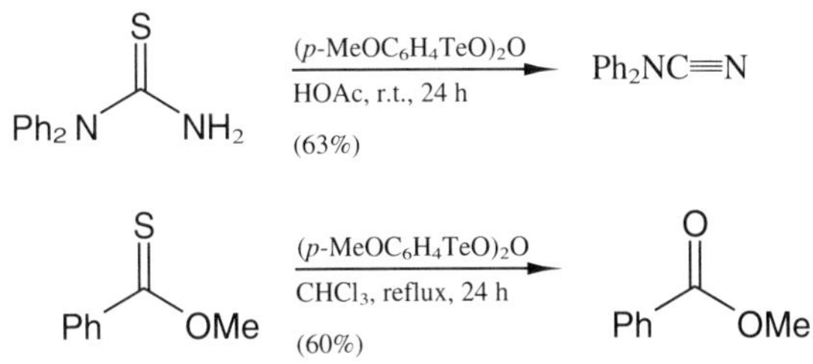

In analogy to tellurones, and in contrast to telluroxides, tellurinic anhydrides oxidize benzylic alcohol to carbonyl compounds.

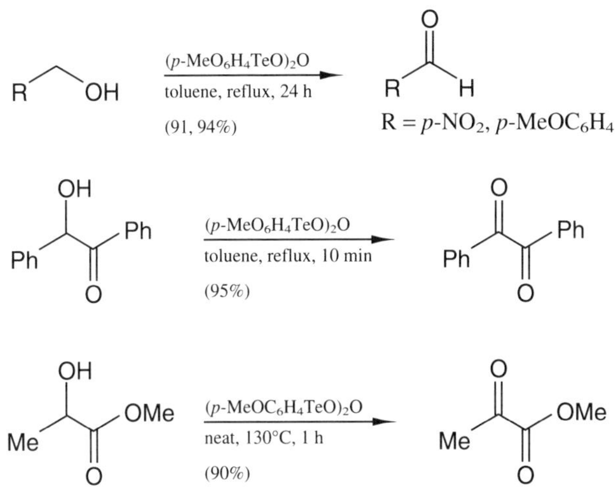

[11]

Oxidation with p-*phenoxyphenyltellurinic anhydride: diphenyl disulphide (typical procedure)* [10] Thiophenol (0.215 g, 2 mmol) is added to a stirred solution of the anhydride (0.210 g, 0.33 mmol) in HOAc (2 ml) under argon. The mixture is stirred for 1 h at room temperature. After evaporation under vacuum, the residue is extracted with CH_2Cl_2. The organic phase is washed with H_2O, 5% aqueous Na_2CO_3 and H_2O. SiO_2 column chromatography yields diphenyl disulphide (0.154 g (86%)) as colourless crystals (from CH_2Cl_2EtOH) (m.p. 58–60°C) and the ditelluride (0.167 g (83%)) as purple crystals (from CH_2Cl_2/EtOH) (m.p. 93–95°C).

Oxidation with p-*methoxyphenyltellurinic anhydride: benzonitrile (typical procedure)* [11] Thiobenzamide (0.15 g, 1.0 mmol) is mixed with the anhydride (0.189 g, 0.356 mmol) in dry CH_2Cl_2 (4 ml). The suspension is stirred at room temperature under N_2 and becomes clear in 0.5 h. After evaporation of the solvent under vacuum, benzonitrile is separated from the residue by Kugelrohr distillation (b.p. 122–123°C/100 torr; 0.107 g (95%)). SiO_2 column chromatography of the residue (eluent: benzene) gives red crystals of An_2Te_2 (0.152 g).

Mixed anhydrides such as benzenetellurinyl acetate (PhTe(O)OAc), benzenetellurinyl trifluoroacetate ($PhTe(O)OCOCF_3$) and benzenetellurinyl trifluoromethanesulphonate ($PhTe(O)OSO_2CF_3$), prepared by reacting benzenetellurinic anhydride with acetic anhydride, trifluoroacetic anhydride and trifluoromethanesulphonic anhydride, respectively, have been recognized as valuable oxidizing reagents, in some cases superior to the parent benzenetellurinic anhydrides [13].

4.5.7 Reaction of oxidizing tellurium reagents with the C=C bond

4.5.7.1 Epoxidation of olefins catalysed by polystyrene-bound tellurinic acid

Anchored aryltellurinic acid has a catalytic activity on the epoxidation of olefins [14]. The reagent is prepared by reaction of tellurium tetrachloride with divinylbenzene–styrene copolymer, followed by hydrolysis of the formed trichloride.

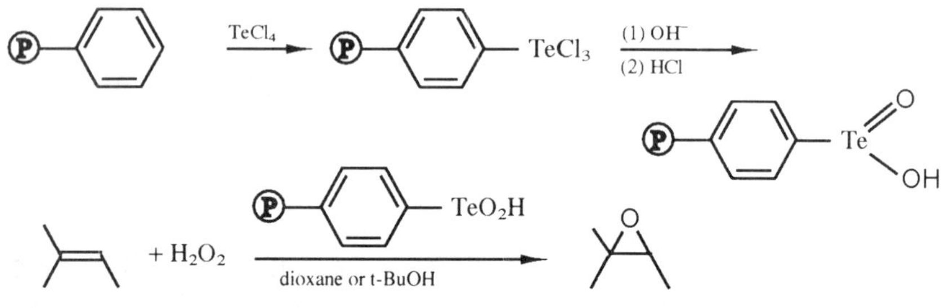

Olefin (relative rates):

1-methyl cyclohexene (54), cyclohexene (21), 3-methyl cyclohexene (21), *trans*-2-butene (15), styrene (10), *cis*-2-octene (4·3), *trans*-2-octene (4·2), 1-octene (1·0), allylchloride (0.66), allyl alcohol (0·13)·

The catalytic activity is dependent on the polymeric structure of the reagent, no epoxidation being observed with free tellurinic acids.

The epoxidation is accelerated by increasing alkyl substitution and is stereospecific as demonstrated by the retained configuration of the olefinic bond.

Catalytic epoxidation of olefins (typical procedure) [14] Solid catalyst (1g) prepared from XAD-2 resin is stirred with cyclohexene (1.6 g, 20 mmol) and H_2O_2 (20 mmol) in t-BuOH or dioxane at 60°C for 24 h. Cyclohexenoxide is obtained in a quantitative yield.

4.5.7.2 Diacetoxylation of olefins

Olefins are diacetoxylated by benzenetellurinic anhydride in boiling acetic acid [15].

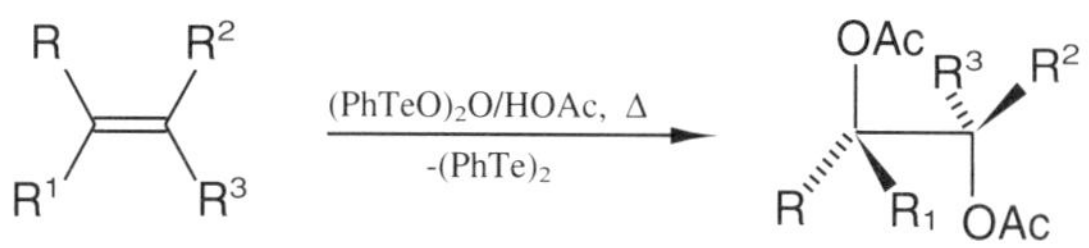

Olefins (% yield):

styrene (82), 1-octene (71), *cis*-2-octene (38), cyclohexene (38)

The reaction, which is accelerated by catalytic amounts of sulphuric acid, exhibits a predominantly *syn* stereochemistry.

The actual reactive species which adds to the C═C bond is a reduction product of the tellurinic anhydride, formed at the expense of the ditelluride produced as the reaction proceeds. Two successive *anti* acetolyses—(a) of the formed telluronium intermediate and (b) of the resulting acetoxytelluride (which is acid catalysed)—give the final diacetate with an overall *syn* stereochemistry.

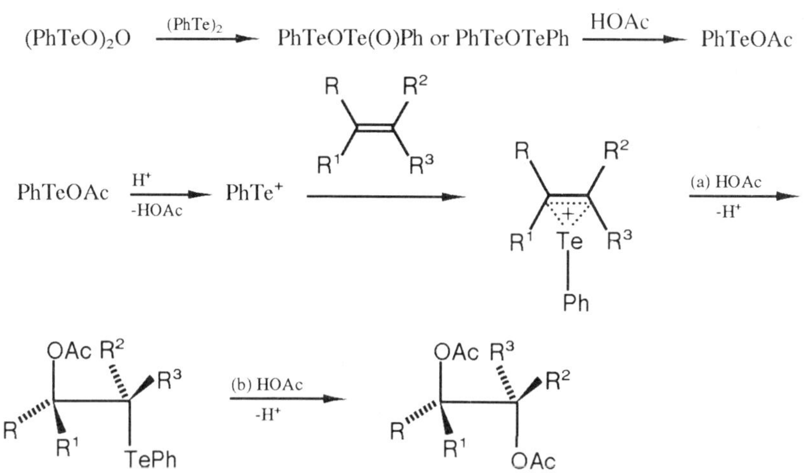

Diacetoxylation of olefins with benzenetellurinic anhydride (typical procedure) [15] To a solution of $(PhTeO)_2O$ (0.95 g, 2.1 mmol) in dry HOAc (15 ml) is added styrene (0.208 g, 2.0 mmol) in HOAc (4 ml) and 98% H_2SO_4 (0.020 g, 0.2 mmol) in HOAc (1 ml). The mixture is gently refluxed for 24 h, which turns red with deposition of a small amount of elemental Te. After removal of Te (0.071 g, 0.56 mmol) by filtration, the solvent is evaporated, the residue extracted with ether and the ether extract dried ($MgSO_4$). Chromatography on SiO_2 gives the *vic*-diacetate (0.363 g (82%) and 1-phenylethyl acetate (0.029 g (9%)).

Other methods achieving the *syn* diacetoxylation of olefins use the TeO_2/LiBr/HOAc [16] or the $TeCl_4$/LiOAc/HOAc [17] systems.

(i) 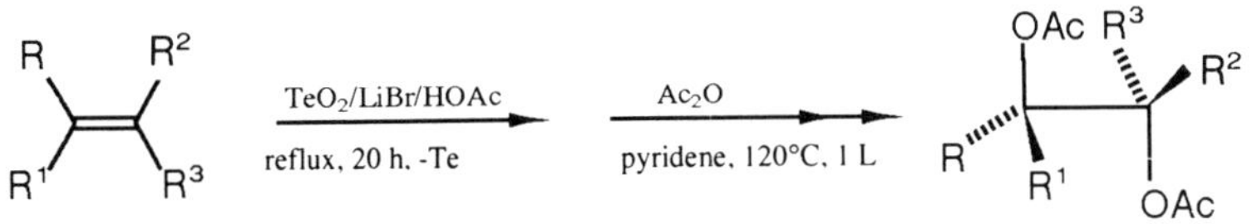

Olefin (% yield, % *erythro/threo* or *meso/dl*):

cyclopentene (68), cyclohexene (41), cycloheptene (59), styrene (42), 1-octene (80), *cis*-2-octene (80·3, 87/13), *trans*-2-octene (91, 42/58),

cis-4-octene (44, 89/11),

trans-4-octene (32·5, 44/56), 1-decene (52·%), allyl acetate (58·5), allylbenzene (61·2), 1,4-cyclohexadiene (51·5)·

The diacetates are accompanied by minor amounts of hydroxyacetates, which are converted to the diacetates by treatment with Ac_2O/pyridine.

As shown for the preceding method employing phenyltellurinic anhydride, the diacetoxylation prefers a *syn* stereochemistry, especially for cyclic alkenes and *cis* linear alkenes, whereas for *trans*-alkenes the preference for the *syn* stereochemistry is decreased. In accordance with the mechanism proposed in the case of the tellurinic anhydride, the reaction can be rationalized as involving the intermediacy of a *trans* adduct,

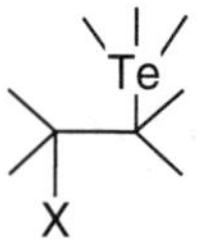

followed by an S_N2-type detellurative acetolysis.

(ii) 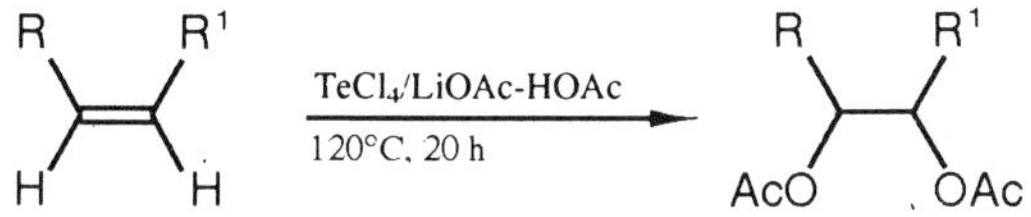

Olefin (% yield, % *meso/dl*): cyclopentene (42), cyclohexene (50), 1-hexene (49)*
styrene (54), *cis*-2-butene (34, 92/8),
trans-2-butene (24, 16/84)

The stereochemistry of the reaction is predominantly *syn*, in accordance with that described for the TeO_2/LiBr/HOAc system, with exception of the reaction with cyclopentene, which gives mainly the *trans* adduct.

Diacetoxylation of olefins with TeO_2/LiBr/HOAc (typical procedure) [16] To a mixture of TeO_2 (1.60 g, 10 mmol), LiBr (0.87 g, 10 mmol) and HOAc (30 ml) is added cyclohexene (1.64 g, 20 mmol) at room temperature. The resulting yellow-orange heterogeneous mixture is stirred and heated under reflux for 20 h, during which time the mixture gradually becomes a black suspension. The mixture is cooled and the solid filtered off. The filtrate is treated with brine, and then extracted with $CHCl_3$ (3×50 ml). The extract is washed successively with aqueous $NaHCO_3$, brine, and then dried ($MgSO_4$). Evaporation of the solvent leaves an oily residue which is treated with $(Ac)_2O$ (3 ml) in pyridene (7 ml) at 120°C for 1 h. The solution is made slightly acidic using dilute HCl and then extracted with $CHCl_3$ (3×30 ml). The extract is washed with brine and dried ($MgSO_4$), and Kugelrohr distillation affords almost pure *cis*-1,2-diacetoxycyclohexane (1.50 g (37.5%); b.p. 160–165°C/3 torr).

Diacetoxylation of olefins with $TeCl_4$/LiOAc/HOAc (typical procedure) [17] A mixture of $TeCl_4$ (1.35 g, 5 mmol), LiOAc (1.32 g, 20 mmol) and HOAc (15 ml) is stirred at 80–90°C for 1 h, during which time the solution becomes homogeneous. To the resulting homogeneous solution is added cyclohexene (0.82 g, 10 mmol), and the solution is heated at 120°C for 20 h. The solution is cooled, and the precipitated black Te is filtered off. The filtrate is extracted with ether (50 ml) and the ether layer washed with H_2O (2×50 ml). The aqueous layer is extracted again with ether (25 ml) and the combined ether extracts washed with aqueous $NaHCO_3$, dried ($MgSO_4$) and evaporated to leave a yellow oil. The oil is treated with Ac_2O (1 ml) in pyridine (3 ml) to acylate any free hydroxy groups. The pyridine solution is diluted with ether (25 ml) and acidified with dilute HCl. The ethereal solution is dried ($MgSO_4$) and evaporated giving a

*$TeBr_4$ used instead of $TeCl_4$.

yellow oil. The oil is chromatographed on a short SiO_2 column (eluting with petroleum ether/ether 1:1), giving *cis*-1,2-diacetoxycyclohexane (0.70 g (70%)).

If the reaction with $TeCl_4$/LiOAc/HOAc is performed at 80°C in the presence of $BF_3{\cdot}Et_2O$, and the mixture treated with sodium thiosulphate after 3 h, acetoxyalkyl ditellurides are obtained, which are converted to the corresponding tribromides by addition of bromine [17].

The stereochemistry of the addition is *anti*, as expected by an attack of the acetate ion at an intermediate telluronium ion.

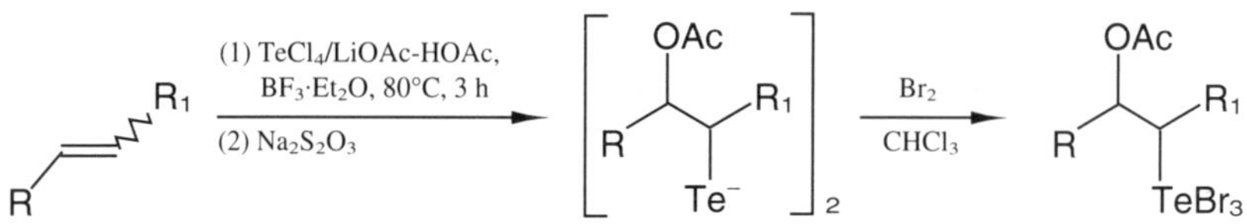

The acetoxytellurium tribromides are converted to the diacetates by the same treatment employed for the diacetoxylation reaction (HOAc, 120°C). The formation of an overall *syn* or *anti* adduct depends on the competition between a rearward attack by the acetate ion at the tellurium atom (as in the case of cyclohexene and 2-butenes), and a front attack by the neighbouring acetate moiety (as in cyclopentenes, where the almost planarity of the five-membered ring makes the conformation of the acetoxytellurium tribromide susceptible to frontal attack).

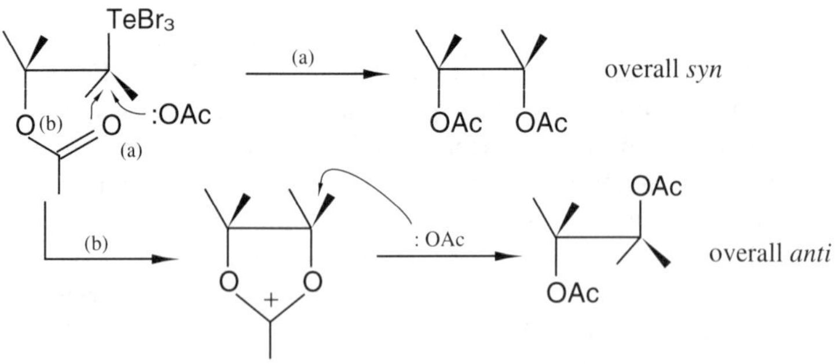

By submitting conjugated dienes to a treatment with the TeO_2/LiBr/HOAc system, a mixture of 1,2- and *cis*- and *trans*-1,4-diacetoxylated adducts is formed [18].

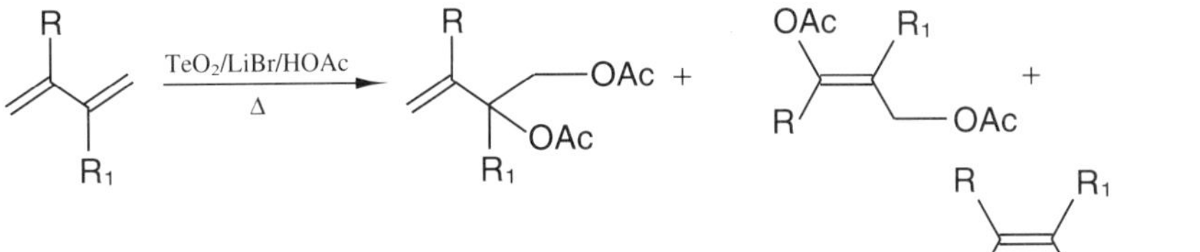

$R = R^1 = H, Me; R = H, R^1 = Me$

The yield and the 1,2/1,4 isomeric ratio is highly dependent on the $LiBr/TeO_2$ ratio, the 1,4 isomer being favoured when the ratio is 5. The *trans* isomer is the major product in the 1,4 addition to 1,3-butadiene.

The reaction can also be carried out catalytically relative to TeO_2 when a reoxidant such as H_2O_2 or t-BuOOH is used.

Unsatisfactory results are obtained with higher and cyclic dienes.

On the basis of the mechanism proposed for the addition of the same TeO_2/LiBr/HOAc system to single olefins, the addition to conjugated dienes probably involves the acetolysis of a homoallylic and an allylic telluroacetate (or tellurohalide) generated by the initial addition of a Te(IV) species and an acetate (or halide) anion to the conjugated system.

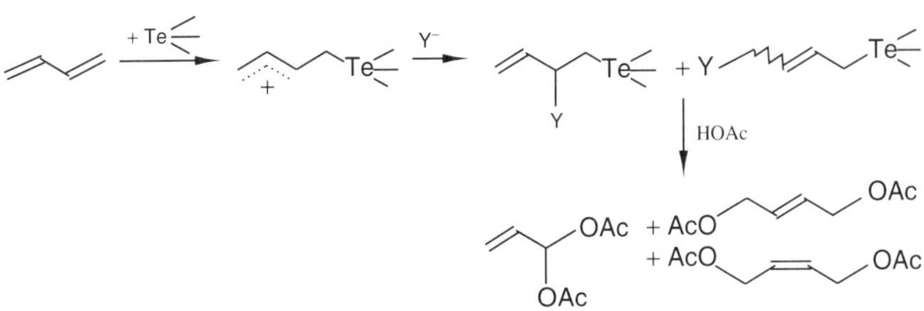

Diacetoxylation of 1,3-butadiene (typical procedure) [18] To a mixture of TeO_2 (0.80 g, 5 mmol), LiBr (2.17 g, 25 mmol), AcOH (18 ml) and Ac_2O (2 ml) in a glass pressure bottle, chilled at −20°C, is added chilled 1,3-butadiene (1.35 g, 25 mmol). The resulting suspension is heated (oil bath) at 120–130°C for 20 h under magnetic stirring, producing a yellow solution which gradually turns to a black suspension. After cooling, the mixture is filtered, and the filtrate is treated with brine, and then extracted with $CHCl_3$ (3 × 50 ml). The extracts are washed successively with aqueous $NaHCO_3$ and brine, and then dried ($MgSO_4$). GC analysis reveals the presence of the three isomeric diacetates. Evaporation of the solvent gives an oil (IR spectroscopy reveals strong OH and AcO absoption bands). After treatment with Ac_2O (3 ml) in pyridine (7 ml) at 80°C for 1 h, GC of the pyridine solution (using ethyl cinnamate as the internal standard) reveals the presence of the 1,2 adduct (0.61 mmol), *cis*-1,4 adduct (0.71 mmol) and *trans*-1,4 adduct (4.78 mmol) (total 1.05 g, 6.10 mmol).

The diacetoxylation of isoprene and 2,3-dimethyl-1,3-butadiene is carried out similarly.

	Reaction time (h)	Temperature (°C)	Total yield (%)	1,2/1,4 ratio
Isoprene	48	95	15.4	1/4
2,3-Dimethyl-1,3-butadiene	72	70	11.5	1/9

4.5.7.3 Methoxytellurenylation and dimethoxylation of olefins

The observation that the previously described diacetoxylation of olefins by means of tellurinic anhydrides produces quantitative yields of the corresponding ditellurides suggested the oxidative functionalization of olefins employing diphenyl ditelluride combined with an oxidizing agent instead of the tellurinic anhydride [19].

Indeed, aliphatic olefins treated with equivalent amounts of diphenyl ditelluride and t-BuOOH in methanol in the presence of sulphuric acid give methoxytellurenylated adducts.*

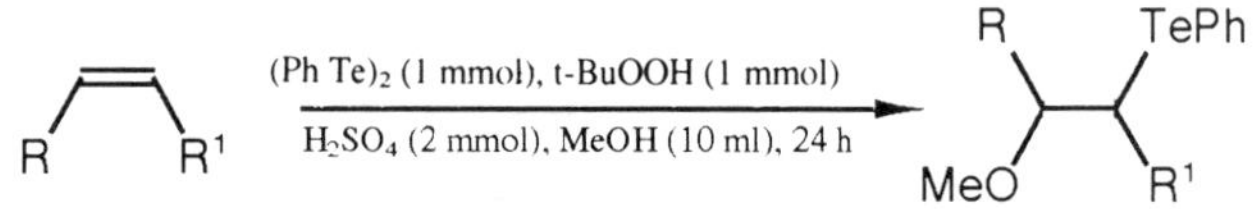

Olefin (mmol, % yield)

$R^1 = H$; $R = n\text{-}C_6H_{13}$ (1, 59), $R = PhCH_2$ (1, 73)

$RR^1 = (CH_2)_4$ (3, 32)

The reaction is regioselective and sterospecific: the tellurium moiety is added exclusively to the terminal carbon of terminal alkenes, and only *trans* adducts are obtained with cyclohexene. The reaction may proceed through the tellurenylation of the olefin (in a rapid equilibrium as demonstrated in separate experiments), forming a telluronium intermediate which is then methanolysed.

*For alkoxydihalotelluration of olefins by means of the $(ArTe)_2/Br_2/ROH$ system, see Section 3.9.3.2.

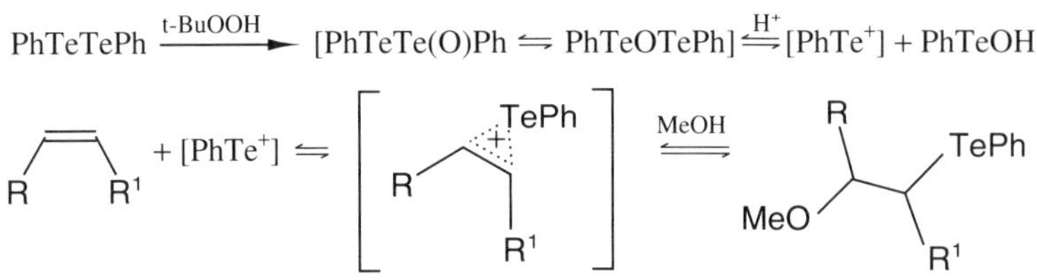

Methoxytellurenylation of olefins (typical procedure) [19] 1-Octene (0.112 g, 1 mmol), t-BuOOH (0.090 g, 1 mmol), $(PhTe)_2$ (0.409 g, 1 mmol), H_2SO_4 (0.196 g, 2 mmol) are heated under reflux in MeOH (10 ml) for 24 h. The phenyl 1-(2-methoxy)octyl telluride is obtained in the pure form by diluting with ether (50 ml), neutralizing with K_2CO_3 and filtration followed by radial layer chromatography (SiO_2, pentane/ether) (0.201 g (59%)). Similar reactions performed using ethanol instead of methanol give the corresponding ethoxy derivatives.

Under identical conditions the aromatic olefins styrene and *p*-methylstyrene give the *vic*-dimethoxy adducts as the sole products. Methoxytellurenylated adducts are formed, however, as minor by-products from different substituted olefins (*p*-chlorostyrene) or exclusively (from styrene) when the amount of H_2SO_4 is reduced.

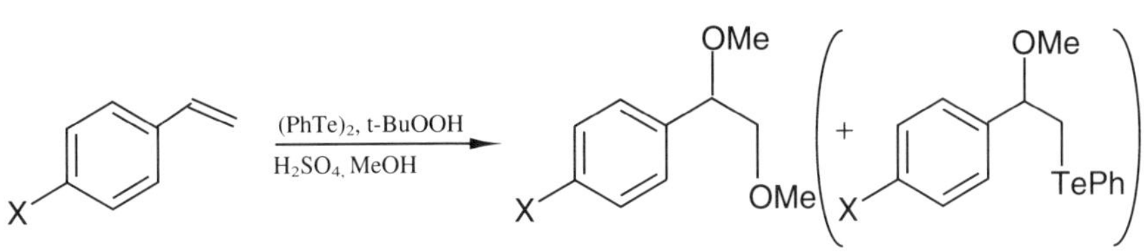

X = H, Me Cl

The dimethoxylation of aromatic olefins can also be performed by employing a catalytic amount of ditelluride, e.g. styrene (5 mmol), $(PhTe)_2$ (0.5 mmol), t-BuOOH (5.5 mmol added in 10 portions), H_2SO_4 (2 mmol), MeOH (10 ml, reflux for 3.8 days) give 1,2-dimethoxy-1-phenylethane in a yield of 88% [19, 20].

A proposed mechanism (on the basis of the identification of the intermediates) for the dimethoxylation involves an equilibrium between the starting olefin and a regioisomeric methoxytellurenylated adduct, the precursor of the dimethoxylated derivatives.

The stereochemistry of the reaction, as demonstrated in the case of indene, involves an *anti*-methoxytellurenylation followed by methanolysis (with inversion) resulting in a net *syn* dimethoxylation

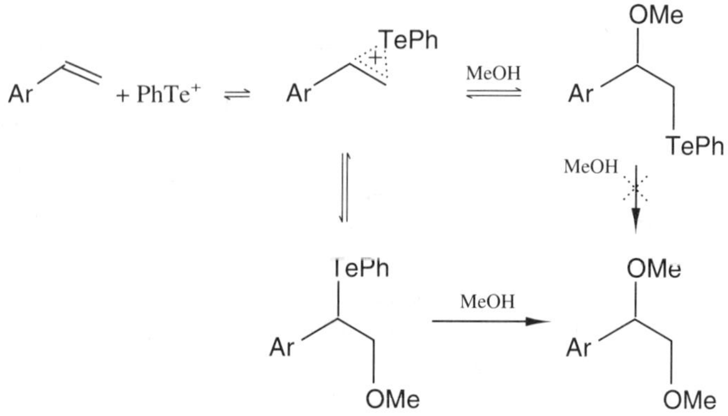

4.5.7.4 Amidotellurinylation of olefins and related reactions

Phenyltellurinyl acetate and trifluoroacetate, prepared *in situ* from phenyltellurinic anhydride and the corresponding carboxylic acids (or anhydrides) [21–23], add to olefins in the presence of excess ethyl carbamate and $BF_3{\cdot}Et_2O$ to give ethyl(2-phenyltellurinyl)alkyl carbamates [12, 21, 23, 24]. Due to difficulties in their purification, the products are reduced with hydrazine hydrate and isolated as tellurides. The following features must be pointed out: reactions with the tellurinyl acetates are carried out in refluxing chloroform, while the more effective trifluoromethanesulphonate reacts in refluxing dichloromethane, even without the aid of the Lewis acid. Markovnikov-type regioselectivity is observed for terminal olefins, and the electrophilic tellurium moiety adds to the terminal carbon, unless electronic factors are operative (as in allyl phenyl ethers) inverting the isomeric ratio. The *trans* stereochemistry of the addition to cyclohexene is clearly consistent with an *anti* addition of ethyl carbamate on an epitelluronium intermediate.

$$(PhTeO)_2O + 2RCO_2H \longrightarrow 2PhTe(O)OCOR + H_2O$$
$$(R = Me,\ CF_3)$$

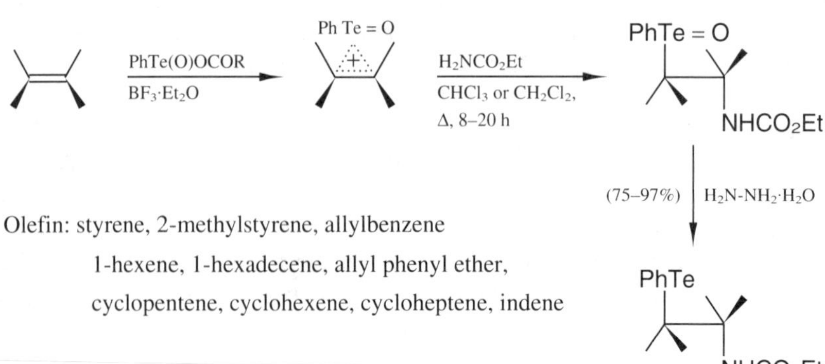

Olefin: styrene, 2-methylstyrene, allylbenzene
1-hexene, 1-hexadecene, allyl phenyl ether,
cyclopentene, cyclohexene, cycloheptene, indene

Amidotellurinylation of olefins (typical procedure) [23] To benzenetellurinyl trifluoroacetate (1.1 mmol) (generated *in situ* by treatment of benzenetellurinic anhydride (0.252 g, 0.55 mmol) with $(F_3CCO)_2O$ (0.116 g, 0.55 mmol) in CH_2Cl_2 (6 ml) at room temperature, for 10 min) are added successively cyclohexene (0.082 g, 1 mmol), ethyl carbamate (0.446, 5 mmol) and $BF_3 \cdot Et_2O$ (0.156 g, 1.1 mmol), and the resulting mixture stirred under reflux for 12 h. The mixture is cooled to room temperature, concentrated under vacuum, and the residue reduced with H_2—$NH_2 \cdot H_2O$ (0.1 g, 2 mmol) in EtOH (6 ml) at 60°C for 10 min. After evaporation of the solvent under vacuum, the mixture is dissolved in CH_2Cl_2 (6 ml), treated with saturated aqueous $NaHCO_3$ (25 ml) and extracted with CH_2Cl_2 (2 × 20 ml). The extract is washed with brine (20 ml) and dried ($MgSO_4$). Evaporation of the solvent under vacuum gives a yellow residue, which is submitted to SiO_2 column chromatography (elution with hexane/ethyl acetate 5:1), giving first a small amount of diphenyl ditelluride and then *trans*-ethyl(2-phenyltelluro)cyclohexyl carbamate. The carbamate is recrystallized from hexane/$CHCl_3$ as colourless needles (0.36 g (96%); m.p. 72–73°C).

The amidotellurinylation of olefins is also achieved by the treatment of olefins with phenyltellurinyl trifluoroacetate, in the presence of $BF_3 \cdot Et_2O$, in acetonitrile, which acts both as a solvent and a nucleophile [25, 26].

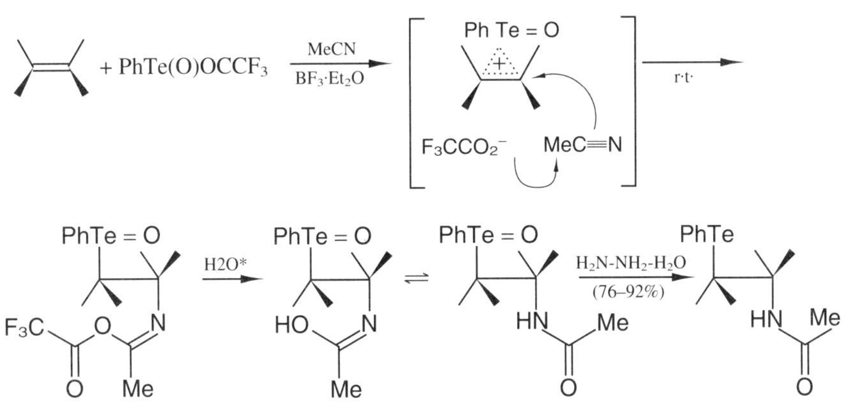

Olefin: cyclopentene, cyclohexene, cycloheptene, 1-hexene

Amidotellurinylation of olefins (typical procedure) [25] To benzenetellurinyl trifluoroacetate (generated *in situ* from benzenetellurinic anhydride (0.27 g, 0.59 mmol) and TFA (0.159 g, 1.4 mmol) in MeCN (6 ml) at room temperature

*Water for hydrolysis is furnished by the reaction forming benzenetellurinyl trifluoroacetate formation [26] (see above).

for 10 min) are added successively cyclohexene (0.082 g, 1.0 mmol) and $BF_3 \cdot Et_2O$ (0.2 g, 1.4 mmol), and the resulting mixture is stirred at room temperature for 12 h. Evaporation of the solvent under vacuum gives a yellow oil, which is reduced with $H_2N—NH_2/H_2O$ (0.1 g, 2 mmol) in EtOH (6 ml) at room temperature for 15 min. The mixture is poured into H_2O and extracted with CH_2Cl_2 (2 × 20 ml). The organic phase is washed with brine, dried ($MgSO_4$) and evaporated. The residue is chromatographed on SiO_2 (eluting with hexane/ethyl acetate 2:3) giving *trans*-2-acetamidocyclohexyl phenyl telluride (0.26 g (76%)). The product is recrystallyzed from hexane/$CHCl_3$ as colourless needles (m.p. 146–146.5°C).

If the two described amidotellurinylation procedures are carried out at a higher temperature, tellurium-free products are obtained in high yields: 2-oxazolidinones [23, 24] and 2-oxazolines [25, 26], respectively. Diphenyl ditelluride is formed as by-product. These heterocycles constitute an important class of compounds with wide applications.

These detellurative cyclizations proceed through a rearward attack of the carbonyl oxygen (a) and of the iminol hydroxyl group (b), respectively, at the carbon bearing the tellurinyl moiety. Taking into consideration the regio- and stereochemistry of the amidotellurinylation step, the entire process is highly regio- and *syn*-stereoselective.

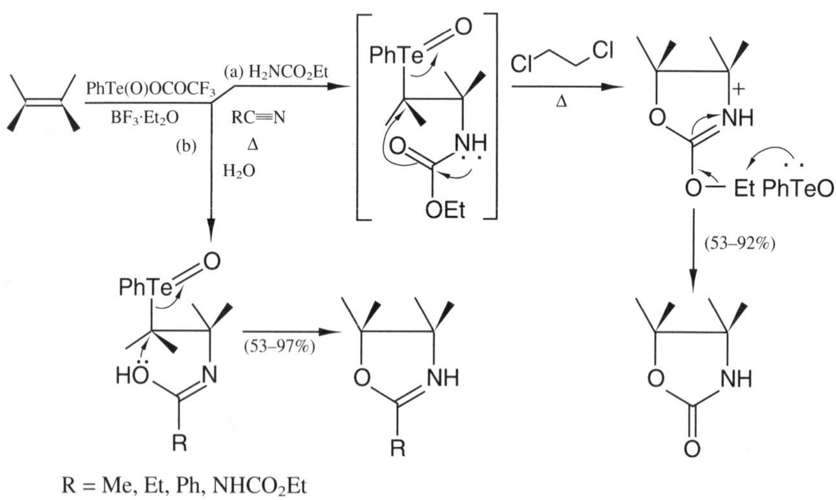

Olefin: styrene, α-methyistyrene, *trans*- β-methylstyrene, 1-hexene, *cis* and *trans*-4-octene, cyclopentene, cyclohexene, cycloheptene, indene, 1,2-dihydronaphthalene

2-Oxazolidinones (typical procedure) [23] To a 1,2-dichloroethane solution (6 ml) of phenyltellurinyl trifluoroacetate (1.1 mmol), generated *in situ* from benzenetellurinic anhydride (0.252 g, 0.55 mmol) and TFA (0.126 g, 1.1 mmol), are added successively cyclohexene (0.082 g, 1 mmol), ethyl carbamate (0.446 g, 5 mmol) and $Et_2O{\cdot}BF_3$ (0.156 g, 1.1 mmol), and the resulting solution is stirred under reflux for 20 h. After cooling, the solution is treated with saturated aqueous $NaHCO_3$ (30 ml) and extracted with $CHCl_3$ (2 × 20 ml). The extracts are dried ($MgSO_4$) and evaporated *in vacuo*. The remaining ethyl carbamate is removed by sublimation (50°C/1 torr) and the residue chromatographed on SiO_2 (elution with EtOAc/hexane 1:1), giving diphenyl ditelluride (0.14 g, 62%) as red needles and then *cis*-4,5-tetramethylene-2-oxazolidinone (0.122 g (86%)) as a white solid. Recrystallization from hexane/acetone gives colourless crystals (m.p. 55–56°C).

2-Oxazolines (typical procedure) [25] To benzenetellurinyl trifluoroacetate (generated *in situ* by treatment of benzenetellurinic anhydride (0.27 g, 0.59 mmol) with TFA (0.159 g, 1.4 mmol) in MeCN (6 ml) at room temperature for 10 min) are added successively cycloheptene (0.096 g, 1 mmol) and $BF_3{\cdot}Et_2O$ (0.27 g, 1.4 mmol). The resulting mixture is heated at 75°C for 3 h, and gradually turns black-red. It is cooled to room temperature, poured into $CHCl_3$ (25 ml), and then extracted with 0.5 M HCl (2 × 25 ml). The aqueous extract is made alkaline by addition of NaOH pellets with ice cooling, and again extracted with ether (2 × 25 ml). The combined organic extracts are dried (K_2CO_3) and evaporated to give pure 2-methyl-4,5-pentamethylene-2-oxazoline (0.149 g (97%)) as a colourless oil (b.p. 130°C/42 torr).

A further demonstration of the synthetic utility of the amidotellurinylation reaction is illustrated by the elimination of phenyltellurinic acid from the tellurinyl carbamates, giving amido olefins as shown in the accompanying example.

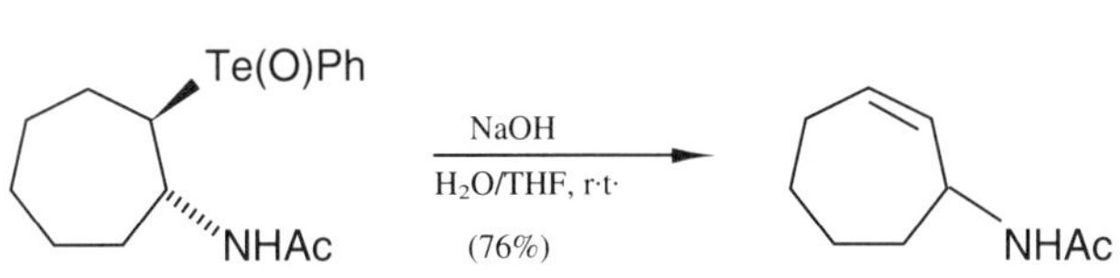

Amido olefins (typical procedure) [25] To *trans*-2-acetamidocycloheptyl phenyl telluroxide in THF (5 ml) (generated from cycloheptene (0.105 g, 1.09 mmol) and benzenetellurinyl trifluoroacetate (1.2 mmol) as previously described) is added aqueous NaOH (0.5 M, 5 ml). The mixture is stirred at room temperature for 3 h, poured into H_2O and extracted with $CHCl_3$ (2 × 20 ml), and the organic phase is dried ($MgSO_4$) and evaporated under vacuum. The residue is purified by SiO_2 column chromatography (eluting with hexane/EtOAc 1:1), giving 3-acetamidocycloheptene (0.127 g (76%)) as colourless crystals (m.p. 74–75°C).

REFERENCES

1. D. H. R. Barton, S. V. Ley and C. A. Merholz, *J. Chem. Soc. Chem. Commun.*, 755 (1979).
2. S. V. Ley, C. A. Merholz and D. H. R. Barton, *Tetrahedron* **37**, 213 (1981).
3. S. V. Ley, C. A. Merholz and D. H. R. Barton, *Tetrahedron Lett.* **21**, 1785 (1980).
4. L. Engman and M. C. Cava, *J. Chem. Soc. Chem. Commun.*, 164 (1982).
5. N. X. Hu, Y. Aso, T. Otsubo and F. Ogura, *Bull Chem. Soc. Jpn* **59**, 879 (1986).
6. M. J. Farrall and J. M. J. Frechet, *J. Org. Chem.* **41**, 3877 (1976).
7. S. J. Falcone and M. P. Cava, *J. Org. Chem.* **45**, 1044 (1980).
8. T. Matsuki, N. K. Hu, Y. Aso, T. Otsubo and F. Ogura, *Bull. Chem. Soc. Jpn* **61**, 2117 (1988).
9. H. Suzuki, S. Kawato and A. Nasu, *Bull. Chem. Soc. Jpn* **65**, 626 (1992).
10. D. H. Barton, J. P. Finet and M. Thomas, *Tetrahedron* **42**, 2319 (1986).
11. N. X. Hu, Y. Aso, T. Otsubo and F. Ogura, *Tetrahedron Lett.* **27**, 6099 (1986).
12. N. X. Hu, Y. Aso, T. Otsubo and F. Ogura, *Phosphorus Sulfur* **38**, 177 (1988).
13. T. Fukumoto, T. Matsuki, N. X. Hu, Y. Aso, T. Otsubo and F. Ogura, *Chem. Lett.*, 2269 (1990).
14. W. F. Brill, *J. Org. Chem.* **51**, 1149 (1986).
15. N. Kambe, T. Tsukamoto, N. Miyoshi, S. Murai and N. Sonoda, *Chem. Lett.*, 269 (1987).
16. S. Uemura, K. Ohe, S. I. Fukuzawa, S. R. Patil and N. Sugita, *J. Organomet. Chem.* **316**, 67 (1986).
17. S. I. Fukuzawa, K. J. Irgolic and D. H. O'Brien, *Organometallics* **9**, 3079 (1990).
18. S. Uemura, S. Fukuzawa, S. R. Patil and M. Okano, *J. Chem. Soc. Perkin Trans I*, 499 (1985).
19. N. Kambe, T. Fujioka, A. Ogawa, N. Miyoshi and N. Sonoda, *Phosphorus Sulfur* **38**, 167 (1988).
20. N. Kambe, T. Fujioka, A. Ogawa, N. Miyoshi and N. Sonoda, *Chem. Lett.*, 2077 (1987).
21. N. X. Hu, Y. Aso, T. Otsubo and F. Ogura, *Chem. Lett.*, 1327 (1987).
22. H. X. Hu, Y. Aso, T. Otsubo and F. Ogura, *Tetrahedron Lett.* **28**, 1281 (1987).
23. N. X. Hu, Y. Aso, T. Otsubo and F. Ogura, *J. Org. Chem.* **54**, 4398 (1989).
24. N. X. Hu, Y. Aso, T. Otsubo and F. Ogura, *J. Chem. Soc. Chem. Commun.*, 1447 (1987).
25. N. X. Hu, Y. Aso, T. Otsubo and F. Ogura, *Tetrahedron Lett.* **29**, 1049 (1988).
26. N. X. Hu, Y. Aso, T. Otsubo and F. Ogura, *J. Chem. Soc. Perkin Trans. I*, 1775 (1989).

4.6 ORGANOTELLURIUM-BASED RING CLOSURE REACTIONS

The addition of an electrophilic tellurium reagent to olefins bearing an effective nucleophilic group at a suitable position, followed by the intramolecular capture of the generated telluronium species, results in the formation of a cyclic structure containing a tellurium moiety linked to the ring. Similar ring closure reactions are well known with selenium electrophilic reagents [1].

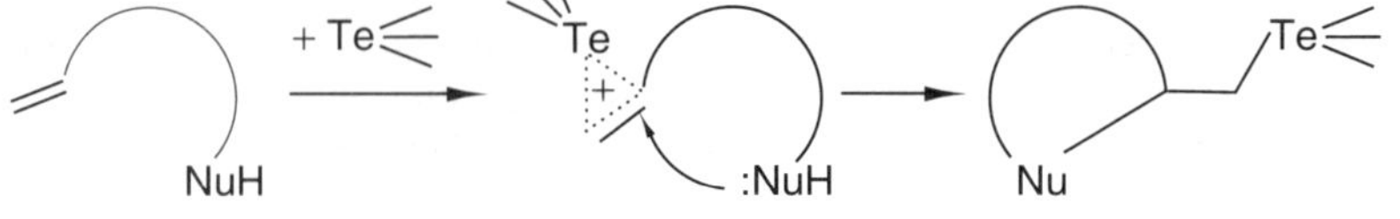

These ring closure reactions, combined with the removal of tellurium by well-established methods, therefore constitute useful procedures for the synthesis of cyclic structures, often present in natural products.

4.6.1 Tellurolactonization of unsaturated carboxylic acids

4.6.1.1 With aryltellurium trichlorides

The reaction of aryltellurium trichlorides with γ,δ-unsaturated carboxylic acids gives rise to aryldichlorotellurium butyrolactones in high yields [2, 3].

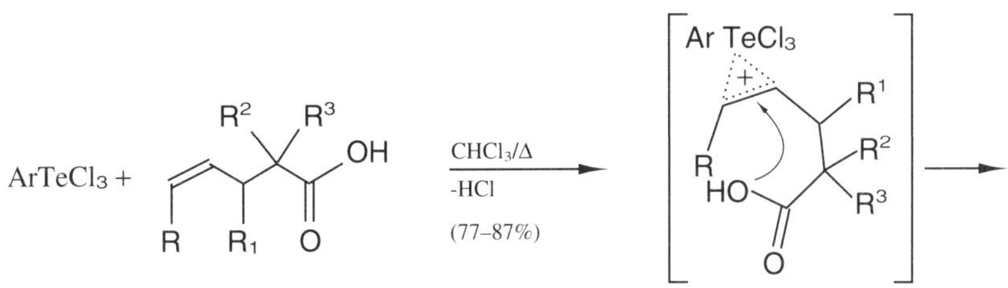

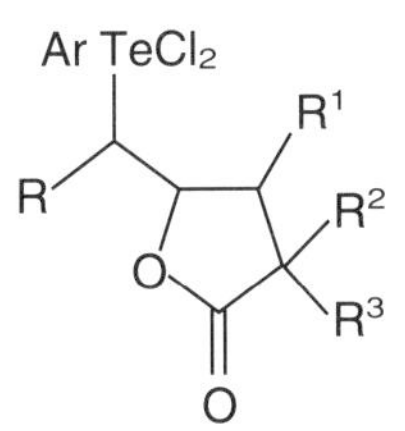

Ar = p-MeOC$_6$H$_4$

$R = R^1 = H$; $R^2 = R^3 = H$, Ph, Me
$R = R^1 = H$; $R^2 = Me$; $R^3 = Ph$
$R,R^1 = (CH_2)_2$, $(CH_2)_3$*; $R^2 = R^3 = H$
$R,R^1 = (CH_2)_3$; $R^2 = H$; $R^3 = Me$
$R,R^1 = (CH_2)_3$; $R^2 = R^3 = Me$

Aryldichlorotellurolactones (general procedure) [3] A solution of the γ,δ-unsaturated acid (5 mmol) and p-methoxyphenyltellurium trichloride (2.0 g, 5.8 mmol) in $CHCl_3$ (80 ml) is heated under reflux (acyclic substrates, 1 h; cyclic substrates, 2.5–7 h). The solution is evaporated and the residue filtered through SiO_2 with the aid of $CHCl_3$. The solution is dried ($MgSO_4$) and evaporated. The residue is recrystallized from $CHCl_3$/petroleum ether 30–60°C, giving the pure product.

Analogous arylselenolactonizations are achieved employing arylselenyl bromides [1, 2].

*As established by X ray [4] analysis, this dichlorotellurolactone exhibits a *cis* ring fusion—in accordance with an *anti* addition mechanism.

4.6.1.2 With benzenetellurenyl nitrobenzenesulphonate

Diphenyl ditelluride reacts with *p*-nitrobenzenesulphonyl peroxide (NBSP), generating the title compound. This efficient electrophilic reagent reacts *in situ* with 4-pentenoic acid, allowing the formation of the corresponding tellurobutyrolactone [5].

$$PhTeTePh + ArSO_2\text{-}OO\text{-}SO_2Ar \longrightarrow 2PhTeOSO_2Ar$$
$$Ar = p\text{-}NO_2C_6H_4$$

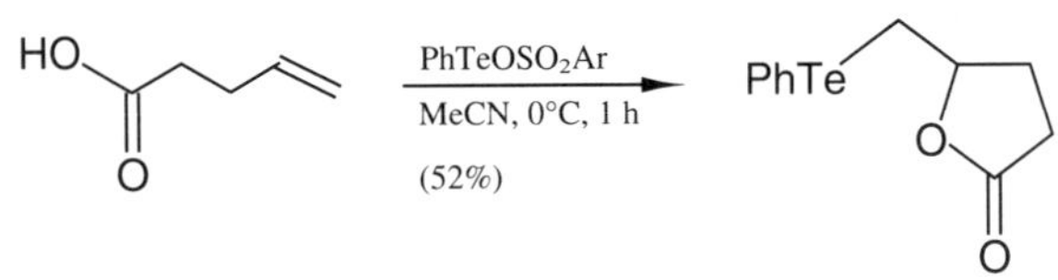

Tellurolactonization of 4-pentenoic acid (typical procedure) [5] To a solution of diphenyl ditelluride (0.102 g, 0.5 mmol) in freshly distilled MeCN (20 ml) is added solid NBSP (0.170 g, 0.5 mmol), in small portions at 0°C, and the solution is stirred for 10 min. 4-Pentenoic acid (0.130 g, 1.1 mmol) is added, the solution is stirred for 2 h at 0°C, then poured into H_2O (10 ml) and extracted with ether (3 × 10 ml). The organic layer is washed with 5% aqueous $NaHCO_3$ (10 ml) and brine (10 ml), and then dried ($MgSO_4$). The solvent is evaporated, and the residue purified by SiO_2 column chromatography (eluting with hexane/$CHCl_3$ 3:1).

5-Hexenoic and 6-heptenoic acids are ineffective for similar lactonizations.

4.6.1.3 Reductive detelluration of tellurolactones

The reductive removal of the tellurium moiety from aryldichlorotellurolactones is achieved by treatment with tributyltin hydride (TBTH) in refluxing toluene [6]. High yields are conditional on a previous careful deoxygenation of the reaction mixture.

If the dichlorotellurolactones are reduced previously to the parent tellurolactones by treatment with $NaBH_4$ [3], the successive detelluration saves 2 eq of TBTH, and the work-up is easier.

Reductive detelluration of aryldichlorotellurolactones (general procedure) [6] To a solution of the dichlorotellurolactone (1 mmol) in deoxygenated toluene (10 ml), under reflux and N_2, is added dropwise Bu_3SnH (4 mmol). The reaction is monitored by TLC. After 4 h the solvent is evaporated and the residue

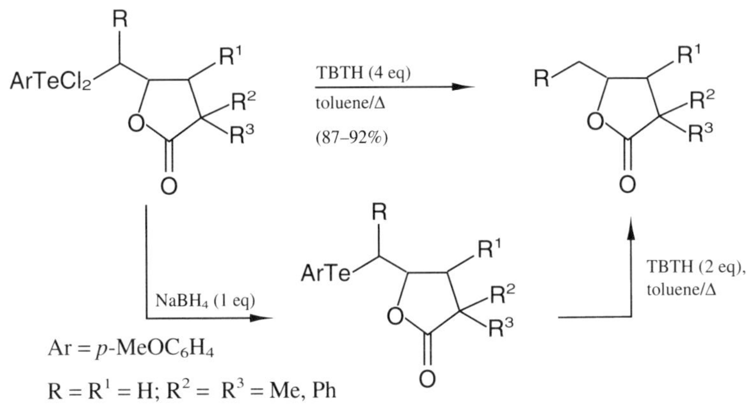

Ar = p-MeOC$_6$H$_4$

R = R^1 = H; R^2 = R^3 = Me, Ph

R = R^1 = R^2 = H; R^3 = Me

R, R^1 = (CH$_2$)$_2$, (CH$_2$)$_3$; R^2 = R^3 = H

chromatographed on SiO_2 (eluted first with petroleum ether to remove Te and Sn by-products, and then with petroleum ether/EtOAc 1:5), giving the tellurium-free lactone.

Reduction of the dichlorotellurolactones to the corresponding tellurolactones (general procedure) [3] A solution of the dichlorotellurolactone (1 mmol) in THF (10 ml) is treated dropwise with $NaBH_4$ (0.04 g, 1.05 mmol) in H_2O (5 ml). An immediate reaction takes place with gas evolution and the reaction mixture turns pale yellow. After stirring for 10 min at room temperature, the reaction mixture is diluted with ether (30 ml) and washed in turn with H_2O, saturated aqueous NH_4Cl and brine. The organic layer is dried ($MgSO_4$) and evaporated, furnishing the tellurolactone as a yellow oil that turns red on standing.

By treatment of the dichlorotellurolactone with excess $NaBH_4$, the fission of the Te—C bond takes place with recovery of the starting γ,δ-unsaturated carboxylic acid [3].

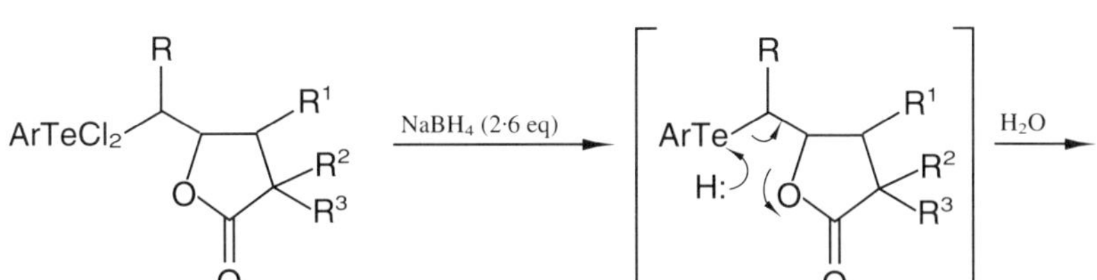

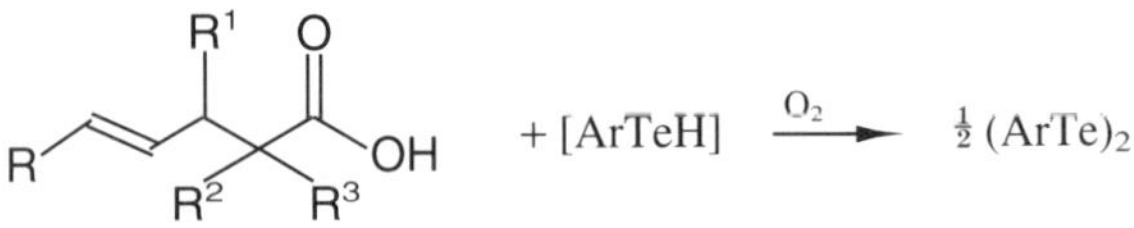

The overall process (lactonization/$NaBH_4$ reduction) therefore constitutes a formal protective method for γ,δ-unsaturated acids.

Reaction of dichlorotellurolactones with excess $NaBH_4$: regeneration of γ,δ-unsaturated acids (general procedure) [3] The dichlorotellurolactone (1 mmol) in THF (10 ml) at room temperature is treated with $NaBH_4$ (0.1 g, 2.6 mmol) in H_2O (5 ml). An immediate reaction takes place and the solution turns dark red. After stirring for 10 min at room temperature, the reaction mixture is diluted with ether (30 ml) and extracted with 2 N NaOH (2 × 5 ml). The organic phase is washed with H_2O, dried ($MgSO_4$) and evaporated, furnishing bis(*p*-methoxyphenyl) ditelluride in a quantitative yield. The cooled aqueous phase is acidified with concentrated HCl, and extracted with ether (3 × 10 ml). The organic extracts are dried ($MgSO_4$), evaporated, and the residue Kugelrohr distilled, furnishing the γ,δ-unsaturated carboxylic acid in a yield of about 90%.

4.6.2 Cyclotelluroetherification of unsaturated alcohols and allylphenols

4.6.2.1 With aryltellurinyl acetates

Aryltellurinyl acetates, generated as previously described from the corresponding tellurinic anhydrides and acetic acid, add to hydroxyolefins, giving tellurinylated cyclic ethers [7–9].

The reaction is carried out in refluxing acetic acid or in chloroform in the presence of $BF_3{\cdot}Et_2O$. Due to their hygroscopicity and intractable nature, the products are reduced with hydrazine hydrate, and isolated as tellurides (as for the amidotellurinylation reactions).

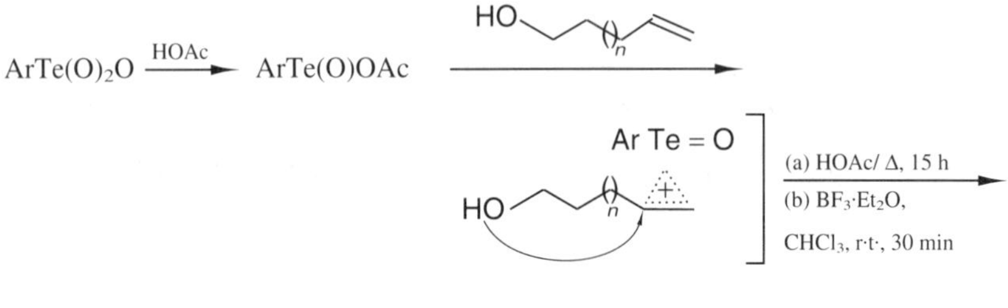

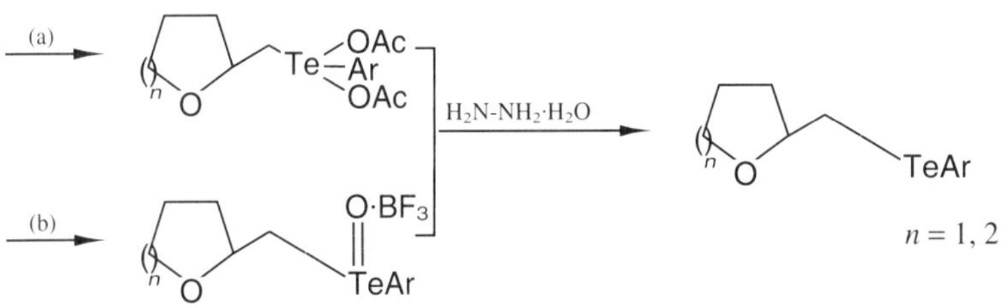

Ar = Ph, *p*-$MeOC_6H_4$, 2-naphthyl

A variety of unsaturated alcohols with a cyclic structure and allyl phenols lead to bicyclic telluroethers.

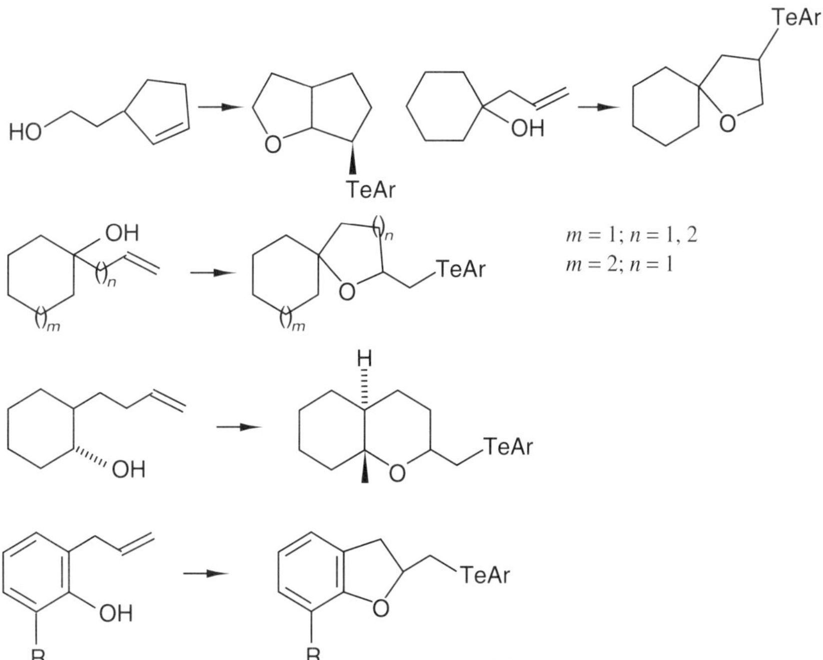

R = H, Me

In accordance with the Baldwin rules, formation of five-membered rings is favoured over the four- or six-membered rings, and the six-membered rings over the seven-membered ones.

Cyclotelluroetherification of allylphenol (typical procedure) [8] To a solution of phenyltellurinyl acetate (1.1 mmol) (prepared *in situ* by treatment of benzenetellurinic anhydride (0.252 g, 0.55 mmol) with Ac_2O (0.056 g, 0.55 mmol) in $CHCl_3$ (6 ml) under reflux for 30 min) is added, after cooling, 2-allylphenol (0.134 g, 1.0 mmol) in $CHCl_3$ (2 ml) and $Et_2O{\cdot}BF_3$ (0.170 g, 1.2 mmol). The resulting mixture is stirred at room temperature for 0.5 h and then concentrated *in vacuo*. The residue is reduced with hydrazine hydrate (0.1 g, 2 mmol) in EtOH (5 ml) at 60°C for 15 min. After cooling to room temperature the mixture is poured into H_2O and extracted with CH_2Cl_2 (2×25 ml). The organic layer is washed with brine, dried ($MgSO_4$), and evaporated *in vacuo*. The residue is chromatographed on SiO_2 (eluting with CH_2Cl_2), giving 2,3-dihydro-2-[(phenyltelluro)methyl]benzofuran as white needles (0.311 g (92%); m.p. 46.5–47°C (from MeOH)). The same reaction carried out in HOAc requires heating under reflux for 15 h.

4.6.2.2 With aryltellurium trichlorides

Aryltellurium trichlorides are suitable reagents for the telluroetherification of olefinic alcohols and allylphenols [10], exhibiting the advantages of stability and easy separability of the formed crystalline dichlorotelluro derivatives compared with the preceding tellurinyl acetate method.

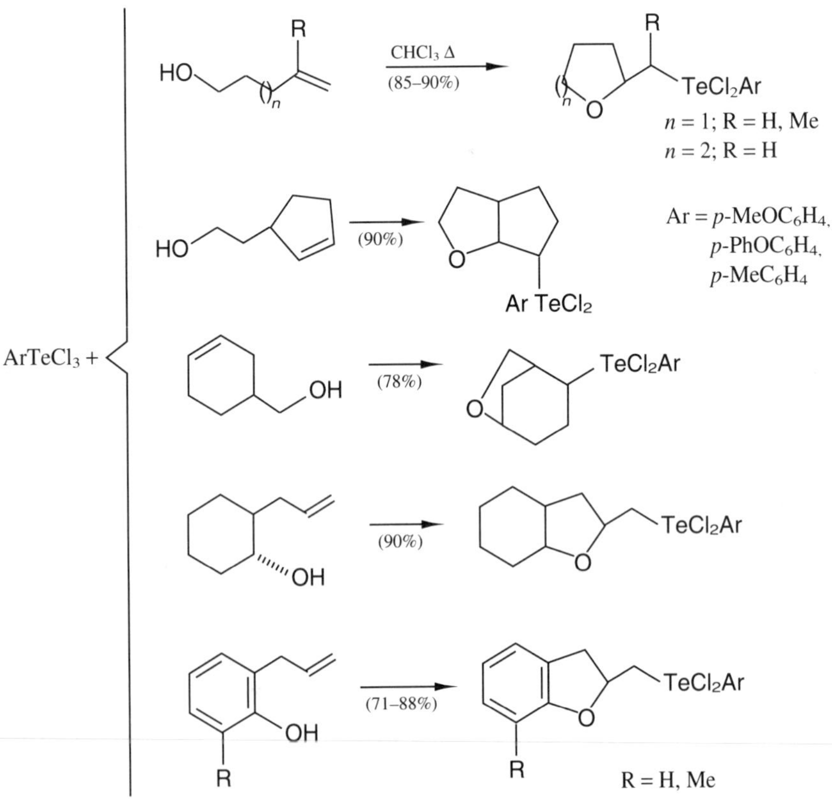

Cyclotelluroetherification (general procedure) [10] The unsaturated alcohol and a slight excess of the aryltellurium trichloride are refluxed in $CHCl_3$ for 15–40 min. The *o*-allylphenols require a longer reaction time (12–24 h). The products are purified by filtration through an SiO_2 column, eluting with $CHCl_3$ to remove the excess tellurium trichlorides. The obtained dichlorotelluroethers are recrystallized from $CHCl_3$/petroleum ether 30–60°C.

Aryltellurium tribromides (Ar = *p*-$MeOC_6H_4$) are also effective in these cycloetherification reactions.

The dichlorotelluroethers are easily reduced to the parent telluroethers by treatment with TUDO [6].

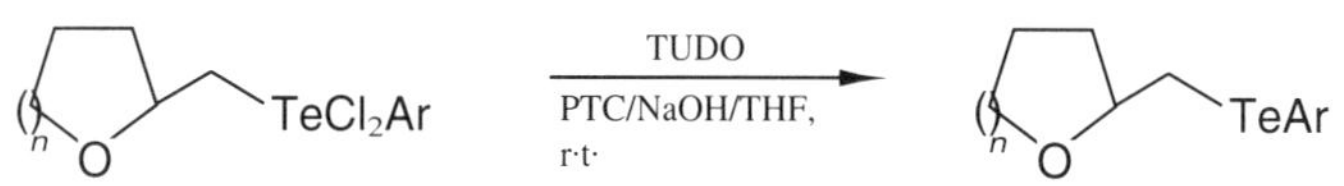

4.6.2.3 With benzenetellurenyl nitrobenzenesulphonate

The title reagent, employed for tellurolactonizations (see Section 4.6.1.2), is also effective for the tellurocyclization of unsaturated alcohols [5].

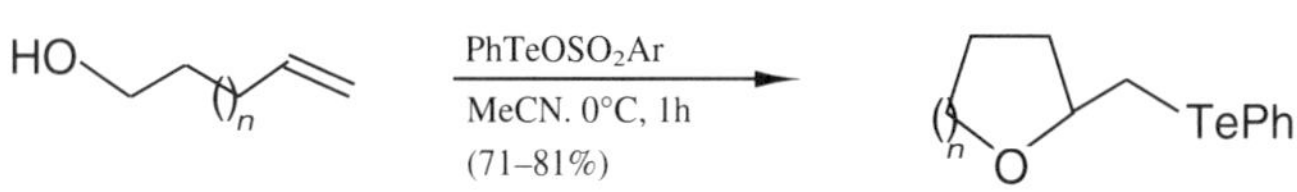

$n = 1, 2$
Ar = p-$NO_2C_6H_4$

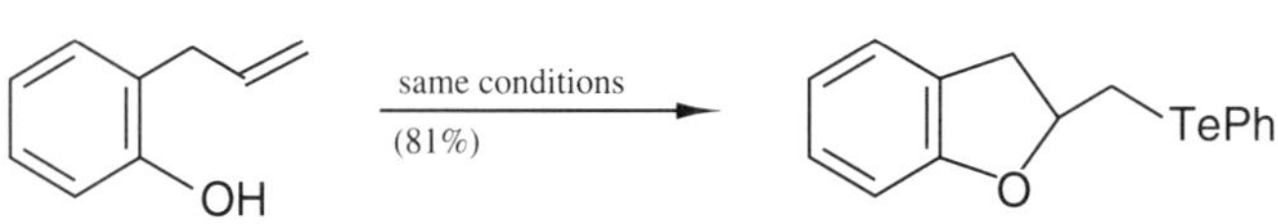

Cyclotelluroetherification (typical procedure) [5] The procedure, starting from $(PhTe)_2$ (0.5 mmol), NBSP (0.5 mmol),4-penten-1-ol or hexen-1-ol (1.1 mmol), is similar to that described for tellurolactonization (see Section 4.6.1.2).

4.6.2.4 With TeO_2/HOAc/LiCl or TeO_2/HCl

Tellurium dioxide in acetic acid/lithium chloride reacts with 2 eq of γ- or δ-hydroxyolefins, giving tetrahydrofuran or tetrahydropyran dichlorotelluro derivatives [11].

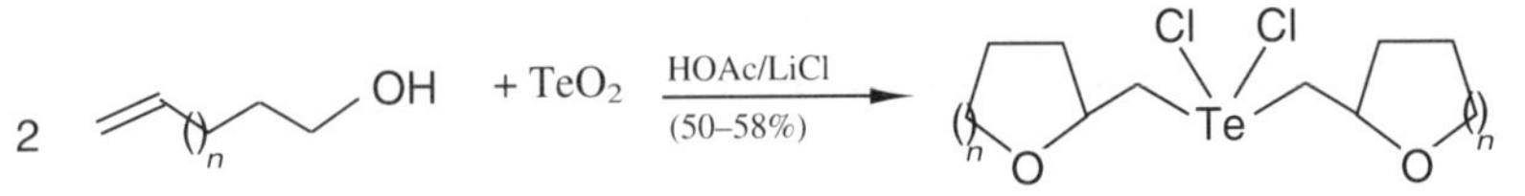

$n = 1, 2$

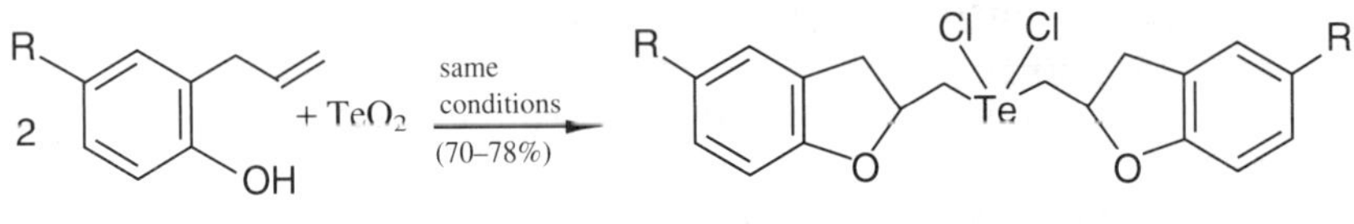

R = H, Me, Cl

The reaction can be carried out in an equimolar ratio, employing tellurium dioxide in alcoholic aqueous hydrochloric acid. The formed trichlorotelluro derivative is reduced without isolation to the corresponding ditelluride (see Section 3.2.3.2) [12].

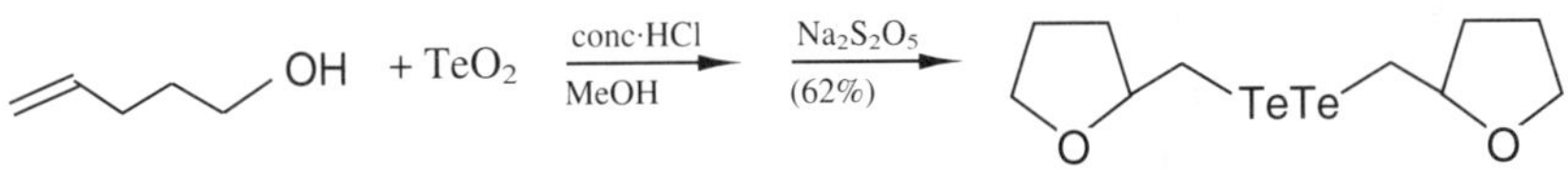

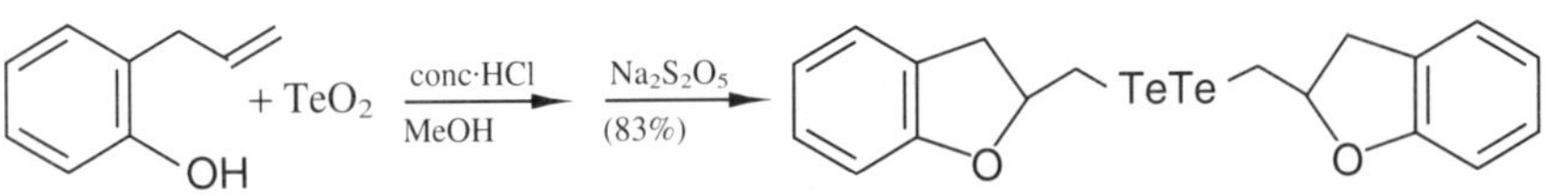

Bis[2-(2,3-dihydrobenzofuranyl)methyl]tellurium dichloride (typical procedure) [11] TeO_2 (2.4 g, 15.0 mmol), 2-allylphenol (4.0 g, 30.0 mmol) and LiCl (3.0 g, 70.8 mmol) are refluxed in HOAc (50 ml) for 1.5 h. The clear solution is then filtered hot from a small amount of elemental Te and, upon cooling (freezing), the product crystallizes as a white solid (5.4 g (78%)). The product is recrystallized from $EtOH/CH_2Cl_2$ (1:1) (m.p. 180–184°C).

Bis[2-(2,3-dihydrobenzofuranyl)methyl] ditelluride (typical procedure) [12] TeO_2 (2.0 g, 12.5 mmol) is dissolved in concentrated HCl (10 ml) and the yellow solution diluted with MeOH (40 ml). 2-Allylphenol (1.98 g, 14.8 mmol) is added and the mixture is heated under reflux for 24 h and then allowed to cool to ambient temperature. The yellowish solution (containing a trace of elemental Te) is then poured into a separating funnel containing $Na_2S_2O_5$ (5 g in 100 ml H_2O) and CH_2Cl_2 (100 ml). After shaking, the red CH_2Cl_2 phase is separated, dried and evaporated, giving a red oil which is purified by SiO_2 flash chromatography (eluting with CH_2Cl_2) (3.23 g (83%)).

4.6.2.5 Synthetic utility of the telluroetherification reactions

The synthetic utility of the described reactions is clearly illustrated by the reductive detelluration of telluroethers with Bu_3SnH, the entire telluration/detelluration process providing a cyclization under mild conditions for unsaturated alcohols.

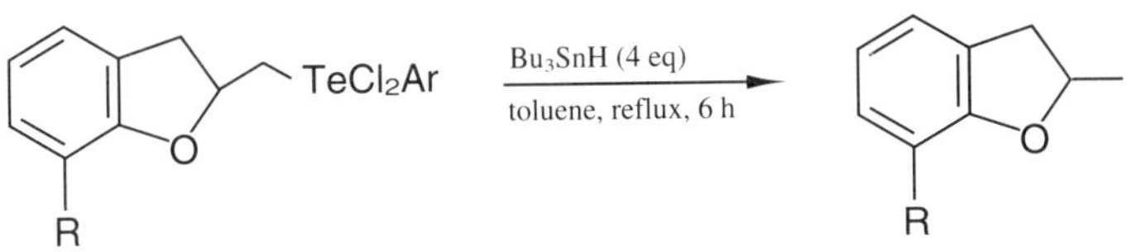

[6]

Ar = *p*-$MeOC_6H_4$; R = H, Me

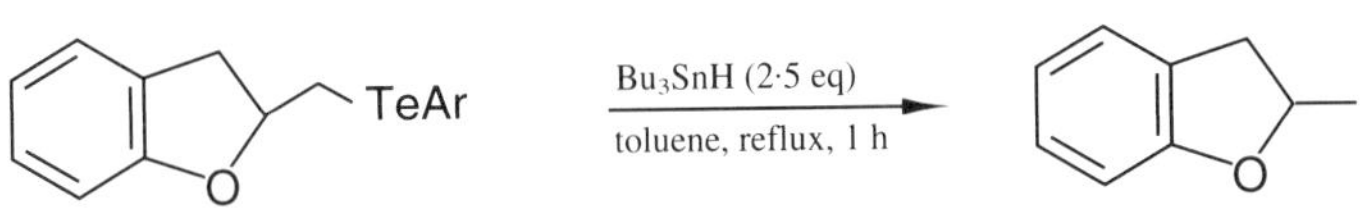

[8]

Ar = Ph

Additional synthetically useful transformations are given in sequence.

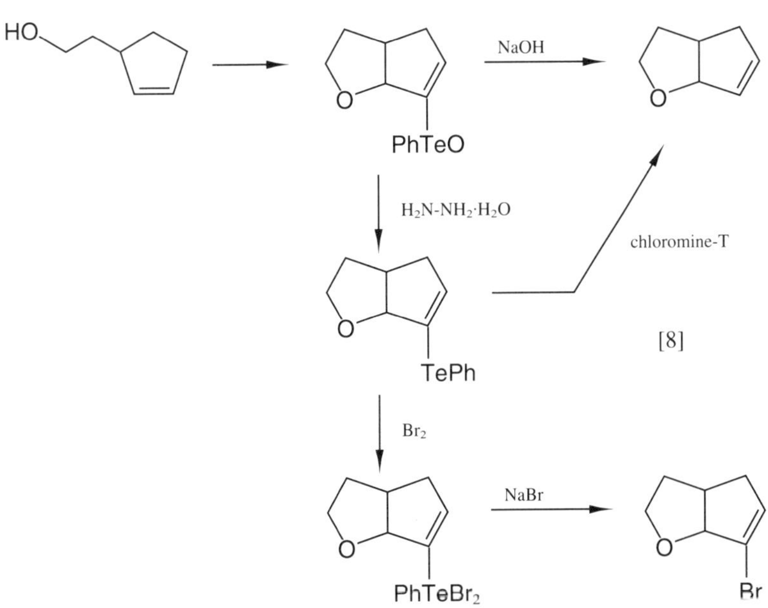

Reductive detelluration of cyclic dichlorotelluroethers (general procedure) [6] This procedure is identical to that described for dichlorotellurolactones (see Section 4.6.1.3).

Reductive detelluration of cyclic telluroethers (typical procedure) [8] To a solution of 2,3-dihydro-2-[(phenyltelluro)methyl]benzofuran (0.338 g, 1 mmol) in toluene (6 ml) is injected TBTH (0.67 ml, 2.5 mmol) at room temperature, and the resulting solution is stirred under reflux for 1 h. The solution is evaporated under vacuum and the residual yellowish oil subjected to SiO_2 column chromatography (eluting with benzene/hexane 3:1) to give pure 2,3-dihydro-2-methylbenzofuran as a colourless oil (0.128 g (95%); b.p. 93–94°C/23 torr).

Telluroxide elimination in cyclic tellurinyl ethers (typical procedure) [8] The cyclofunctionalization of 2-cyclopent-2-enylethanol (0.152 g, 1.35 mmol) with benzenetellurinyl acetate (1.48 mmol) in $CHCl_3$ (6 ml) is carried out by the general method described in Section 4.6.2.1. The reaction solution containing the crude telluroxide is diluted with $CHCl_3$ (25 ml), washed with brine, and evaporated under reduced pressure. To the residue are added successively THF (5 ml) and NaOH (0.5 M, 10 ml). The resulting mixture is stirred under reflux for 5 h, poured into H_2O and extracted with ether (2 × 20 ml). The extract is dried ($MgSO_4$) and concentrated under reduced pressure. The residue is chromatographed on SiO_2, eluting with CH_2Cl_2/hexene (5:2), to give 8-oxabicyclo[3.3.0]oct-2-ene as a colourless oil (0.079 g (53%)).

4.6.3 Tellurocyclization of olefinic carbamates

When olefins bearing a carbamate group at a suitable position are submitted to treatment with phenyltellurinyl acetate in the presence of $BF_3 \cdot Et_2O$ (previously described for amidotellurinylation of olefins—see Section 4.5.7.4) an intramolecular cyclization takes place to give pyrrolidine and piperidine derivatives [13].

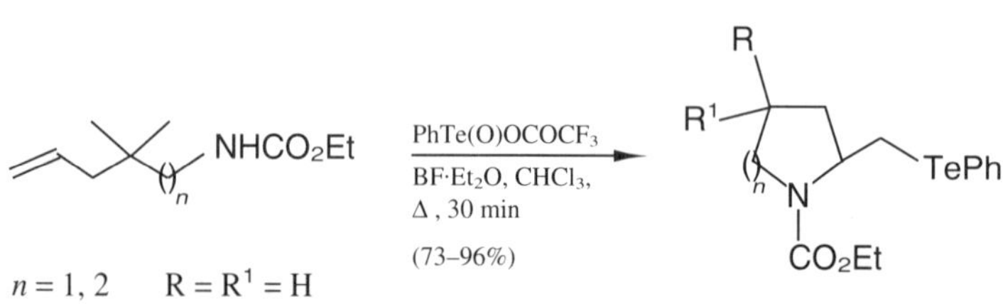

$n = 1, 2$ $R = R^1 = H$
$n = 1$ $R = H; R^1 = Me$
$n = 1$ $R = Me; R^1 = allyl$

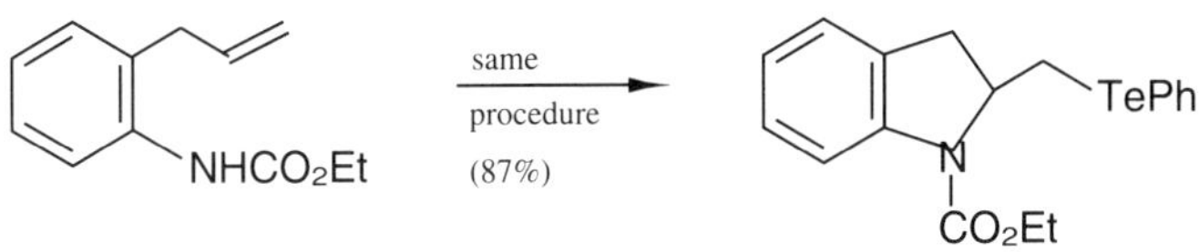

The cyclization is much faster (30 min) than the intermolecular addition (8–20 h).

REFERENCES

1. (a) C. Paulmier, *Selenium Reagents and Intermediates in Organic Synthesis*, Chapt. VIII. Pergamon Press, Oxford, 1986; (b) K. C. Nicolan, N. A. Petasis and D. A. Claremon, in *Organoselenium Chemistry* (ed. D. Liotta), Chapt. 2. Wiley, Chichester, 1987.
2. M. Moura Campos and N. Petragnani, *Chem. Ber.* **93**, 317 (1960).
3. J. V. Comasseto and N. Petragnani, *Synth. Commun.* **13**, 889 (1983).
4. S. Husebye, E. A. Meyers, R. A. Zingaro, J. V. Comasseto and N. Petragnani, *Acta Cryst.* **C43**, 1147 (1987).
5. M. Yoshida, T. Suzuki and N. Kamigata, *J. Org. Chem.* **57**, 383 (1992).
6. J. V. Comasseto, H. M. C. Ferraz, C. A. Brandt and K. K. Gaeta, *Tetrahedron Lett.* **30**, 1209 (1989).
7. N. X. Hu, Y. Aso, T. Otsuba and F. Ogura, *Tetrahedron Lett.* **28**, 1281 (1987).
8. N. X. Hu, Y. Aso, T. Otsuba and F. Ogura, *J. Org. Chem.* **54**, 4391 (1989).
9. N. X. Hu, Y. Aso, T. Otsuba and F. Ogura, *Phosphorus Sulfur*, **38**, 177 (1988).
10. J. V. Comasseto, H. M. C. Ferraz and N. Petragnani, *Tetrahedron Lett.* **28**, 5611 (1987).
11. J. Bergman and L. Engman, *J. Am. Chem. Soc.* **103**, 5196 (1981).
12. L. Engman, *Organometallics* **8**, 1997 (1989).
13. N. X. Hu, Y. Aso, T. Otsuba and F. Ogura, *Chem. Lett.*, 1327 (1987).

4.7 CONVERSION OF ORGANOTELLURIUM COMPOUNDS INTO TELLURIUM-FREE ORGANIC COMPOUNDS

4.7.1 Detelluration of organotellurium compounds with the formation of new C—C bonds (carbodetelluration)

4.7.1.1 Synthesis of biaryls by tellurium extrusion from diaryltellurium dichlorides and aryltellurium trichlorides with Raney nickel

Diaryltellurium dichlorides are detellurated by heating with Raney nickel in appropriate solvents, giving the corresponding biaryls [1].

$$\mathrm{ArTe(Cl)_2Ar} \xrightarrow[\text{diglyme, } \Delta, \text{ 8 h}]{\text{Raney nickel}} \mathrm{Ar\text{-}Ar}$$

(60–91%)

Ar = Ph, *p*-MeO, *p*-EtO, *p*-Me, *p*-Br, *p*-$Me_2NC_6H_4$,
p-MeO-*m*-MeC_6H_3, *p,m*-$(MeO)_2C_6H_3$, *o,p*-$(MeO)_2C_6H_3$,
2-naphthyl

Synthesis of biaryls (general procedure) [1] The diaryltellurium dichloride (50 mmol) is heated with degassed Raney nickel [2] (60 g) in diglyme (500 ml) for 8 h. The mixture is filtered while still hot, and the solvent evaporated (water pump). The residue is recrystallized from ethanol or toluene.

2,2′-Binaphthyl is obtained by submitting 2-naphthyltellurium trichloride to a similar treatment [3].

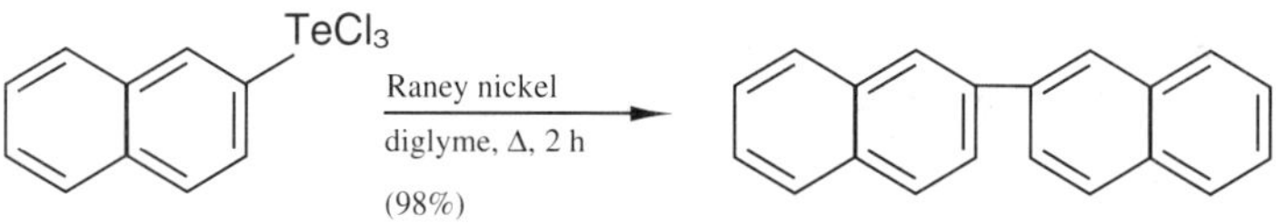

4.7.1.2 Conversion of diorganyl tellurides (and ditellurides) into coupled products by reaction with palladium(0) and palladium(II)

Palladium(0) converts symmetrical and unsymmetrical tellurides into coupled compounds under mild conditions [4].

$$RTeR^1 \xrightarrow[\text{MeCN, }\Delta]{\text{Pd(O)}} R\text{-}R^1 + Te$$

(76–100%)

R = R^1 = *p*-$MeOC_6H_4$, $PhCH_2CH_2$

R = $PhCH_2CH_2$; R^1 = $PhCH_2$

R = *p*-$MeOC_6H_4$; R^1 = n-$C_{15}H_{31}$, t-adamantyl

The reaction is carried out by treating the telluride (1 mmol) with the Pd(0) catalyst, prepared *in situ* from $Pd(OAc)_2$ (1 mmol) and Et_3N (2 mmol) in MeCN at 65°C, during 5–10 h under an argon atmosphere.

Diaryl ditellurides are converted to biaryls (Ar = *p*-$MeOC_6H_4$) by treatment with an additional eq of Pd(0).

Styryl tellurides, on treatment with lithium chloropalladate (2 eq), are detellurated, giving stereoisomeric mixtures of the coupled 1,3-butadiene derivatives [5].

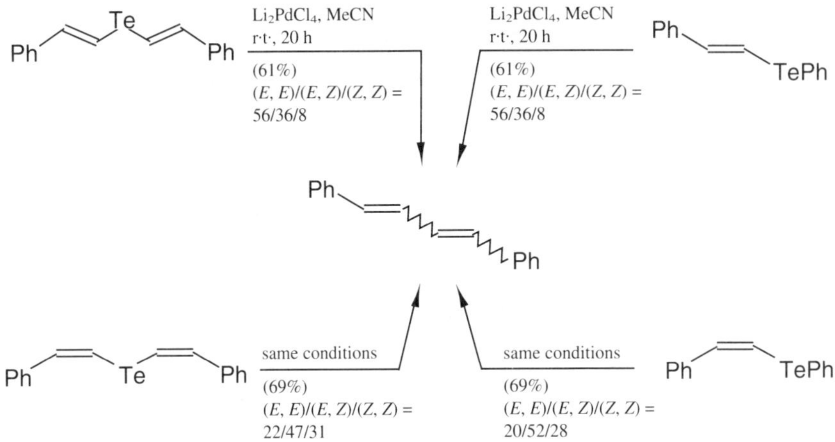

On treatment with Pd(II) acetate under the same conditions, distyryl telluride furnishes styryl acetate.

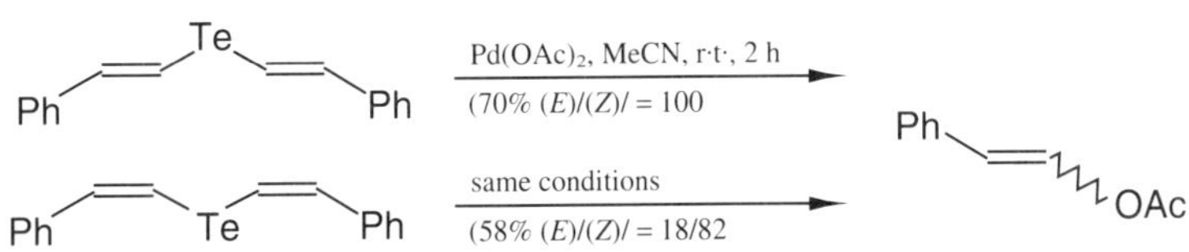

4.7.1.3 Correlate reactions

Some particular carbocyclic systems are obtained by pyrolytic tellurium extrusion [6, 7].

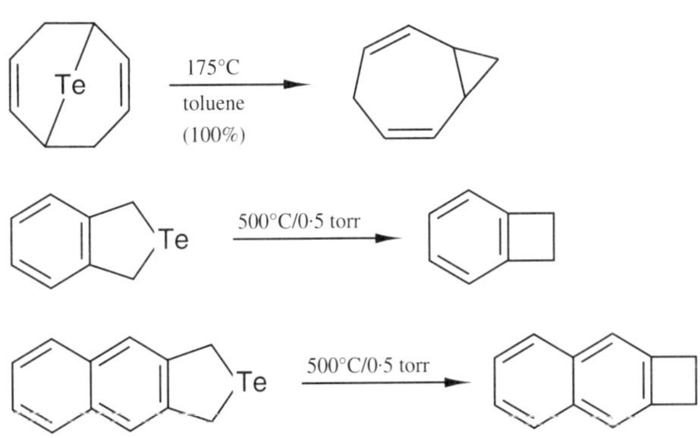

4.7.1.4 Olefin arylation by carbodetelluration of aryltellurium compounds under Pd(II) catalysis

The phenylation of ethyl acrylate by diphenyl telluride in the presence of $Pd(OAc)_2$ was first described as part of a general study of the reactions of phenyl derivatives of the group V and VI elements with olefins, catalysed by Pd(II) salts [8].

The Pd(II)-catalysed arylation of olefins by means of different aryltellurium (IV) chlorides has been more extensively investigated [9].

The reactions are carried out treating diaryltellurium dichlorides or aryltellurium trichlorides with excess olefin (5–10 mol eq) in the presence of Pd(II) chloride–sodium acetate in refluxing acetic acid or acetonitrile. The yields are moderate to good with the dichlorides and poor with the trichlorides. Minor amounts of biaryls are the usual by-products beyond acetoxylated arylalkenes (formed by the addition of acetic acid to the olefins). The reactions can also be performed catalytically using Pd(II) chloride if a suitable oxidant such as t-butyl hydroperoxide or Cu(II) chloride is added to the system.

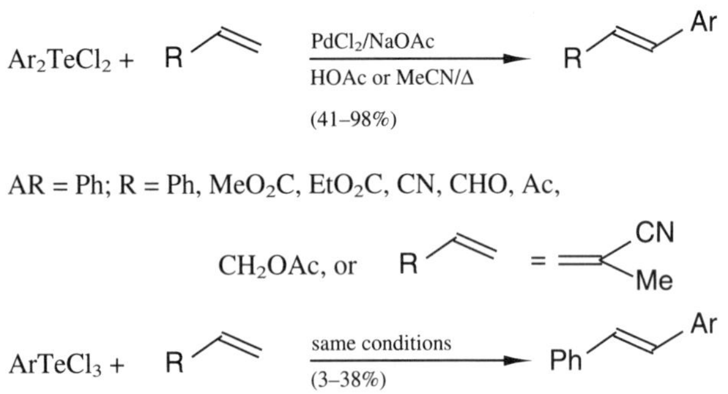

Ar = Ph, *p*-Me, *p*-$MeOC_6H_4$

The products exhibit the (*E*) configuration, with the exception of the acrylonitrile and methacrylonitrile derivatives.

In the absence of the olefin, the corresponding biaryls are formed in moderate yields (in analogy to the Raney nickel-promoted detellurations (see Section 4.7.1.1).

$$Ar_2TeCl_2 \xrightarrow[\text{HOAc, }\Delta]{\text{PdCl}_2\text{/NaOAc}} \text{Ar-Ar} \quad (10\text{–}51\%)$$

Ar = Ph, *p*-Me, *p*-MeO, *p*-BrC_6H_4

$$ArTeCl_3 \xrightarrow[(3\%)]{\text{same conditions}} \text{Ar-Ar}$$

Ar = *p*-$MeOC_6H_4$

Arylation of styrene (catalytic in Pd(II); typical procedure) [9] A mixture of Ph_2TeCl_2 (0.35 g, 1 mmol), styrene (1.04 g, 10 mmol), $PdCl_2$ (0.03 g, 0.17 mmol) and $CuCl_2$ (0.23 g, 1.7 mmol) in HOAc (20 ml) is refluxed for 2.5 h with stirring. After cooling to room temperature the black precipitate is filtered off, and the pale green filtrate is treated with brine (200 ml) and extracted with benzene. The extract is washed with aqueous $NaHCO_3$ and then with brine, and dried (Na_2SO_4). GLC analysis using ethyl cinnamate as the internal standard shows the presence of *trans*-stilbene (35.1% based on Ph_2TeCl_2; 41.3% based on $PdCl_2$) and a trace of biphenyl and unidentified compounds.

4.7.1.5 Cross-coupling of Grignard reagents with organic tellurides catalysed by Ni(II) or Co(II) complexes

Some organic tellurides react with Grignard reagents in the presence of Ni(II) or Co(II). Phosphine complexes give mixtures of cross-coupling products together with the homo-coupling products of the telluride in good-to-moderate yields [10]. Some illustrative examples are shown.

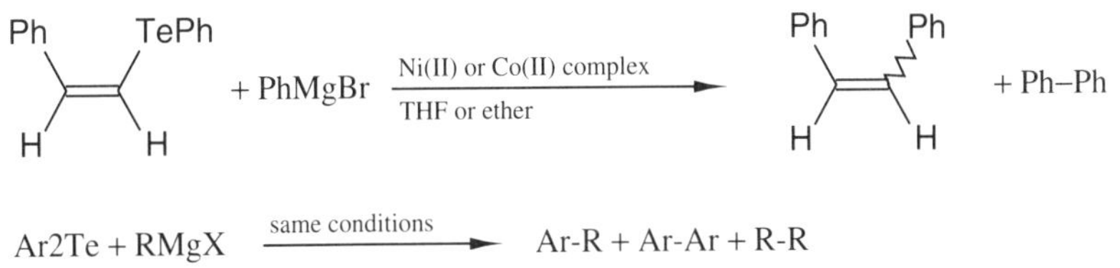

Coupling reaction (typical procedure) [10] To a solution of *cis*-PhCH$=$CHTePh (0.307 g, 1.0 mmol) in THF (10 mol) containing $NiCl_2(PPh_3)_2$ (0.032 g, 0.05 mmol) is added a THF solution of PhMgBr (1 M, 2.5 ml, 2.5 mmol) under N_2 at room temperature. The mixture is stirred for 10 min. A black precipitate of elemental Te is formed. The mixture is then decomposed with dilute HCl and filtered from the precipitate, which is washed, leaving black Te (0.11 g). The filtrate is treated with brine and extracted with ether (3 × 30 ml). The extract is dried ($MgSO_4$). GLC of the extract reveals the presence of stilbene (0.108 g (60%, *cis*/*trans* = 90/10)) and biphenyl (0.142 g (92%)), dibenzyl being used as the internal standard.

By submitting tellurophene to a similar treatment, 1,4-disubstituted (*Z,Z*)(1,3)-butadienes are formed in good yields [11].

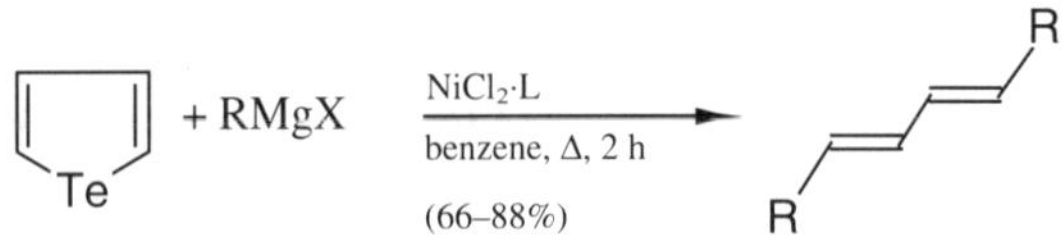

R = Ph; L = 2(Ph_3P)
R = Me, n-Bu; L = 

molar ratio: tellurophene/RMgX/cat. = 1:2·1:0·1

Furans, thiophenes and selenophenes react similarly, but with the reactivity decreasing from tellurophene to furans.

4.7.1.6 Detellurative carbonylation of organotellurium compounds: preparation of carboxylic acids

Aryltellurium trichlorides and diaryltellurium dichlorides react with nickel tetracarbonyl in DMF, giving benzoic acids. Small amounts of diaryl tellurides and diaryl ketones are by-products of these reactions [12].

$$ArTeCl_3 + 2Ni(CO)_4 \xrightarrow[\text{(2) } H_2O \text{ (52, 35\%)}]{\text{(1) DMF, 70°C, 24 h}} ArCO_2H + NiCl_2 + NiTe + 7CO + HCl$$

Ar = *p*-$MeOC_6H_4$, 2-naphthyl

$$Ar_2TeCl_2 + Ni(CO)_4 \xrightarrow[\text{(71, 58\%)}]{\text{same conditions}} 2ArCO_2H + NiTe + 2CO + 2HCl$$

Ar = *p*-$MeOC_6H_4$, Ph

Detellurative carbonylation of aryltellurium trichloride: $Ni(CO)_4$ method (typical procedure) [12] $Ni(CO)_4$ (2.6 ml, 20 mmol) is added to *p*-methoxyphenyltellurium trichloride (2.0 g, 5.9 mmol) in dry DMF (50 ml) under N_2. After 24 h at 70°C the excess $Ni(CO)_4$ is evaporated off in a stream of N_2. The mixture is then poured into H_2O (200 ml) containing 48% HBr (10 ml). The product is extracted with ether (3 × 75 ml), and the ether solution extracted with 2 M NaOH. The ether solution containing non-acid products is dried ($CaCl_2$) and evaporated, giving (*p*-$MeOC_6H_4)_2Te$ (0.15 g (15%)) and a trace amount of (*p*-$MeOC_6H_4)_2CO$. The alkaline extract is acidified and extracted with ether. Evaporation of this ether extract gives *p*-$MeOC_6H_4CO_2H$ (0.45 g (52%)).

Detellurative carbonylation of diaryltellurium dichlorides (typical procedure) [12] Following the above procedure, $Ni(CO)_4$ (1.5 ml, 11.5 mmol) and di-*p*-methoxyphenyltellurium dichloride (2.0 g, 4.8 mmol) furnish *p*-$MeOC_6H_4CO_2H$ (1.05 g (71%)), (*p*-$MeOC_6H_4)_2Te$ (0.25 g (15%)) and a trace amount of (*p*-$MeOC_6H_4)_2CO$.

Carboxylic acids are also produced through the detellurative carbonylation of several types of telluride by treatment with carbon monoxide at atmospheric pressure and room temperature in the presence of Pd(II) salts, in various solvents [13, 14].

These reactions are of particular interest since the corresponding sulphur and selenium compounds give unsatisfactory results.

Among the systems examined, the $PdCl_2/Et_3N/MeOH$ system provides the best results (furnishing the methyl esters directly) [15, 16].

$$\mathrm{RTeAr} \xrightarrow[\mathrm{Et_3N/MeOH}]{\mathrm{CO\ (1\ atm)/PdCl_2}} \mathrm{RCO_2Me + ArCO_2Me}$$

Ar	R	Percentage yield	
Ph	*cis*-Ph /=	96	30
Ph	*trans*-Ph /=	71	24
Ph	*cis*-EtO_2C /=	72	32
Ph	Ph—≡	67	17
Ph	n-$C_{12}H_{25}$*	22	98
Ph	Ph	92	
p-$MeOC_6H$	*p*-$MeOC_6H_4$	99	

In all cases a black precipitate is deposited during the reaction. Complete retention of the configuration is observed with alkenyl tellurides. This type of carbonylation can be extended to the synthesis of butenolides by the use of hydroxyvinyl phenyl tellurides as substrates. The use of a non-polar solvent such as dichloromethane is essential for the success of the reaction.

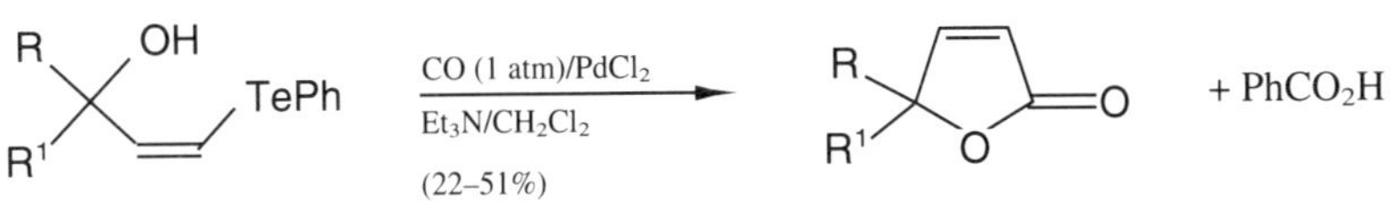

$R = R^1 = Me, (CH_2)_5$

$R = Me; R^1 = i\text{-}Bu$

When R and/or R^1 is hydrogen, no reaction occurs at all.

These carbonylation reactions can be carried out catalytically using Pd(II) when performed in the presence of a reoxidant such as Cu(II) chloride. In the

*A considerable amount of dodecene is formed as a by-product.

catalytic procedure, in contrast to the stoichiometric one, the formation of methyl benzoates is minimized in the case of alkenyl phenyl tellurides and alkynyl phenyl tellurides. The catalytic procedure is especially suitable for the stereoselective synthesis of various ring-substituted methyl *cis*-cinnamates from the corresponding acetylenes, in view of the facile regio- and stereoselective addition of phenyltellurol to acetylenes (see Section 3.11).

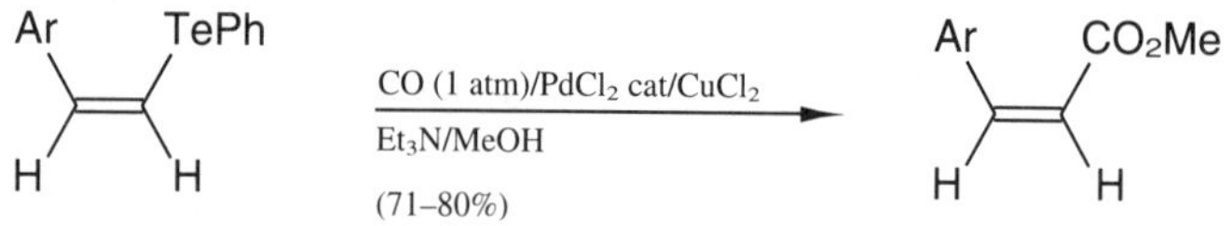

Ar = *p*-Me, *p*-Cl, *p*-MeO, *o*-ClC_6H_4

The catalytic procedure is unsuitable for the conversion of hydroxyvinyl tellurides into butenolides.

These reactions have been rationalized as involving a key step migration of an organic moiety from tellurium to palladium in telluride–$PdCl_2$ complexes, probably formed *in situ*, affording organopalladium compounds.

Detellurative methoxycarbonylation of diorganyl tellurides (a) *$PdCl_2$ stoichiometric method (general procedure)* [15]. $PdCl_2$ (0.177 g, 1.0 mmol) and the telluride (1.0 mmol) are flushed with CO from a CO balloon connected to the reaction flask at 25°C, to which dry MeOH (10 ml) and Et_3N (0.202 g, 2.0 mmol) are added by a syringe. After stirring for 5 h at 25°C the black precipitate formed is filtered off, and the filtrate is poured into aqueous NH_4Cl and extracted with ether (3 × 30 ml). The products are determined by GLC (using an EGSS-X 3% (1 m) column).

(b) *$PdCl_2$ catalytic procedure.* $PdCl_2$ (0.017 g, 0.1 mmol), $CuCl_2$ (0.270 g, 2.0 mmol) and the telluride (1.0 mmol) are flushed with CO from a CO balloon connected to the reaction flask at 25°C, to which dry MeOH (10 ml) and Et_3N (0.303 g, 3.0 mmol) are added by a syringe. After stirring for 75 h at 25°C, during which time the green colour turns dark brown, the brown solid is filtered off. The filtrate is poured into aqueous NH_4Cl and extracted with ether (3 × 30 ml). The products are isolated by preparative TLC or Kugelrohr distillation, or determined by GLC (using an EGSS-X 3% (1 m) column with an appropriate inner standard).

4.7.2 Replacement of the tellurium moiety by other functionalities

4.7.2.1 By halogens

The transformation of a carbon–tellurium bond into a carbon–halogen bond has been achieved in several types of organotellurium compound.

(a) Halogenodetelluration of vinylic tellurium trichlorides

By treatment of (*Z*)-2-chlorovinyltellurium trichlorides, easily obtained by the addition of tellurium tetrachloride to phenylacetylenes (see Section 3.11), with 1–2 mol eq of iodine or *N*-bromosucinimide–aluminium trichloride, a halogeno-detelluration occurs, generating the corresponding (*Z*)-iodo- or (*Z*)-bromo-chloroalkenes [17].

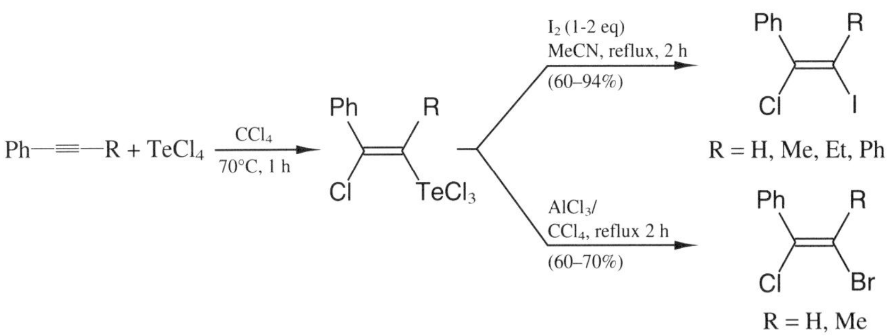

The chlorotelluration–iododetelluration of propargyl alcohol proceeds analogously.

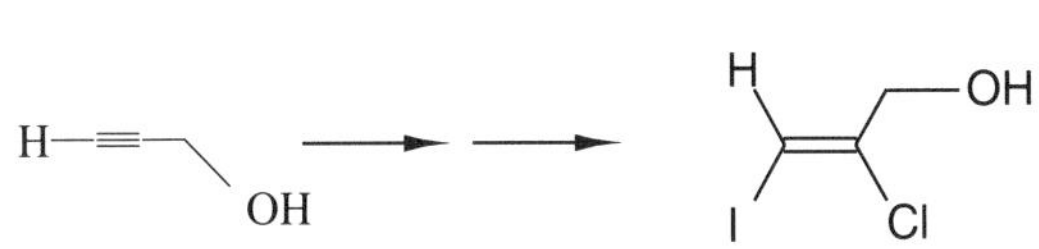

This two-step procedure therefore offers a useful method for the *syn*-iodo- and *syn*-bromochlorination of acetylenes.

(b) Halogenodetelluration of aryltellurium(IV) halides

Aryltellurium trichlorides bearing a *para* electron-releasing group undergo substitution of the trichlorotelluro group by iodine, furnishing the corresponding iodoarenes [18]. The reaction is greatly enhanced by the addition of potassium, caesium or ammonium fluoride. Diaryltellurium dichlorides (and dicarboxylates) react similarly but with lower yields.

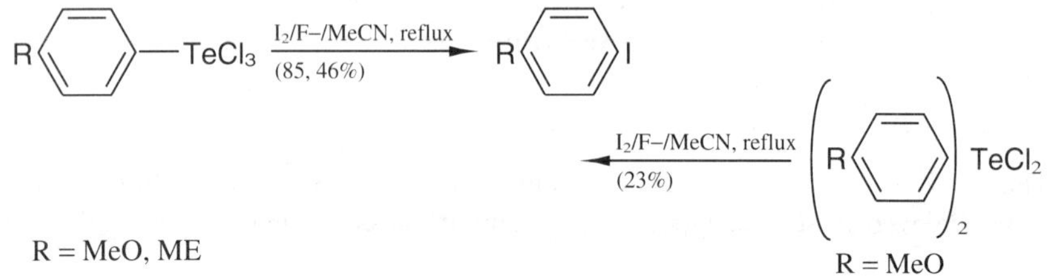

Similar bromodetellurations afford *o,p*-dibromo derivatives as the main product (formed by a further bromination of the primary *p*-bromo derivative under Te(IV) catalysis). The addition of F^- ions is unnecessary.

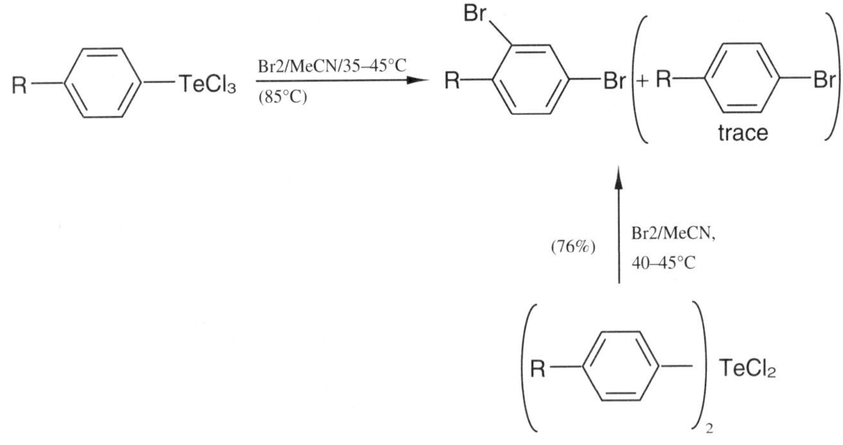

An electrophilic mechanism is plausible for these reactions, on the basis of the dependence on *para* electron-releasing groups and the enhancement by F^- ions (forming $ArTeCl_3F_2^{2-}$ or $Ar_2TeCl_2F_2^{2-}$ species which make the electrophilic attack by halogens easier).

Iododetelluration of aryltellurium trichlorides (typical procedure) [18] A mixture of *p*-methoxyphenyltellurium trichloride (0.34 g, 1 mmol), iodine (1.01 g, 4 mmol) and NH_4F (0.15 g, 4 mmol) in CH_3CN (20 ml) is stirred under reflux for 5 h. After cooling to room temperature the mixture is treated with brine (200 ml) and extracted with benzene (3 × 50 ml). The extracts are washed with 10% aqueous $Na_2S_2O_3$ and then with brine, and dried (Na_2SO_4). GLC analysis, using *p*-iodotoluene as the internal standard, shows the presence of *p*-methoxyiodobenzene (85%).

Bromodetelluration of aryltellurium trichlorides (typical procedure) [18] A mixture of *p*-methoxyphenyltellurium trichloride (0.68 g, 2 mmol) and bromine

(0.64 g, 4 mmol) in MeCN (20 ml) is stirred at 35–45°C for 20 h. After work-up as described above, GLC analysis using ethyl cinnamate as the internal standard shows the presence of *o,p*-dibromo-*p*-methoxybenzene (69.5%) and a trace of *p*-methoxybromobenzene.

(c) α elimination of organic halides from organotellurium(IV) halides

Organotellurium tri- and dihalides undergo α elimination by oxidative, photolytic or thermal routes, giving the corresponding halides with a selective transference of the halogen at the position where the tellurium moiety was originally attached (*ipso* substitution).

The oxidative procedure A [19, 20] is based on the treatment with oxidants, preferably t-butyl hydroperoxide in refluxing dioxane, acetic acid or acetonitrile.

The photolytic procedure B [20, 21] is carried out by irradiation in benzene with a high-pressure mercury lamp, at room temperature, in the presence of atmospheric oxygen. The reaction is accompanied by deposition of elementary tellurium.

In the thermal procedure C the substrate is heated at 230–250°C under reduced pressure (in a Kugelrohr distillation apparatus) (method C^1) [20] or heated in DMF at 70–100°C in the presence of an additional alkali metal halide (method C^2) [22]. The tellurium moiety is isolated as diphenyl ditelluride, in spite of the formal formation of phenyltellurenyl halide.

α elimination of organic halides from organotellurium(IV) halides

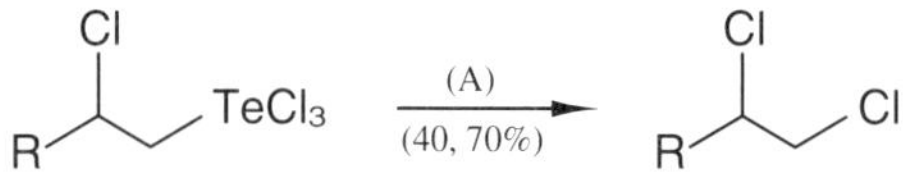

(1)

$R = n\text{-}C_6H_{17}$, $PhCH_2$

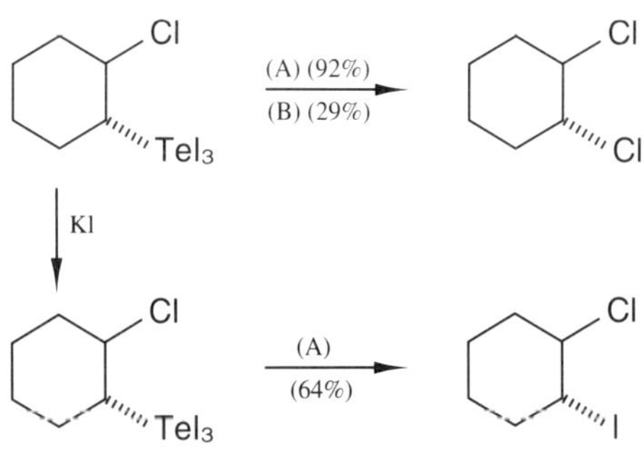

(2)

(3)

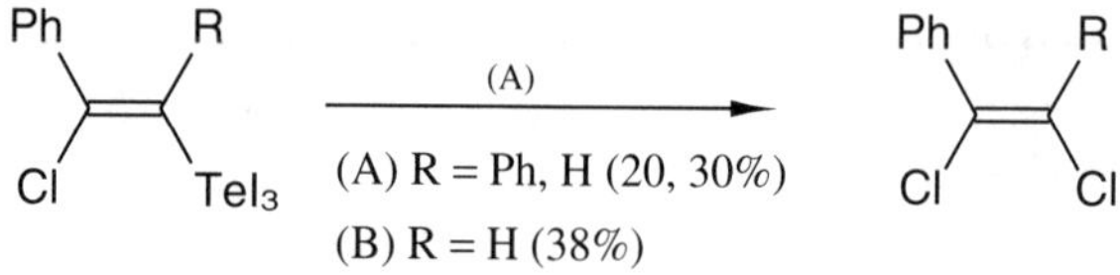
Ph
R
(A)
Cl
TeI3
(A) R = Ph, H (20, 30%)
(B) R = H (38%)
Ph
R
Cl
Cl

(4)

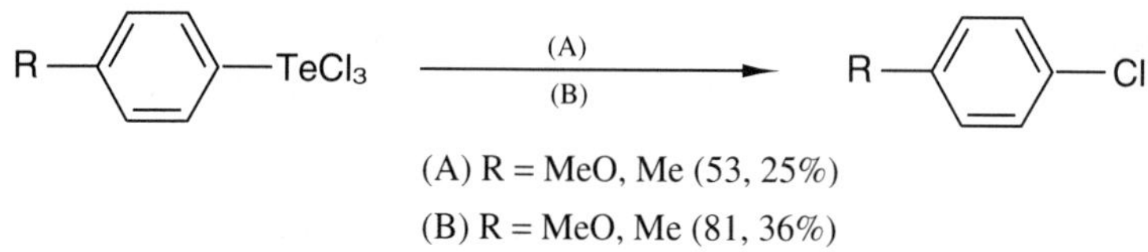
R
TeCl3
(A)
(B)
R
Cl
(A) R = MeO, Me (53, 25%)
(B) R = MeO, Me (81, 36%)

(5)

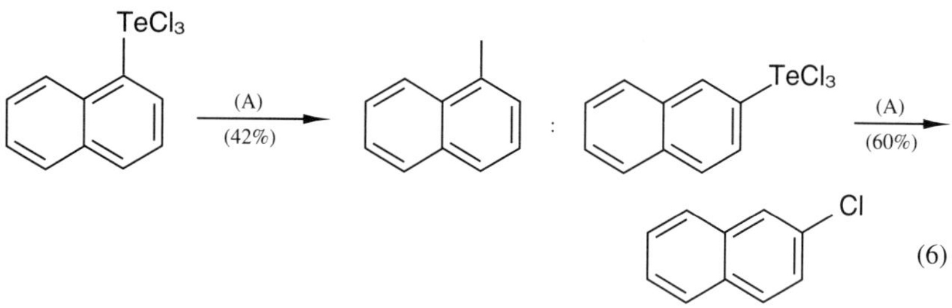
TeCl3
(A)
(42%)
TeCl3
(A)
(60%)
Cl

(6)

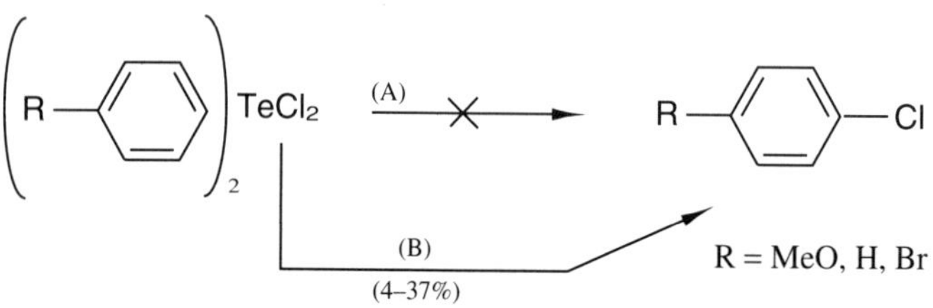
R
TeCl2
2
(A)
R
Cl
(B)
(4–37%)
R = MeO, H, Br

(7)

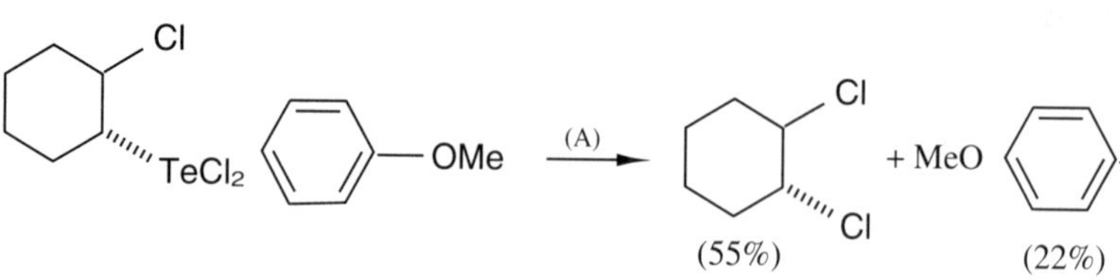
Cl
TeCl2
OMe
(A)
Cl
Cl
+ MeO
Cl
(55%)
(22%)

(8)

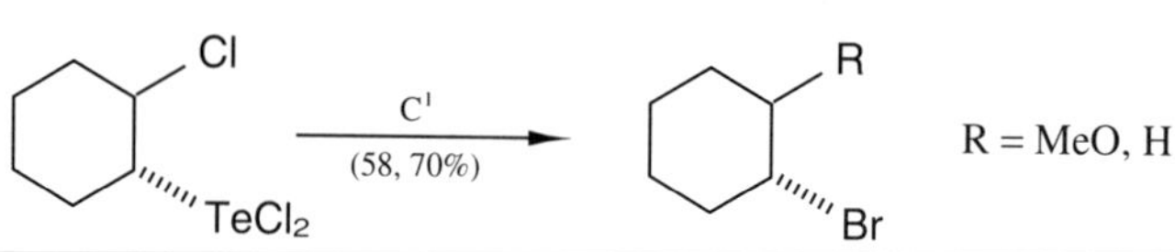
Cl
TeCl2
C[1]
(58, 70%)
R
Br
R = MeO, H

(9)

$$PhTeX_2CHRR^1 \xrightarrow[(83\text{–}98\%)]{C^1} RR^1CHX + [PhTeX] \qquad (10)$$

C^2 (59–95%): $PhTeX_2CHRR^1 \rightarrow RR^1CHX$

X = Br R = H; R1 = $C_{12}H_{15}$, $PhCH_2$
R = Me; $R^1 = C_6H_{13}$, $C_{12}H_{25}$

X = Cl, Br, I R = H; R1 = $C_{10}H_{21}$, $C_{11}H_{23}$, $C_{13}H_{27}$, $C_{15}H_{31}$, $PhCH_2$, $PhCH_2CH_2$
R = Me; $R^1 = C_{11}H_{23}$

Methods A, B, C^1 and C^2 are described fully in the text.

The following remarks are pertinent:

- Retention of the configuration is observed in all the reactions involving the cyclohexyl and alkenyl derivatives.
- By comparison of the yields of the reactions shown in the scheme, the following order for the trichlorides has been assessed for method A: alkyl > aryl > alkenyl.
- Diaryltellurium dichlorides are unreactive by method A, and by method B furnish lower yields than aryltellurium trichlorides (equation (7)).
- Equation (8) shows that the cycloalkyl group is more reactive towards the oxidative α elimination than the aryl group.
- By method B, the reactivity order of the trichlorides is aryl > alkenyl > alkyl (equations (5), (2) and (4)).

The described oxidative α elimination reactions have been rationalized as proceeding through a 1,2 tellurium–halogen shift.

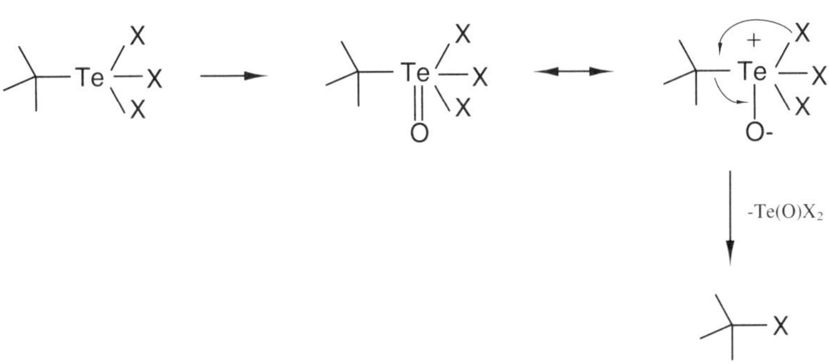

Organic halides from organotellurium(IV) halides (typical procedures) Method A [20]. To 1-chlorocyclohexyltellurium trichloride (0.35 g, 1.0 mmol) in dioxane (5 ml) is added t-BuOOH (70%, 0.26 g, 2.0 mmol) and the mixture stirred under reflux for 30 min, during which period a pale yellow solid precipitates. After cooling at room temperature, the precipitate is filtered off, and the filtrate poured into brine and then extracted with ether

(3 × 30 ml). The ether extract is dried ($MgSO_4$), and GLC analysis reveals the presence of 1,2-dichlorocyclohexane (0.14 g (92%), *cis/trans* = 7/93) (using phenyl acetate as the internal standard).

Method B [20]. A solution of *p*-methoxyphenyltellurium trichloride (0.34 g, 1.0 mmol) in benzene (200 ml) is irradiated with a high-pressure mercury lamp for 1 h (formation of a black precipitate). The mixture is filtered and the filtrate is washed with brine and then dried ($MgSO_4$). GLC analysis shows the presence of *p*-chloromethoxybenzene (0.14 g (70%)) (and minor amounts of *o,p*-dichloromethoxybenzene and *p*-methoxybiphenyl) using *p*-chlorotoluene, 2,4-dichlorotoluene and *p*-methylbiphenyl as the internal standard, respectively.

Method C^1 [20]. Neat n-dodecylphenyltellurium dibromide (0.53 g, 1.0 mmol) is heated at 230–250°C at 3 torr using a Kugelrohr distillation apparatus. A colourless liquid distills which is identical to n-dodecyl bromide (^1H NMR and GLC analysis) (0.25 g (90%)). (A red-brown residue seems to be mainly PhTeBr and on standing exposed to the ambient atmosphere solidifies to a grey solid—PhTe(O)Br and the starting dibromide by IR and combustion analytical data.)

Method C^2 [22]. Hexadecylphenyltellurium dichloride (0.199 g, 0.40 mmol) and NaCl (0.030 g, 0.51 mmol) are heated in DMF (2 ml) under N_2, with stirring at 100°C for 1 h. The mixture is quenched with H_2O and extracted with hexane. The extract is dried ($MgSO_4$) and filtered through a short column of SiO_2 to give 1-chlorohexadecane as a colourless oil (0.096 g (92%)).

Since alkyl phenyl tellurides are easily prepared from alkyl halides and phenyltelluromethyllithium, procedure C above constitutes a homologation of alkyl halides.

$$\mathrm{RX} \xrightarrow{\mathrm{PhTeCH_2Li}} \mathrm{PhTeCH_2R} \xrightarrow{\mathrm{X_2}} \longrightarrow \mathrm{RCH_2X}$$

Similarly, 1,1-dibromoalkanes can be prepared from lithium diphenyltelluromethane, whereas homologated aldehydes are obtained via iodination (probably through the hydrolysis of an intermediate diiodide) [22].

$$\mathrm{RX} \xrightarrow{\mathrm{(PhTe)_2CHLi}} \mathrm{(PhTe)_2CHR} \xrightarrow[\text{(76–85\%)}]{\mathrm{Br_2/NaBr}} \mathrm{RCHBr_2}$$

$$\mathrm{(PhTe)_2CHR} \xrightarrow[\text{(77–93\%)}]{\mathrm{L_2/NaI}} \mathrm{RCHO}$$

$$\mathrm{R} = \text{n-}C_{11}H_{23},\ \text{n-}C_{12}H_{25},\ \text{n-}C_{14}H_{29},\ \text{n-}C_{16}H_{33},\ Ph(CH_2)_3$$

An alternative iodination method is provided by treatment of the telluride with methyl iodide and sodium iodide in DMF [22].

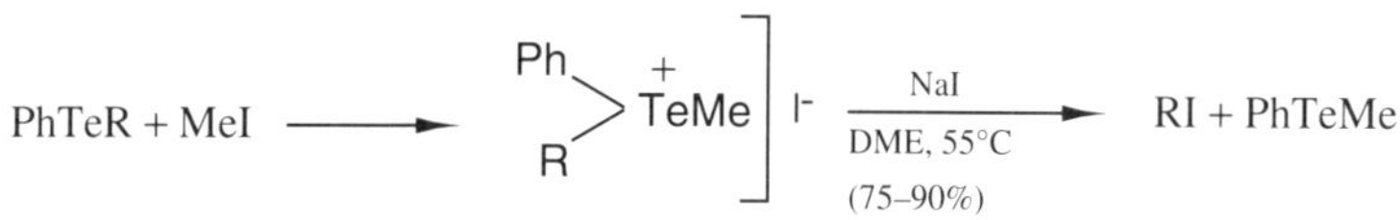

R = n-$C_{11}H_{13}$, n-$C_{12}H_{25}$, n-$C_{14}H_{29}$, n-$C_{16}H_{33}$, $Ph(CH_2)_2$, $Ph(CH_2)_3$
$C_{12}H_{25}CHCH_3$

Iodination of alkyl phenyl tellurides with MeI/NaI (typical procedure) [22] MeI (0.20 ml, 3.21 mmol) and NaI (75 mg, 0.50 mmol) are successively added to a solution of 1-phenyltellurododecane (149 mg, 0.40 mmol) in DMF (2 ml) under N_2. The mixture is stirred for 10 min at room temperature and then for 2 h at 55°C. It is quenched with H_2O and extracted with hexane. The extract is dried ($MgSO_4$) and filtered through a short column of SiO_2 to give 1-iodododecane as a colourless oil (103 mg (87%)).

(d) Halogenodetelluration of α-dichloroaryltelluroketones

Reaction of α-dichloroaryltelluroketones (see Section 3.9.3) with equimolar amounts of chlorine in dichloromethane at room temperature leads to the corresponding α-chloroketones in good yields, accompanied by the corresponding aryltellurium trichlorides [23].

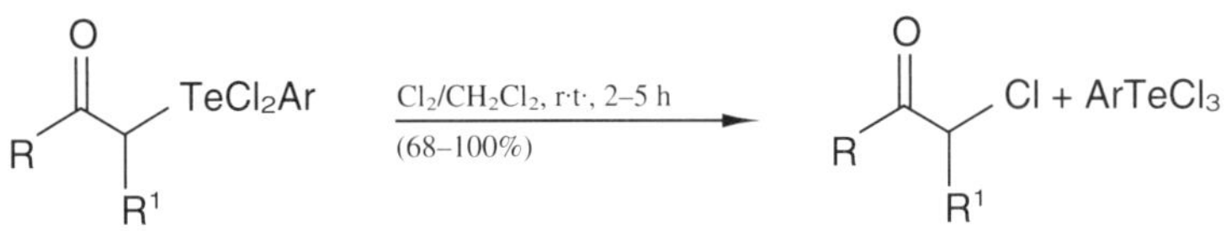

R^1 = H; R + Ph, t-Bu
R, R^1 = $(CH_2)_3$, $(CH_2)_4$, $(Ch_2)_5$

Ar = *p*-$MeOC_6H_4$

Similar reactions with bromine produce a mixture of α-chloro- and α-bromoketones.

Reaction of α-dichloroaryltelluroketones with chlorine: α-chloroketones (general procedure) [23] To a solution of the α-(dichloroaryltelluro)ketone (2 mmol) in CH_2Cl_2 (10 ml) is added a solution of Cl_2 (2 mmol) in the same solvent. The mixture is vigorously stirred until precipitation of the aryltellurium trichloride, which is then filtered off. The filtrate is washed with H_2O (25 ml) and then with brine (2 × 30 ml). The organic phase is dried ($MgSO_4$) and evaporated in a rotary evaporator. The residue is purified by preparative TLC on SiO_2 (eluting with petroleum ether 30–60°C).

(e) Halogenodetelluration of acetylenic tellurides

Acetylenic tellurides (see Section 3.12), on treatment with 3 mol eq of bromine or iodine, undergo halogenolysis, accompanied by addition of halogen to the acetylenic carbon–carbon bond, giving trihaloalkenes and organyltellurium trihalides [24].

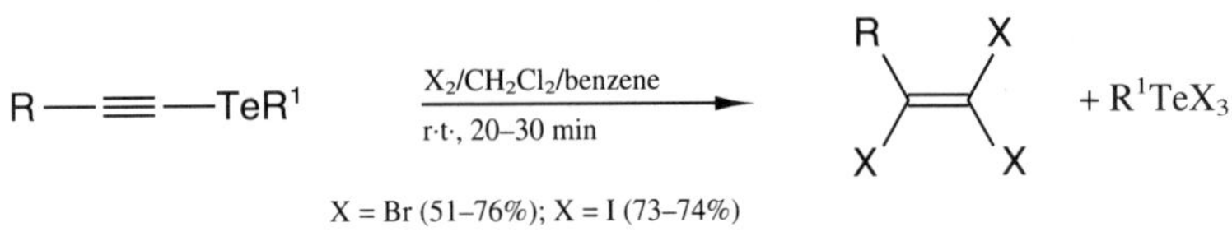

X = Br; R = n-C_5H_{11}, Ph, p-BrC_6H_4; R^1 = n-C_4H_9

R = Ph; R^1 = *p*-$MeOC_6H_4$

X = I; R = n-C_5H_{11}, Ph; R^1 = n-C_4H_9

Reaction of acetylenic tellurides with iodine (typical procedure) [24] To a solution of 1-(butyltelluro)heptyne (1.40 g, 5 mmol) in CH_2Cl_2 (10 ml) is added I_2 (3.81 g, 15 mmol). The pale yellow colour of the solution turns dark red and a precipitate is formed. The mixture is stirred at room temperature for 30 min and then the solvent is evaporated. The residue is filtered through an SiO_2 column (eluting with hexane), giving 1,1,2-triiodoheptene (1.76 g (74%)).

(f) Correlate reactions

Dibenzyl ditelluride, benzyl telluride [25, 26] and benzyl tellurocyanate [27] generate benzyl bromide on treatment with excess bromine.

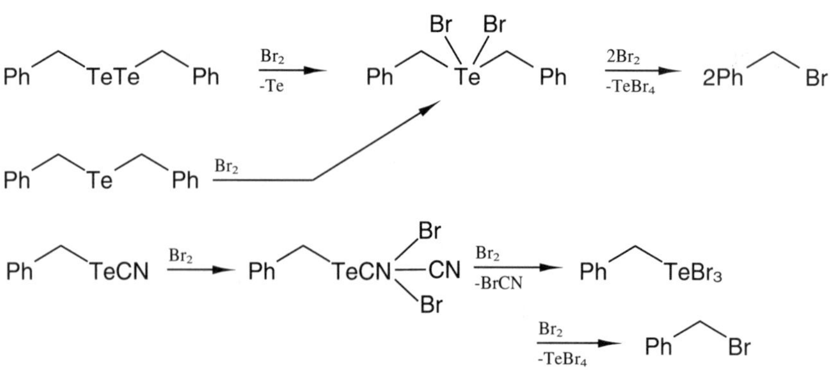

4.7.2.2 By the methoxy group

(a) Detellurative methoxylation

Oxidation of alkyl phenyl telluride with excess MCPBA (3–5 eq) in methanol affords the replacement of the phenyltellurium moiety by a methoxy group, giving the corresponding methyl ethers [28, 29] (method A). This reaction, which also proceeds by using alkyl phenyl selenides, therefore contrasts sharply with the well-known selenoxide (and with the less familiar telluroxide) elimination leading to olefins.

Improved yields are obtained by the similar treatment of telluroxides (prepared by hydrolysis of the parent dibromides [30]), and different alcohols such as ethanol, propanol and isopropanol can be employed to give the corresponding ethers (method B).

The detellurative methoxylation of the telluroxides also proceeds in high yields by treatment with trifluoroperoxyacetic acid (generated *in situ* from TFA and H_2O_2) [29] (method C).

Finally, alkyl phenyl tellurones (prepared by oxidation of telluroxides with $NaIO_4$ [31] are susceptible to similar conversions [29] (method D).

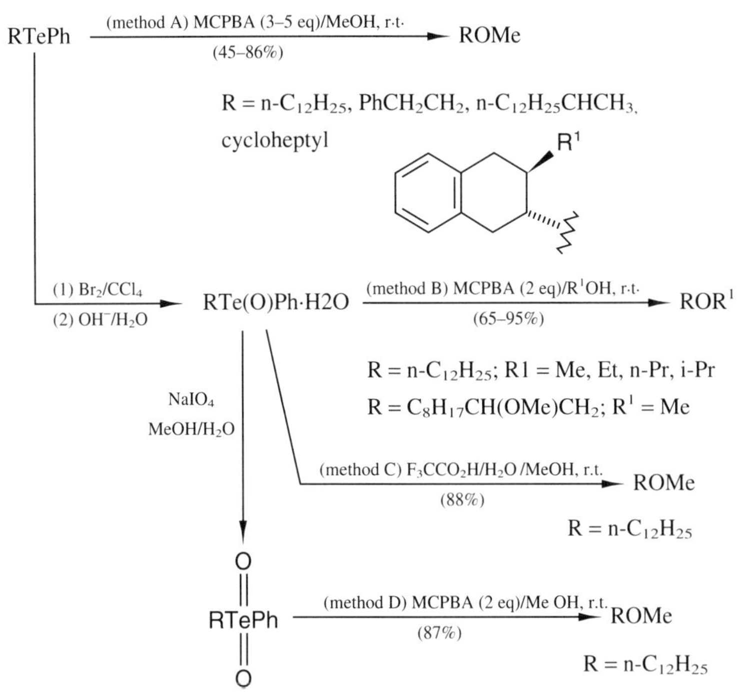

In the above procedures, the telluroxide *syn* elimination (see Section 4.8.1) was observed as a competitive interference only in the case of cycloheptyl phenyl telluride, where cycloheptene is formed in a yield of 45%.

When a phenyl group is linked at a vicinal position to tellurium, the replacement of tellurium by the methoxy group is accompanied by phenyl migration [28, 29].

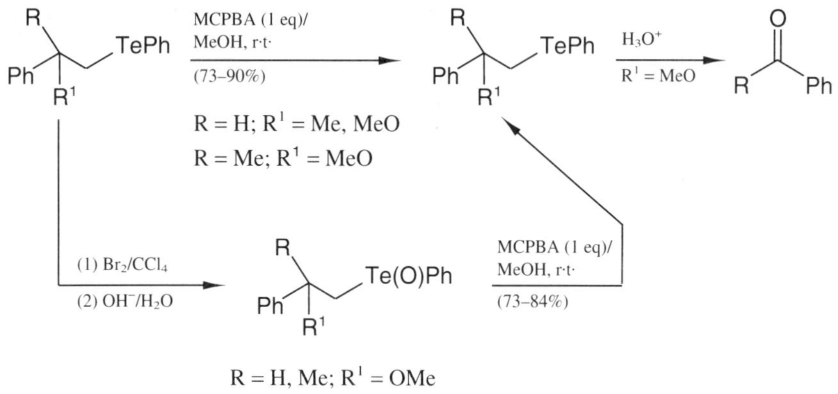

(2-Methoxy)cyclohexyl phenyl telluride and cycloheptyl phenyl telluride (easily accessible by methoxy telluration of the corresponding cycloalkenes, see Section 4.5.7.3) and (2-hydroxy)cyclohexyl phenyl telluride (prepared by the opening of cyclohexenoxide with the phenyl tellurolate ion, see Section 3.1.3.2) are converted by the above-described procedure into the dimethyl acetals of the ring-contracted aldehydes, in contrast, therefore, to the detellurative methoxylation of the tetrahydronaphthalene derivative (see above).

The difference in stability between the cyclic six- and seven-membered methoxy telluroxides is noteworthy. Indeed, while the cyclohexane derivative is stable and isolable, giving ring contraction on treatment with 1 eq of MCPBA, the cycloheptane derivative is unstable, suffering telluroxide elimination (in analogy to cycloheptene formation from cycloheptyl phenyl telluride, as shown previously).

Detellurative methoxylation of alkyl phenyl tellurides (typical procedure) [29] 2-Methoxy-3-phenyltelluro-1,2,3,4-tetrahydronaphthalene (1.85 g, 5 mmol) is treated with MCPBA (3.3 g, 15 mmol) in MeOH (30 ml) at 25°C with stirring. After 1 h the mixture is treated with aqueous $Na_2S_2O_3$ followed by aqueous $NaHCO_3$ and extracted with ether (3 × 30 ml). The extract is dried ($MgSO_4$) and evaporated, to leave a residue which is purified by SiO_2 TLC

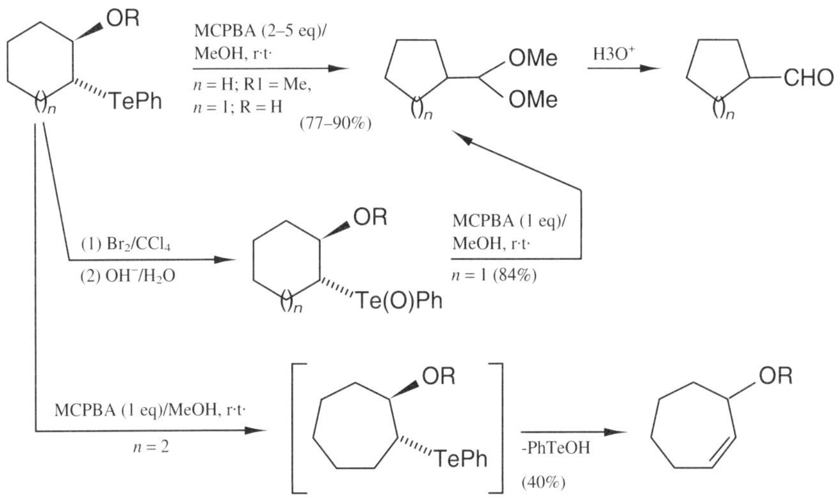

chromatography (eluting with hexane/EtOAc 4:1) to afford pure *trans*-2,3-dimethoxy-1,2,3,4-tetrahydronaphthalene (0.37 g, 1.94 mmol) and the *cis* isomer (0.15 g, 1.76 mmol (54%)).

Detellurative methoxylation of alkyl phenyl telluroxides (typical procedure) [29] To a heterogeneous solution of dodecyl phenyl telluroxide (0.26 g, 2 mmol) in MeOH (10 ml) is added solid MCPBA (80% purity, 0.88 g, 4 mmol) at 25°C with stirring. The resulting solution becomes homogeneous after 1 h, when it is treated with aqueous $Na_2S_2O_3$ followed by aqueous $NaHCO_3$. The mixture is treated with ether (3 × 30 ml), and the extract dried ($MgSO_4$) and evaporated to leave a pale yellow oily residue which is purified by SiO_2 column chromatography (eluting with hexane/EtOAc 5:1) to afford pure dodecyl methyl ether (0.38 g (95%)).

(b) Correlate reaction—synthesis of α-arylpropanoic acids

The title compounds, important *anti*-inflammatory pharmaceuticals, can be synthesized, starting from aromatic ketones, as depicted in the accompanying scheme, involving the above-described detellurative methoxylation in the key step [32, 33].

In an analogous sequence using the corresponding selenides, the yield of the substitution of the bromine for the phenylseleno group is higher. This result depends on the self-oxidation of the telluride to the telluroxide, followed by elimination to the vinylic derivative.

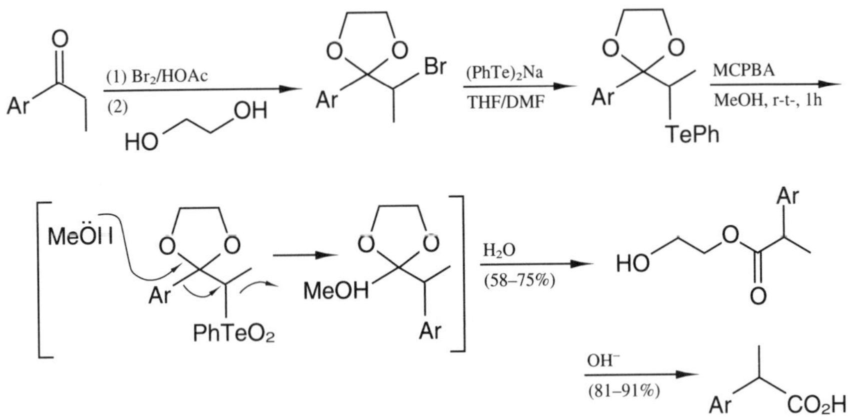

Ar = Ph, *p*-Me, *p*-i-Bu, *p*-Ph, *p*-BrC_6H_4,

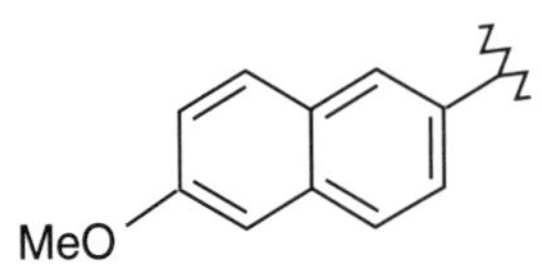

4.7.2.3 Reductive detelluration of tellurides by triphenyltin hydride

Phenyl alkyl tellurides are reduced by triphenyltin hydride under mild conditions, giving the corresponding hydrocarbons in high yields [34]. An excess of the reagent is necessary (2–2.8 eq), but a radical initiator need not be employed.

$$\mathrm{RTePh} \xrightarrow{\mathrm{Ph_3SnH(2\text{–}2.8eq),\ 25\text{–}80°C}} \mathrm{RH + Ph_3SnTePh}$$

$$\mathrm{R = C_{12}H_{25},\ CH_3CHC_{10}H_{21},\ C_8H_{17}CH(OH)CH_2,}$$
$$\mathrm{C_8H_{17}CH(OAc)CH_2,\ C_3H_7CH(OH)CHC_3H_7,}$$
$$\text{5-}\alpha\text{-cholestanyl}$$

The corresponding tellurium dichlorides can also be employed as starting materials, and although a greater excess of the reagent is needed (3.5–5 eq) the reactions are faster.

These reductive cleavages can be rationalized through a radical mechanism.

$$\mathrm{Ph_3SnH \longrightarrow Ph_3Sn\cdot}$$

$$\mathrm{Ph_3Sn\cdot + RTePh \longrightarrow RSnTePh + R\cdot}$$

$$\mathrm{R\cdot + Ph_3SnH \longrightarrow RH + PhSn\cdot}$$

Since the starting tellurides are easily prepared from alkyl halides or epoxides by displacement with tellurolate anions (see Section 3.1.3.2), the overall sequence constitutes a mild reduction of these substrates and is advantageous over the analogous reductions of selenides, which require more severe conditions (temperature of 120°C necessary).

The example sequence, where a carbonyl function remains untouched, clearly demonstrates the value of this method.

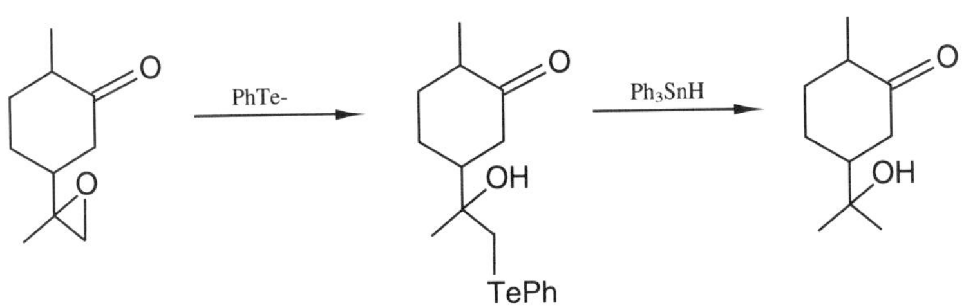

Reduction of alkyl phenyl tellurides to alkanes (general procedure) [34] All the operations are executed under red light using a one-piece apparatus (condenser fused to the flask) to avoid loss of material, closed with septum through which are inserted inlet and exit needles for N_2. The telluride is weighed directly in the flask, benzene is introduced from a syringe, and N_2 is swept for about 5 min through the system, which is then closed (by removing the exit needles) to ensure a slight static pressure. Ph_3SnH is injected into the mixture and the reaction mixture maintained at room temperature or heated under reflux. The course of the reaction is monitored by TLC. The product is isolated by chromatography and/or distillation.

In a typical experiment, n-dodecyl phenyl telluride (0.160 g, 0.32 mmol) in benzene (1 ml) is treated with Ph_3SnH (0.361 g, 1.03 mmol) during 5 h at room temperature, furnishing n-dodecane (0.080 g (87%)).

Further examples of reductive detelluration of tellurium compounds have been shown in connection with tellurolactonization and telluroetherification reaction (see Section 4.6).

REFERENCES

1. J. Bergman, *Tetrahedron* **28**, 3323 (1972); J. Bergman, *Org. Synth.* **57**, 18 (1977).
2. G. R. Petit and E. E. Van Tamelen, *Org. React.* **12**, 409 (1962).
3. J. Bergman and L. Engman, *Tetrahedron* **36**, 1275 (1980).

4. D. H. R. Barton, N. Ozbalik and M. Ramesh, *Tetrahedron Lett.* **29**, 3533 (1988).
5. S. Uemura, H. Takahashi and K. Ohe, *J. Organomet. Chem.* **423**, C9 (1992).
6. E. Cuthbertson and D. D. MacNicol, *J. Chem. Soc. Chem. Commun.*, 498 (1974).
7. E. Cuthbertson and D. D. MacNicol, *Tetrahedron Lett.*, 1893 (1975).
8. T. Kawamura, K. Kikukawa, M. Takagi and T. Matsuda, *Bull. Chem. Soc. Jpn* **50**, 2022 (1977).
9. S. Uemura, M. Wakasugi and M. Okano, *J. Organomet. Chem.* **194**, 277 (1980).
10. S. Uemura, S. I. Fukuzawa and S. R. Patil, *J. Organomet. Chem.* **243**, 9 (1983).
11. E. Wenkert, M. H. Leftin and E. L. Michelotti, *J. Chem. Soc. Chem. Commun.*, 617 (1984).
12. J. Bergman and L. Engman, *J. Organomet. Chem.* **175**, 233 (1979).
13. S. Uemura, K. Ohe, J. R. Kim, K. Kudo and N. Sugita, *J. Chem. Soc. Chem. Commun.*, 271 (1985).
14. K. Ohe, H. Takahashi, S. Uemura and N. Sugita, *J. Organomet. Chem.* **326**, 35 (1987).
15. K. Ohe, H. Takahashi, S. Uemura and N. Sugita, *J. Org. Chem.* **52**, 4859 (1987).
16. K. Ohe, H. Takahashi, S. Uemura and N. Sugita, *Nippon Kagaku Kaishi*, 1469 (1987).
17. S. Uemura, H. Miyoshi and M. Okano, *Chem. Lett.*, 1357 (1979).
18. S. Uemura, S. I. Fukuzawa, M. Wakasugi and M. Okano, *J. Organomet. Chem.* **214**, 319 (1981).
19. S. Uemura and S. I. Fukuzawa, *J. Chem. Soc. Chem. Commun.*, 1033 (1980).
20. S. Uemura and S. I. Fukuzawa, *J. Organomet. Chem.* **268**, 223 (1984).
21. S. Uemura and S. I. Fukuzawa, *Chem. Lett.*, 943 (1980).
22. K. Chikamatsu, T. Otsubo, F. Ogura and H. Yamaguchi, *Chem. Lett.*, 1081 (1982).
23. H. A. Stefani, A. Chieff and J. V. Comasseto, *Organometallics* **10**, 1178 (1991).
24. M. J. Dabdoub, J. V. Comasseto, S. M. Barros and F. Moussa, *Synth. Commun.* **20**, 2181 (1990).
25. G. Vicentini, *Chem. Ber.* **91**, 801 (1956).
26. H. K. Spencer and M. P. Cava, *J. Org. Chem.* **42**, 2937 (1977).
27. H. K. Spencer, M. V. Lakshmikantham and M. P. Cava, *J. Am. Chem. Soc.* **99**, 1470 (1977).
28. S. Uemura and S. I. Fukuzawa, *Tetrahedron Lett.* **24**, 4347 (1983).
29. S. Uemura and S. I. Fukuzawa, *J. Chem. Soc. Perkin Trans I*, 471 (1985).
30. M. R. Detty, *J. Org. Chem.* **45**, 274 (see also Section 3.10.2) (1980).
31. H. Lee and M. P. Cava, *J. Chem. Soc. Chem. Commun.*, 277 (1981); L. Engman and M. P. Cava, *J. Chem. Soc. Chem. Commun.*, 164 (1982).
32. S. Uemura, S. I. Fukuzawa, T. Yamauchi, K. Hattori, S. Mizutaki and T. Tamaki, *J. Chem. Soc. Chem. Commun.*, 426 (1984).
33. S. Uemura and S. I. Fukuzawa, *J. Chem. Soc. Perkin Trans. I*, 1983 (1986).
34. D. L. J. Clive, G. J. Chittattu, V. Farina, W. A. Kiel, S. M. Menchen, C. G. Russel, A. Singh, C. K. Wong and N. J. Curtis, *J. Am. Chem. Soc.* **102**, 4438 (1980).

4.8 SYNTHESIS OF OLEFINS

4.8.1 By telluroxide elimination

The telluroxide elimination appeared initially to be of little value in comparison with the popularity of the analogous selenoxide elimination. However, after the previous results which established the *syn* stereochemistry of the elimination [1], more detailed investigations clearly demonstrated the synthetic utility of this methodology [2–4].

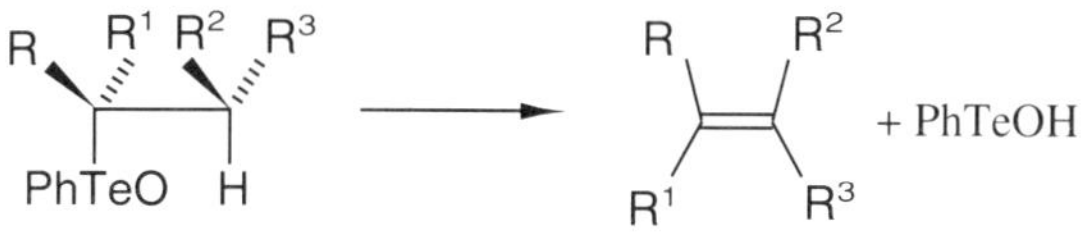

The starting telluroxides (see Section 3.10) are prepared by hydrolysis of the corresponding tellurium dibromides [3] or by direct oxidation of the tellurides (MCPBA is the more appropriate oxidizing agent [4]). By the first method the telluroxides are obtained as hydrates and normally give higher elimination yields.

s-Alkyl phenyl telluroxides and cycloalkyl phenyl telluroxides, with the exception of cyclohexyl derivatives, are unstable compounds, suffering the elimination reaction at room temperature. In contrast, n-alkyl phenyl telluroxides and cyclohexyl phenyl telluroxide are stable compounds and undergo elimination only by prolonged heating in toluene or THF, or by pyrolysis at 200–240°C.

Since the starting tellurides are easily prepared (see Section 3.1.3.2) from the corresponding alkyl bromides and tellurolate ions, and β-hydroxyalkyl tellurides by the opening of epoxides with the same reagents, the combined procedures furnish a method for the dehydrobromination of alkyl bromides and for the conversion of epoxides into allylic alcohols. Moreover, combining the telluroxide elimination with the methoxytelluration of olefins (see Sections 3.9.3.2 and 4.5.7.3), allylic and vinylic ethers are easily prepared. The accompanying scheme illustrates these useful synthetic transformations.

The following important features should be noted.

The telluroxide elimination shows a preference towards the less substituted carbon (as observed by the 2.48–2.50:1 ratio assessed in the elimination of 2-bromooctyl phenyl telluroxide and 2-bromodecyl phenyl telluroxide) that is more marked than that observed for the selenoxide and sulphoxide eliminations. By procedure b (MCPBA oxidation), the ratio is decreased to 1.5–1.7:1.

In all cases, small amounts of the corresponding alcohols and ketones are detected.

The by-product of the elimination, phenyltellurenic acid, is detected, at least partially, as diphenyl ditelluride, and is probably formed by disproportionation and oxidation reactions.

$$2PhTeOH \longrightarrow PhTe(O)OH + PhTeH$$

$$2PhTeH \xrightarrow{[O]} (PhTe)_2 + H_2O$$

The telluroxide elimination proceeds much more slowly than selenoxide elimination.

Telluroxide eliminations

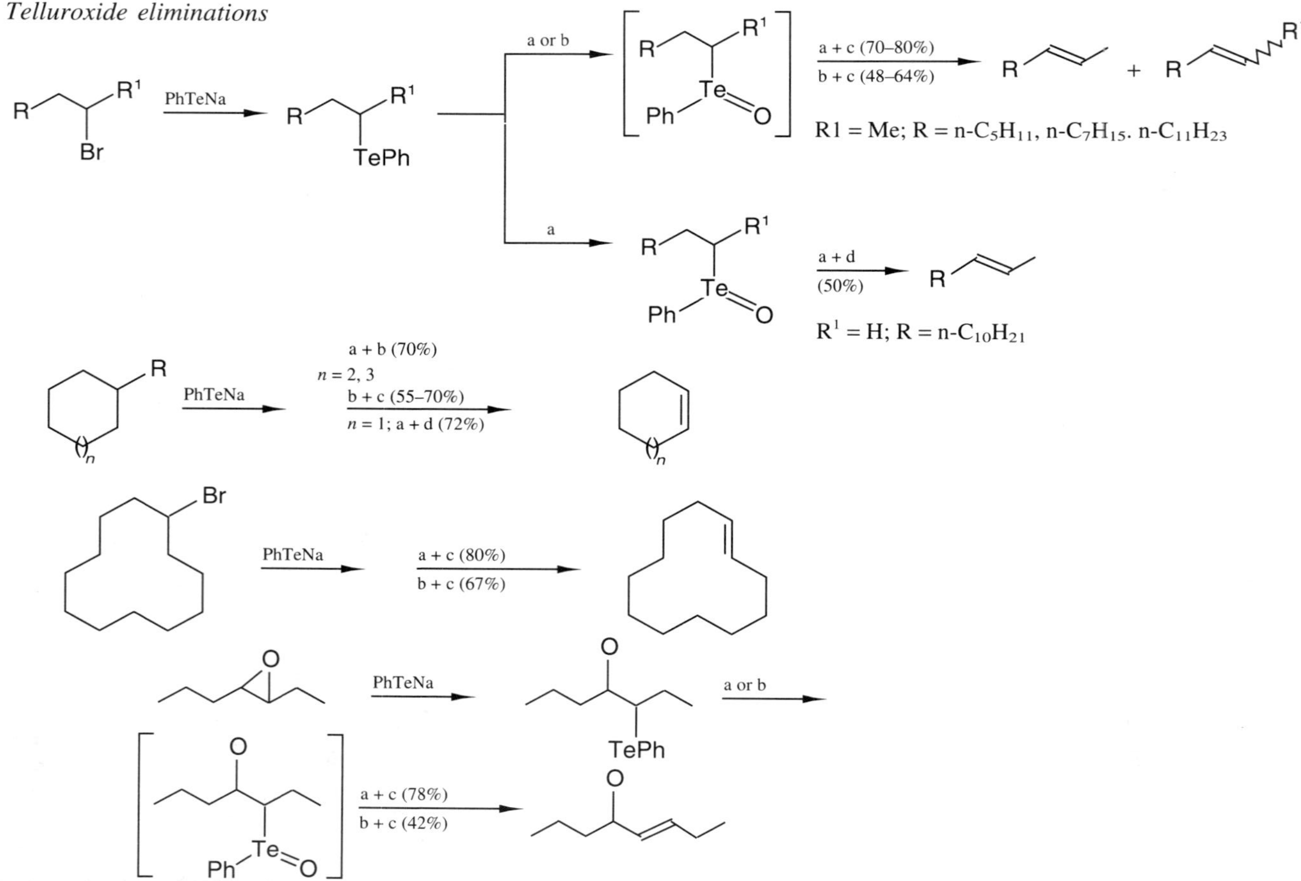

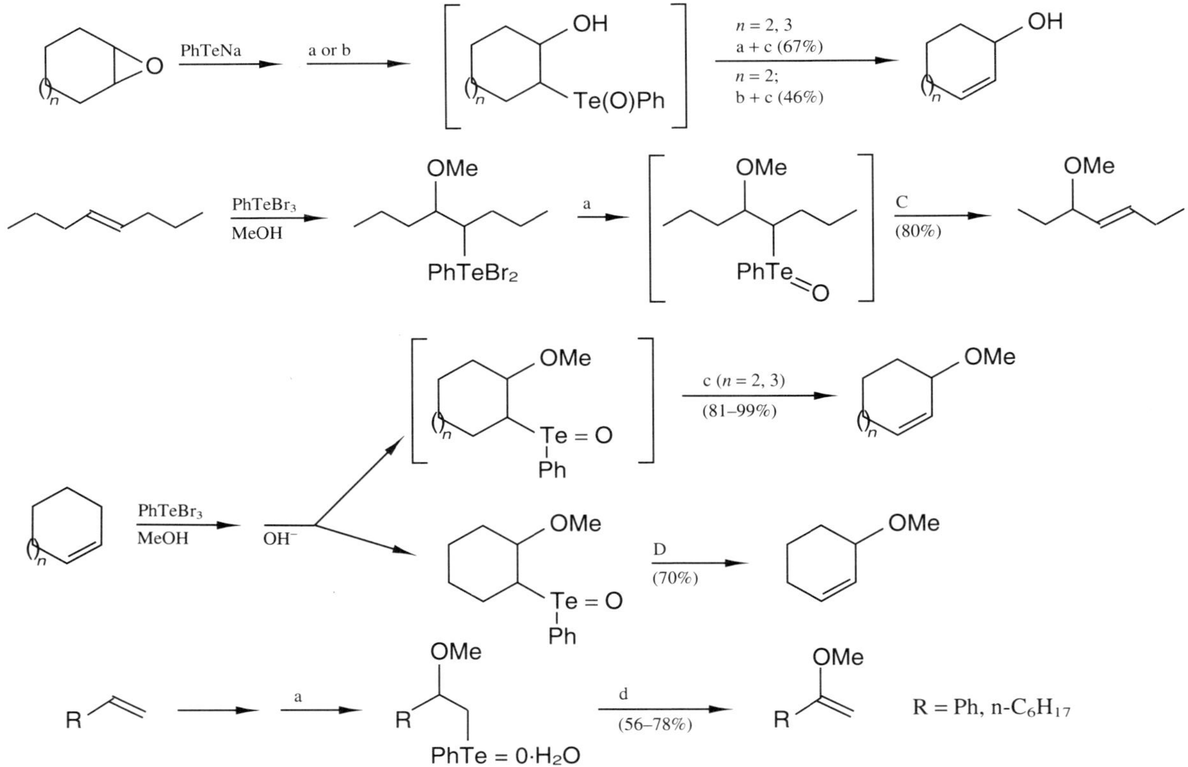

a: (1) Br_2/CCl_4, (2) OH^-/H_2O (telluroxide as hydrate).

b: MCPBA (1·5 eq) ether.

c: 20–25°C, 1–3 h (for a) or 2 h (for b).

d: Kugelrohr distillation, 200–240°C/20–760 torr.

In contrast to selenides, in the case of tellurides the double bond geometry of the formed alkenes is markedly dependent upon the amount of oxidant employed. Thus, *trans*-cyclododecene is nearly the sole product of the oxidation of cyclododecyl phenyl telluride with 1 eq of oxidant, whereas with excess oxidant (even 2 eq) a mixture of *cis* and *trans* isomers is formed, in some cases the *cis* isomer being the major product [5].

Olefins and allylic alcohols from alkyl or cycloalkyl bromides and from epoxides (general procedures) [3] The s-alkyl or cycloalkyl bromide is treated with sodium phenyl tellurolate (1 eq, from diphenyl ditelluride and $NaBH_4$, see Section 4.1.3.2). The crude telluride is purified by SiO_2 chromatography (elution with hexane) and converted to the corresponding dibromide by addition of bromine (1 eq) in CCl_4 at 0°C. Epoxides are converted to β-hydroxylalkyl and β-hydroxycycloalkyl phenyl tellurium dibromide by a similar procedure, except that the intermediate tellurides are chromatographed on SiO_2 using hexane/EtOAc (5:1) as the eluent.

Method a + c. The tellurium dibromide (1.0 mmol) in THF (20 ml) is treated with 0.5 N NaOH (10 ml) at room temperature with stirring, and the resulting solution is stirred for 1 h, the colour of the solution turning to orange. The mixture is diluted with brine and extracted with ether. The extracts are dried ($MgSO_4$), evaporated under vacuum and the residue purified from diphenyl ditelluride by SiO_2 chromatography (elution with hexane). The pure olefin is isolated by distillation of the residue.

Method a + d. The tellurium dibromide is treated with 0.5 N NaOH at room temperature for 1 h. The mixture is diluted with H_2O and extracted with ether. The extracts are dried ($MgSO_4$), and evaporated under vacuum, furnishing the telluroxide hydrate in a quantitative yield. Pyrolysis of the telluroxide is carried out in a Kugelrohr apparatus at 200–240°C/760 torr, giving the olefin as an oil.

Allylic and vinylic ethers via methoxytelluration of olefins (general procedure) [3] To a solution of phenyltellurium tribromide (2.223 g, 5 mmol), prepared from diphenyl ditelluride and bromine in methanol (5 ml), is added the olefin (10 mmol) and the mixture refluxed for 1 h. On cooling, the (β-methoxy)alkyl- or cycloalkylphenyltellurium dibromide precipitates and is separated by filtration.

Method c. The dibromide is converted to the corresponding allylic methyl ethers by treatment with NaOH as described previously.

Method d. The dibromide is converted to the stable telluroxide hydrate and then pyrolysed to the allylic and vinylic methyl ether as described previously.

4.8.2 Correlate method: reaction of alkyl phenyl tellurides with chloramine-T

The reaction of alkyl phenyl tellurides with excess chloramine-T (*N*-chloro-*N*-sodio-*p*-tolysulphonamide) in refluxing THF leads to olefins, presumably through a tellurosulphimino intermediate [6]. Due to the high yields obtained, this method seems to be highly competitive with the telluroxide elimination.

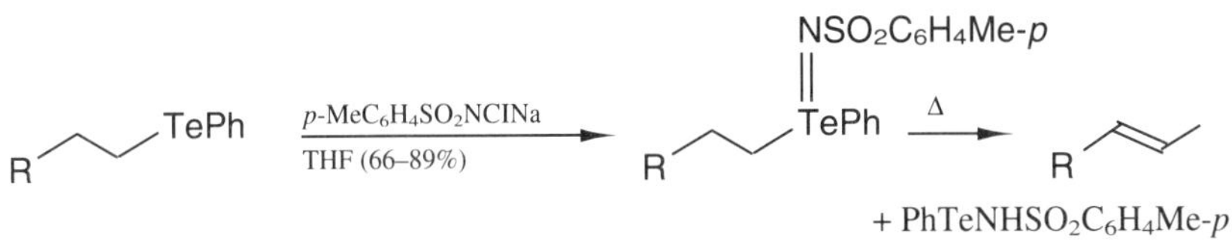

R = n-C_8H_{17}, n-$C_{10}H_{21}$, n-$C_{12}H_{25}$, n-$C_{13}H_{27}$

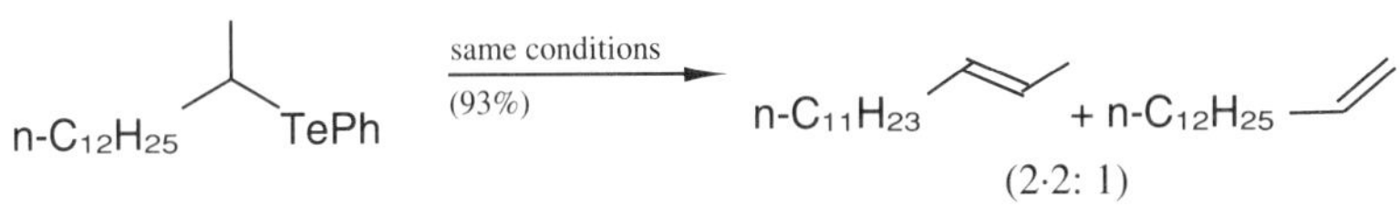

Olefins by reaction of alkyl phenyl tellurides with chloramine-T (typical procedure) [6] A solution of 1-dodecyl phenyl telluride (0.198 g, 0.52 mmol) and commercial chloramine-T trihydrate (0.300 g, 1.0 mmol) in THF (5 ml) is refluxed for 40 min under N_2. After evaporation of the solvent, the residue is treated with hexane (20 ml), and then filtered. The filtrate is passed through a short SiO_2 column with hexane, giving, after evaporation, 1-dodecene as a colourless oil (0.068 g (78%)).

Vinylsilanes, compounds of great synthetic utility [7], can be synthesized by this procedure, starting from 1-phenyltelluro-1-trimethylsilylalkanes [8] (thio and seleno analogues react similarly).

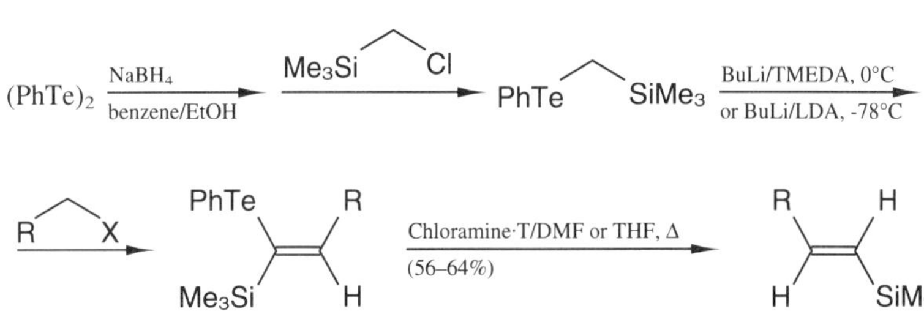

R = n-C_9H_{19}, n-$C_{11}H_{23}$, n-$C_{13}H_{27}$, n-$C_{15}H_{31}$

4.8.3 From telluronium ylides

4.8.3.1 Stabilized telluronium ylides

Stabilized telluronium ylides, such as dibutyltelluronium carbethoxy [9], phenacyl [10], cyano- [10] and carbamoyl [11] methylide (easily prepared by the reaction of dibutyl tellurides with the appropriately substituted methyl halides, followed by treatment with a base), undergo Wittig-type olefination reactions with a variety of carbonyl compounds, giving the expected olefins in satisfactory yields (method A). This behaviour is in sharp contrast to that of stabilized sulphonium ylides, which are inert towards carbonyl compounds.

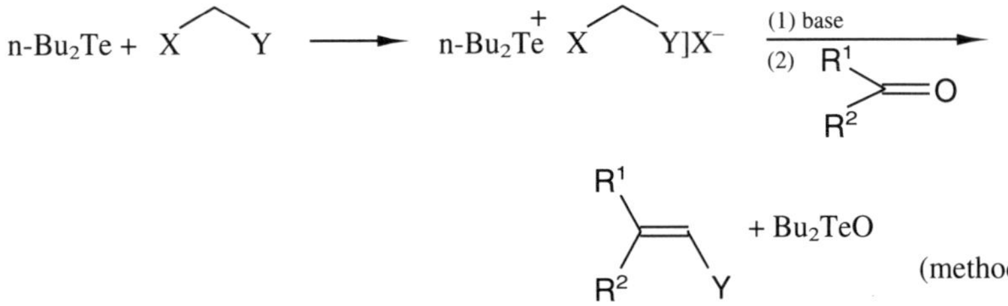

(method A)

Y	X	Base/solvent/temperature	R^1R^1CO	Percentage yield
CO_2Et	Br	KOt-Bu/THF/–20°C	aliphatic and aromatic aldehydes aliphatic, cycloaliphatic and aromatic ketone, α,β-unsaturated aldehydes and ketones	52–90
COPh	Br	KOt-Bu/THF/–20°C	aromatic aldehydes	74–81
CN	Cl	KOt-Bu/THF/–20°C	aromatic aldehydes	60–83
			aliphatic and cycloaliphatic ketones	36–49
CONHBu-i	Br	NaH/THF/HMPA/–50°C,r.t.	aliphatic and cycloaliphatic aldehydes α,β-unsaturated aldehydes	40–71

Noteworthy features of these reactions are the high predominance of (*E*)stereochemistry in the case of aldehydes and the good results obtained even with highly enolizable ketones (such as cyclopentanone), α,β-epoxy ketones (isophorone oxide) and α,β-unsaturated compounds (benzalacetophenone and cinnamaldehyde).

Dibutyltelluronium benzylide, generated by treatment of the telluronium salt with potassium t-butoxide, behaves similarly to the above stabilized ylides, undergoing Wittig-type olefinations with aromatic aldehydes [12].

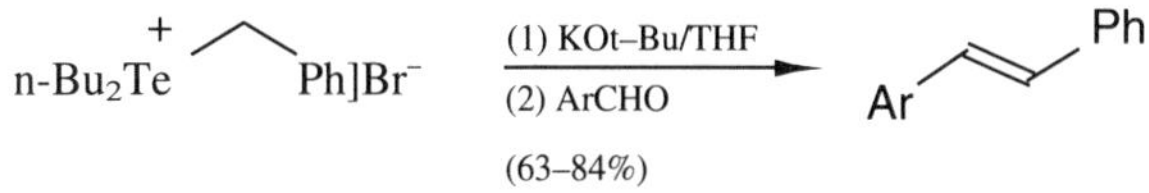

Ar = Ph, *p*-Cl, *p*-MeC_6H_4, PhCH = CH, 2-naphthyl

In the case of carbamoyl derivatives (Y = CONHBu-i) the reactions can be carried out even under phase transfer catalysis, by simply treating the aldehyde with the telluronium salt (itself behaving as a phase transfer catalyst) in THF/trace H_2O, in the presence of K_2CO_3 at 50–60°C (method B) [13].

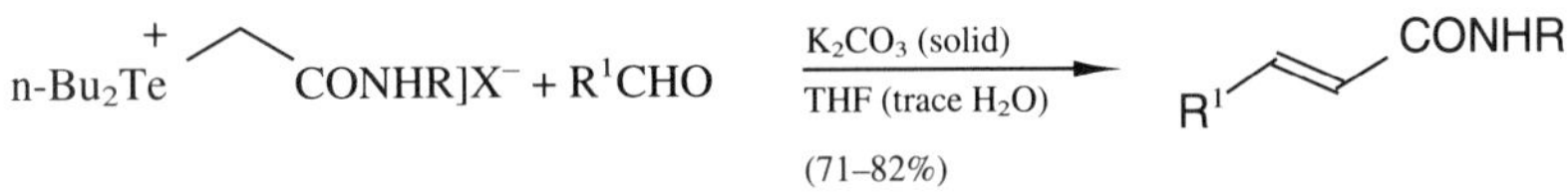

R = i-Pr, i-Bu; R^1 = Ph, *p*-Cl, *p*-Br, *p*-Me, *p*-$MeOC_6H_4$

X = Cl, Br (method B)

(a) Simplified procedures

The low energy and high polarity of the Te—C bond allows the reactions to be performed starting from the telluronium salts, instead of the ylides, under neutral conditions (method C) [10, 14].

Finally, the preparation of the telluronium salts and the olefination reaction can be combined in a simplified and very practical one-pot procedure (method D) [10, 14].

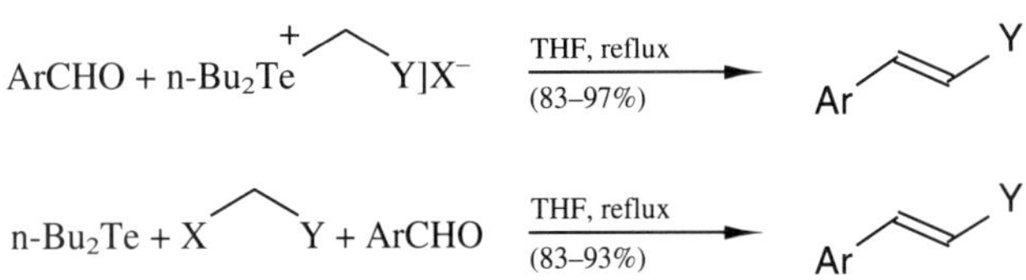

(method C)

(method D)

Y = CO_2Et, CO_2Me, COPh; X = Br

Y = CN; X = Cl

Ar = *p*-NO_2, *m*-NO_2, *p*-Br, *p*-ClC_6H_4

Catalytic one-pot procedure. Since in the described telluronium ylide olefination, telluroxide is formed as a by-product, and the telluroxide is susceptible to be reduced by triphenyl phosphite, a catalytic procedure can be employed, providing a practical one-pot synthesis of α,β-unsaturated esters and ketones (method E) [15]. By this procedure, a catalytic amount of n-dibutyl telluride reacts with the α-bromoester or α-bromoketone, and the formed telluronium salt is converted *in situ* under phase transfer conditions (solid K_2CO_3/trace H_2O) into the ylide, which reacts in turn with the aldehyde, giving the olefin. Since the reaction is carried out in the presence of triphenyl phosphite, the formed dibutyl telluroxide is reduced back to the dibutyl telluride, which is then recycled.

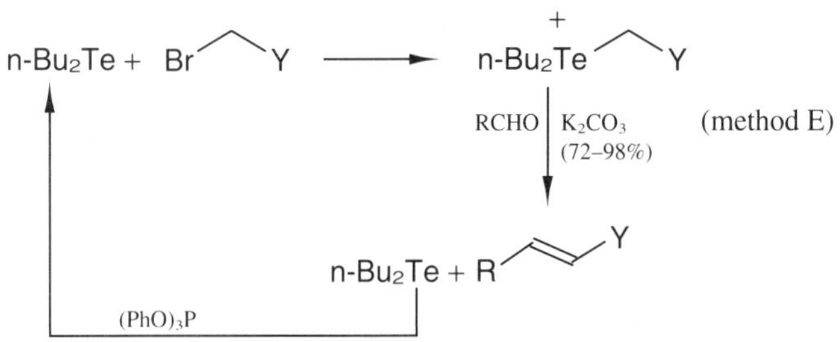

Y = CO_2Me, COPh, COPr-i

R = Ph, *p*-Cl, *p*-MeC_6H_4, styrl, 2-pyridyl, 2-furyl, cyclohexyl, n-Bu, n-C_9H_{19}

Reaction of stabilized telluronium ylide with aldehydes Method A—using telluronium ylide (typical procedure) [9]. A solution of carbethoxymethyldibutyltelluronium bromide (1.23 g, 3 mmol) in dry THF (3 ml) is added dropwise to a solution of KOt-Bu (0.337 g, 3 mmol) in THF (4 ml) at −20°C. After a few minutes, *trans*-cinnamaldehyde (0.264 g, 2 mmol) is added over 1 min. The reaction mixture is then stirred for 1 h at −20°C. After the usual work-up, the product is purified by SiO_2 column chromatography, to give ethyl 5-phenylpenta-(2*E*,4*E*)-dienoate (0.291 g (72%)). GC and ^1H NMR analysis of the product reveals a purity of 98%.

Method B—using telluronium ylide under PTC (general procedure) [13]. A mixture of dibutyl(alkylcarbamoylmethyl)telluronium halide (2.5 mmol), aldehyde (2.5 mmol) and K_2CO_3 (2.5 mmol) in THF (25 ml) and a trace of H_2O is stirred at 50–60°C for about 6 h. The end-point of the reaction is monitored by TLC (GF_{254}, petroleum ether/acetone 4:1). Brine (15 ml) is added, and the reaction mixture is extracted with ether (3 × 20 ml). The combined ether extracts are dried ($MgSO_4$) and concentrated under vacuum to give the crude α,β-unsaturated amide, which is purified by SiO_2 chromatography.

Method C—using telluronium salts (typical procedure) [10]. A mixture of carbethoxymethyldibutyltelluronium bromide (0.99 g, 2.5 mmol) and 4-nitrobenzaldehyde (0.38 g, 2.5 mmol) is refluxed in THF (30 ml). After 6 h, it is quenched with H_2O and extracted with ether. The organic extract is purified by recrystallization to afford pure (*E*)-methyl 3-(4′-nitrophenyl)propenoate (0.491 g 95%)).

Method D—using dibutyl telluride (typical procedure) [10]. A mixture of 3-nitrobenzaldehyde (0.38 g, 2.5 mol), methyl bromoacetate (0.38 g, 2.5 mmol), and dibutyl telluride (0.61 g, 2.5 mmol) is refluxed in THF. After 6 h, the mixture is worked up as for method C to give pure (*E*)-methyl 3-(3′-nitrophenyl)propenoate (0.460 g (89%)).

Method E—catalytic procedure (typical procedure) [15]. Benzaldehyde (106 mg, 1.0 mmol), methyl bromoacetate (165 mg, 1.1 mmol), triphenyl phosphite (356 mg, 1.2 mmol), dibutyl telluride (48 mg, 0.2 mmol), K_2CO_3 (179 mg, 1.3 mmol) and THF (4 ml) are mixed and stirred at 50°C for 13 h (monitored by TLC). The reaction mixture is filtered rapidly through a small amount of SiO_2 with EtOAc as the eluent to remove inorganic salts and dibutyltellurium oxide. Preparative TLC with EtOAC/petroleum ether 60–90°C (1:9) as the eluent yields 3-phenylpropenoate (160 mg (98%)).

4.8.3.2 Semi-stabilized and non-stabilized telluronium ylides

In sharp contrast to the olefination reaction employing stabilized telluronium ylides, semi- and non-stabilized ylides react with carbonyl compounds to give epoxides (in analogy to non-stabilized sulphonium and selenonium ylides).

(a) Dialkyltelluronium allylides

Moderate-to-good yields of α,β-unsaturated epoxides are obtained, allowing aromatic and aliphatic aldehydes to react with dialkyltelluronium allylide (the diisobutyl derivatives are the reagent of choice) [16].

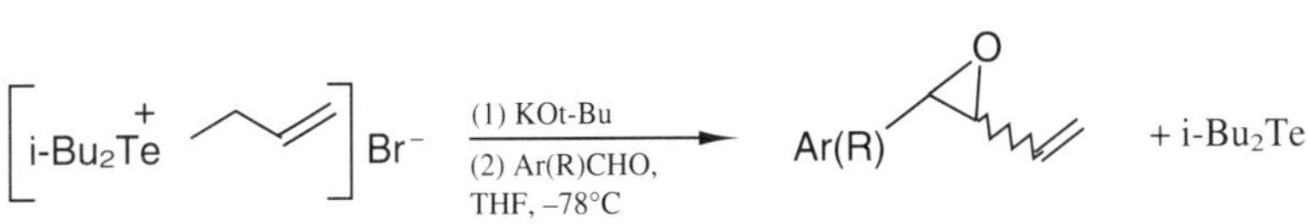

Ar(R) = Ph and (methyl, alkoxy, halo, nitro) derivatives, 3-pyridyl, 1-naphthyl, Me_2CPh, 4-cyclohexenyl, cyclohexyl (55–98%)

R = $PhCH_2CH_2$, n-C_6H_{13}, n-C_9H_{19} (30–38%)

The moderate *cis* selectivity, observed for aromatic and aliphatic aldehydes, decreases if an NO_2 group or *ortho* substituents are linked to the aromatic ring.

Reaction of telluronium allylides with aldehydes (typical procedure) [16] Allyldiisobutyltelluronium bromide (0.87 g, 2.4 mmol) and KOtBu (0.269 g, 2.4 mmol) are placed in a reaction vessel under N_2, and dry THF (4 ml) is added with stirring at −78°C. After a few minutes, a solution of benzaldehyde (0.216 g, 2 mmol) in THF (2 ml) is added dropwise, and the reaction mixture is slowly allowed to warm to room temperature. Normal work-up followed by SiO_2 column chromatography gives diisobutyl telluride (0.55 g (95% based on the used salt); eluted by hexane) and 2-phenyl-3-vinyloxirane (0.356 g (82%); *cis/trans* = 85:15; eluted by 5% ether/hexane).

(b) Diphenyltelluronium methylide

This was the first-described non-stabilized ylide, obtained by treatment of the corresponding telluronium tetrafluoroborate with LiTMP. Epoxides are obtained by reaction with both aldehydes and ketones [17].

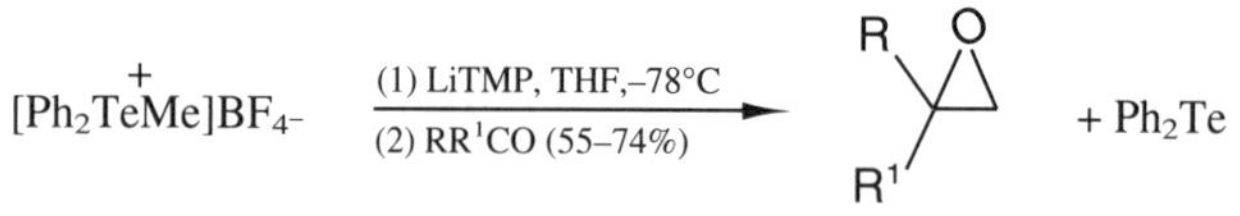

R = H; R[1] = Ph, *p*-Cl, *p*-Br, *p*-FC_6H_4, *p*-biphenyl, cyclohexyl

R = Ph; R[1] = Me, Ph

Reaction of telluronium methylide with carbonyl compounds (general procedure) [17] A solution of LiTMP (1.2 mmol) in THF is added to a solution of diphenyltelluronium methyl tetrafluoroborate (0.459 g, 1.2 mmol) in THF (8 ml) at −78°C under N_2. The mixture is warmed to −70°C and stirred for 30 min. After cooling to −78°C, the carbonyl compound (1.0 mmol) in THF (2 ml) is added. The reaction mixture is then allowed to warm at room temperature. After the reaction is complete (monitored by TLC), the usual work-up and flash chromatography give the product.

(c) Diisobutyltelluronium trimethylsilylpropynylide

As in the preceding case, the ylide is obtained using LiTMP as the base and gives epoxides with aldehydes and ketones [18, 19].

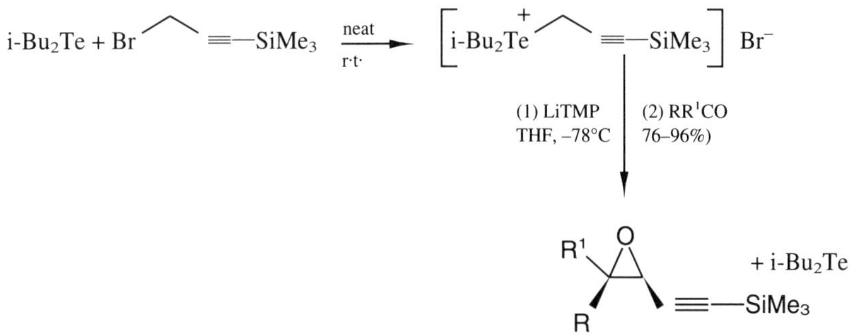

R = H; R' = Ph, *p*-Cl, *p*-BrC_6H_4, 2-naphthyl, *p*-biphenyl,
cyclohexyl, n-Bu, n-C_5H_{11}((*Z*)/(*E*): 99/1-81/19)
R = Ph; R' = Me

As shown by the obtained products, the reaction is highly *cis*-stereospecific. The utility of this method is evident in view of the extensive use of unsaturated silicon reagents in organic synthesis [20].

Reaction of telluronium trimethylsilylpropynylide with carbonyl compounds (general procedure) [18] A solution of LiTMP (1.2 mmol) in THF (2 ml) is added dropwise to a solution of [i-Bu_2Te^+ ⁄╲ ≡ $SiMe_3$]Br^- in THF (8 ml) at −78°C under N_2. The mixture is stirred for 30 min and then the carbonyl compound (1.0 mmol) in THF (2 ml) is added. The reaction mixture is allowed to warm at room temperature, and when the reaction is complete (monitored by TLC) the usual work-up and flash chromatography give the pure (trimethylsilyl)ethynyl epoxide.

4.8.3.3 Correlate reaction: the reaction of telluronium salts with carbonyl compounds mediated by organolithium reagents—formation of secondary alcohols

Telluronium salts, precursors of non-stabilized, semi-stabilized and stabilized telluronium ylides, react with carbonyl compounds on treatment with organolithium reagents to give secondary alcohols instead of alkenes [12, 19, 21, 22]. This different pathway arises due to the conversion of the telluronium salts into unstable tetraorganyltellurium intermediates (instead of the deprotonation giving ylides), which, through a cleavage of one of the Te—C bonds, generates a nucleophilic species suitable to be added to the carbonyl group.

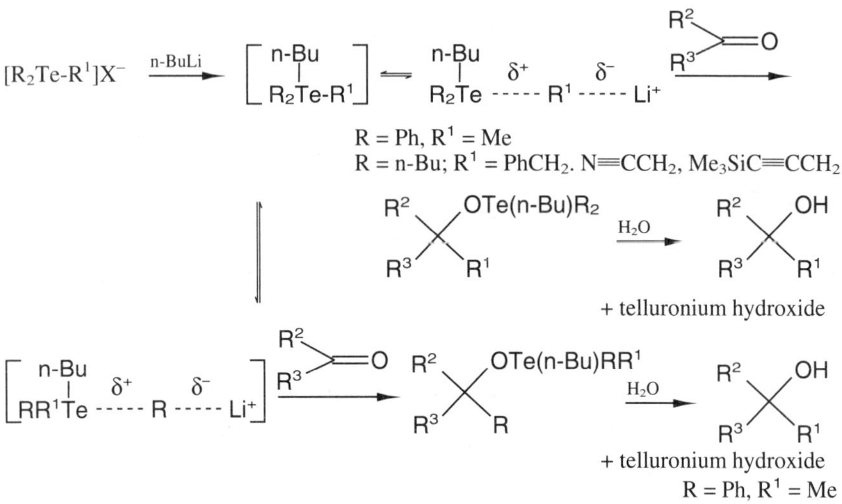

Telluronium salt	Carbonyl compound	Percentage yield
(a1) $[Me_3Te^+]I^-$ [21], (a2) $[Ph_2Te^+Me]BF_4^-$ [21]	benzaldehyde and derivatives, 2-naphthyl-CHO, 2-pyridyl-CHO	56–85
(b) $[n\text{-}Bu_2Te^+CH2Ph]Br^-$ [12]	benzaldehyde and derivatives, cinnamaldehyde, 2-pyridyl-pCHO, cyclohexanone	58–96
(c) $[n\text{-}Bu_2Te^+CH_2CN]Cl^-$ [22]	benzaldehyde and derivatives, aliphatic aldehydes, benzophenone, acetophenone	85–98
(d) $[i\text{-}Bu_2Te^+CH_2C{\equiv}CSiMe_3]Br^-$ [19]	benzaldehyde and derivatives, 2-naphthyl-CHO, 2-pyridyl-CHO, cyclohexyl-CHO, acetophenone, cyclohexanone, 2-cyclohexenone	67–93

In the reactions with trimethyltelluronium salts (a1), alcohols with the structure RCH(-Bu)OH are formed as minor by-products (6–23%), resulting from the transference of the n-butyl group.

With the telluronium salts (a2), (b), (c) and (d), phenyl, benzyl, cyanomethyl and (trimethylsilyl)propargyl groups, respectively, are transferred in preference to the other groups.

The β-hydroxynitriles obtained from telluronium salt (c) form an important class of compounds since the cyano group can be submitted to several transformations [23].

The importance of the propargylic anions in synthetic chemistry [24] enhances the utility of the reaction involving telluronium salt (d).

Reaction of telluronium salts/n-BuLi with carbonyl compounds: synthesis of alcohols *With telluronium salt (a2) (typical procedure)* [21]. A solution of n-BuLi (1.1 mmol) in hexane is added to a solution of the telluronium salt (0.459 g, 1.2 mmol) in THF (6 ml) at −78°C under N_2. The mixture is warmed at −60°C and stirred for 10 min. After cooling again at −78°C, *p*-chlorobenzaldehyde (0.140 g, 1.0 mmol) in THF (2 ml) is added. The mixture is then allowed to warm at room temperature. After the reaction is complete (monitored by TLC), the usual work-up and flash chromatography yield pure *p*-chlorophenyl(phenyl)carbinol.

With telluronium salt (b) (general procedure) [12]. A solution of n-BuLi (1.2 mmol) in hexane (0.6 ml) is added to a solution of the telluronium salt (0.494 g, 1.2 mmol) in dry THF (4 ml) at −70°C under N_2. After 30 min, a solution of a carbonyl compound (1.0 mmol) in THF (2 ml) is added. The mixture is then allowed to warm at room temperature. After the reaction is complete (monitored by TLC), the usual work-up and flash chromatography yield pure benzylcarbinol.

With telluronium salt (c) (general procedure) [22]. A solution of n-BuLi (1.5 mmol) in hexane is added to a solution of the telluronium salt (0.475 g 1.5 mmol) in dry THF at −70°C under N_2. After 1 h, a solution of the carbonyl compound (1.0 mmol) in THF (2 ml) is added. After the reaction is complete (monitored by TLC), AcOEt is added. The resulting mixture is filtered through a short SiO_2 column. The solvent is evaporated under vacuum, and flash chromatography gives pure cyanomethylcarbinol.

With telluronium salt (d) (typical procedure) [19]. A solution of n-BuLi (1.5 mmol) in hexane is added to a solution of the telluronium salt (0.65 g, 1.5 mmol) in dry THF (10 ml) at −78°C under N_2. After 30 min, a solution of *p*-chlorobenzaldehyde (0.168 g, 1.2 mmol) in THF (2 ml) is added dropwise at −78°C and the reaction mixture is allowed to warm at room temperature. After the reaction is complete (monitored by TLC), H_2O (1 ml) is added and the solution is stirred for another 1 h. The mixture is then extracted with ether (3 × 5 ml). The combined extracts are washed with brine, dried (Na_2SO_4), filtered and concentrated under vacuum. Flash chromatography on an SiO_2 column gives 1-(*p*-chlorophenyl)-4-(trimethylsilyl)-3-butyn-1-ol (0.265 g (87%) (98% GC purity)).

4.8.4 Tellurium-catalysed decomposition of α-lithiated benzylic sulphones into 1,2-diarylethylenes

α-Lithiated benzyl sulphones slowly decompose to give 1,2-diarylethylenes. In the presence of catalytic amounts of elemental tellurium this reaction acquires synthetic utility since the decomposition time is decreased to 3 h from 2–10 days [25]. In contrast to the uncatalysed reaction which furnishes pure *trans*-olefin, in the catalytic process substantial amounts of *cis* isomers (15–35%) are formed. However, *cis*/*trans* isomerization is promoted by tellurium tetrachloride in chloroform under reflux, and pure *trans*-stilbenes can, therefore, be obtained.

Sulphur and selenium are less effective as catalysts.

Since the required starting sulphones are easily prepared by the reaction of sodium benzene sulphinate with benzyl bromides, the described method comprises the formal dimerization of a benzylic fragment.

The reaction has been rationalized as involving an attack of lithium tellurolate on a lithiated sulphone, promoting the elimination of lithium phenyl sulphinate and the formation of a labile epitelluride that readily collapses into stilbene and elemental tellurium.

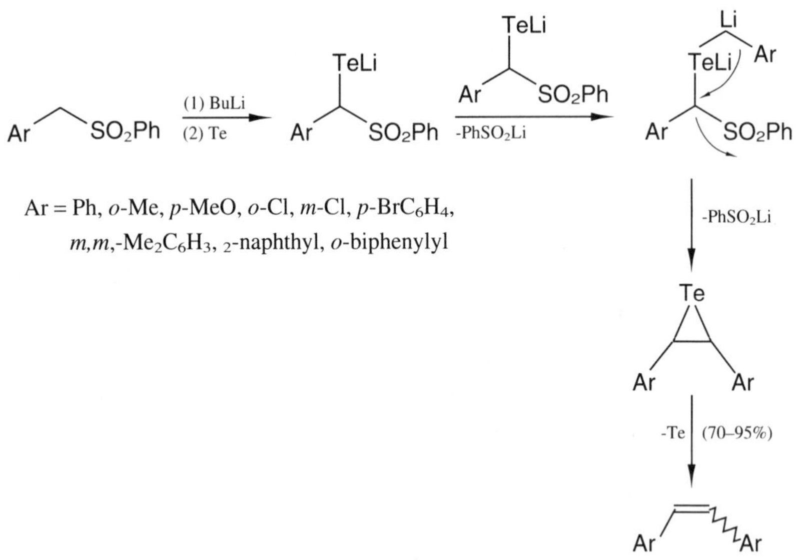

Ar = Ph, *o*-Me, *p*-MeO, *o*-Cl, *m*-Cl, *p*-BrC_6H_4, *m*,*m*,-$Me_2C_6H_3$, 2-naphthyl, *o*-biphenylyl

Tellurium-catalysed conversion of aryl phenyl sulphones into stilbenes (typical procedure) [25] To a solution of benzyl phenyl sulphone (1.0 g, 4.3 mmol) in dry THF (50 ml) is added n-BuLi (1.6 M, 2.7 ml, 4.3 mmol) at 0°C with stirring. Finely ground elemental Te (0.10 g, 0.78 mmol) is then added to the orange-yellow solution, and stirring continued at room temperature for 3 h. During this period the colour of the solution slowly fades. Protonation of any resulting anion is effected by the addition of aqueous HOAc (50%, 2 ml). After evaporation of the solvent, the product is extracted with CH_2Cl_2 and washed once with 5% aqueous Na_2CO_3. The extract is dried, evaporated and chromatographed on SiO_2 (eluting with CH_2Cl_2/petroleum ether 40–60°C 3:1), giving stilbene (0.35 g (90%)) as a 1:4 *cis/trans* mixture (determined by integration of ^1H NMR spectrum).

The product is heated under reflux in $CHCl_3$ (15 ml) containing $TeCl_4$ (0.10 g, 0.37 mmol) for 1 h. The cooled solution is then shaken with 5% aqueous Na_2CO_3 in a separating funnel. The organic layer is dried and evaporated, giving pure *trans*-stilbene (0.33 g (85%); m.p. 124°C).

REFERENCES

1. K. B. Sharpless, K. M. Gordon, R. F. Lauer, D. W. Patrick, S. P. Singer and M. W. Young, *Chem. Soc.*, 8A, 9 (1975).
2. H. Lee and M. P. Cava, *J. Chem. Soc. Chem. Commun.*, 277 (1981).
3. S. Uemura and S. I. Fukuzawa, *J. Am. Chem. Soc.* **105**, 2748 (1983).
4. S. Uemura, K. Ohe and S. I. Fukuzawa, *Tetrahedron Lett.* **26**, 895 (1985).
5. S. Uemura, Y. Hirai, K. Ohe and N. Sugita, *J. Chem. Soc. Chem. Commun.*, 1037 (1985).
6. T. Otsuba, F. Ogura and H. Yamaguchi, *Chem. Lett.*, 447 (1981).
7. E. W. Colvin, *Silicon Reagents in Organic Synthesis*, p. 7. Academic Press, London, 1988.
8. F. Ogura, T. Otsubo and N. Ohira, *Synthesis*, 1006 (1983).
9. A. Osuka, Y. Mori, H. Shimizu and H. Suzuki, *Tetrahedron Lett.* **24**, 2599 (1983).
10. X. Huang, L. Xie and H. Wu, *J. Org. Chem.* **53**, 4862 (1988).
11. A. Osuka, Y. Hanasaki and H. Suzuki, *Nippon Kagaku Kaishi*, 1505 (1987).
12. S. W. Li, Z. L. Zhou, Y. Z. Huang and L. L. Shi, *J. Chem. Soc. Perkin Trans. I*, 1099 (1991).
13. Z. Z. Huang, L. W. Wen and X. Huang, *Synth. Commun.* **20**, 2579 (1990).
14. X. Huang, L. Xie and H. Wu, *Tetrahedron Lett.* **28**, 801 (1987).
15. Y. Z. Huang, L. L. Shi, S. W. Li and X. O. Wen, *J. Chem. Soc. Perkin Trans. I*, 2397 (1989).
16. A. Osuka and H. Suzuki, *Tetrahedron Lett.* **24**, 5109 (1983).
17. L. L. Shi, Z. L. Zhou and Y. Z. Huang, *Tetrahedron Lett.* **31**, 4173 (1990).
18. Z. L. Zhou, Y. Z. Huang and L. L. Shi, *J. Chem. Soc. Chem. Commun.*, 986 (1992).
19. Z. L. Zhou, Y. Z. Huang, L. L. Shi and J. Hu, *J. Org. Chem.* **57**, 6598 (1992).
20. E. W. Colvin, *Silicon Reagents in Organic Synthesis*. Academic Press, London, 1988.
21. L. L. Shi, Z. L. Zhou and Y. Z. Huang, *J. Chem. Soc. Perkin Trans. I*, 2847 (1990).
22. Z. L. Zhou, L. L. Shi and Y. Z. Huang, *J. Chem. Soc. Perkin Trans. I*, 1931 (1991).
23. *Houben–Weyl—Methoden der Organischen Chemie*, 4th edn, Vol. 8, pp. 427 and 661. Georg Thieme, Stuttgart, 1952.
24. S. Patai (ed.), *The Chemistry of the Carbon–Carbon Triple Bond*. Wiley, New York, 1978; L. Brandsma and H. D. Verkruijsse, *Synthesis of Acetylenes, Allenes and Cumulenes*. Elsevier, Amsterdam, 1981.
25. L. Engman, *J. Org. Chem.* **49**, 3559 (1984).

4.9 TRANSMETALLATION REACTIONS OF ORGANOTELLURIUM COMPOUNDS

4.9.1 Lithium–tellurium exchange: generation of organolithium reagents

4.9.1.1 Alkyl-, aryl-, allyl-, benzyl-, vinyl- and ethynyllithium compounds

By treatment of diorganyl tellurides with alkyllithiums in THF at −78°C, a lithium–tellurium exchange occurs, generating the more stable organolithium compounds. By this procedure alkyl-, aryl-, vinyl- and ethynyllithiums, and even allyl- and benzyllithiums (barely accessible by conventional lithium–halogen exchange, due to the interference of Wurtz-type coupling), can easily be obtained and trapped *in situ* with selected electrophiles [1–3].

Depending on whether 1 or 2 eq of the starting organolithium reagent is used, only one or both organic groups, respectively, are displaced from the telluride (methods A and B).

Moreover, since organic tellurides are easily accessible from lithium tellurolates and organic halides, a one-pot procedure can be carried out, overcoming the isolation of the telluride (method C).

$$\mathrm{RTeR^1 + R^2Li \xrightarrow[\text{-}R^1TeR^2]{\text{THF, }-78^\circ C} RLi \xrightarrow{E} RE} \quad \text{(method A)}$$

R^2 = n-Bu

$$\mathrm{RTeR + 2R^1Li \xrightarrow[\text{-}R^1TeR^1]{\text{THF, }-78^\circ C} 2RLi \xrightarrow{E} 2RE} \quad \text{(method B)}$$

$$\mathrm{RX + R^1TeLi \xrightarrow[\text{-}LiX]{\text{THF, }-78^\circ C} RTeR^1 \xrightarrow[\text{-}R^1TeR^1]{R^1Li} RLi \xrightarrow{E} RE} \quad \text{(method C)}$$

Transmetallations of vinylic tellurides deserve particular attention. These tellurides (prepared by *anti* addition of tellurols to acetylenes, see Section 3.11) exhibit the (*Z*)-configuration and therefore generate (*Z*)-vinyllithiums. These results are in sharp contrast to the earlier tin–lithium exchange performed with vinylstannanes (characterized by the (*E*)-configuration), giving (*E*)-vinyllithiums [4].

Lithium–tellurium exchange: generation of vinyllithium (typical procedure) *Method A* [2]. To a solution of butyl vinyl telluride (0.211 g, 1 mmol) in THF (4 ml) at −78°C under N_2 is added dropwise a solution of n-BuLi (1.5 M in hexane, 0.67 ml, 1 mmol). After stirring for 40 min at −78°C, benzaldehyde (0.016 g, 1 mmol) is added. The mixture is allowed to react at room temperature for 30 min, diluted with EtOAc (40 ml), and washed with brine. After normal work-up of the organic layer, the crude product is chromatographed on SiO_2. Elution with hexane removes dibutyl telluride, and elution with EtOAc removes the product (PhCH(OH)Bu-n) (0.115 g (70%)).

Method C [1]. To a solution of lithium phenyl tellurolate (prepared *in situ* from Te (2 mmol) and PhLi (2.0 M, 1 ml, 2 mmol)) in THF (5 ml) is added benzyl bromide (0.342 g, 2 mmol) at –78°C. The mixture is warmed at 25°C and stirred for 30 min. At this stage the conversion to phenyl benzyl telluride is complete. The subsequent Li–Te exchange and reaction with an electrophile are the same as described for method A.

It has been reported that sodium–, potassium–, calcium– and magnesium–tellurium exchanges are achieved by similar procedures with alkyl, aryl, ethynyl, vinyl, allyl and benzyl tellurides [5].

$$RTeR^1 + R^2Li \xrightarrow[-R^1TeR^2]{THF,\ -78^\circ C} RLi \xrightarrow{E} RE \text{ (method A)}$$

R^2 = n –Bu

Telluride	E	Product	Percentage Yield%
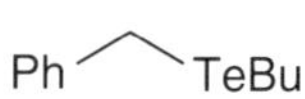	PhCHO	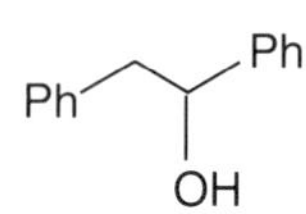	74
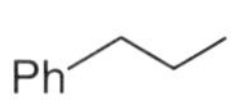	PhCHO	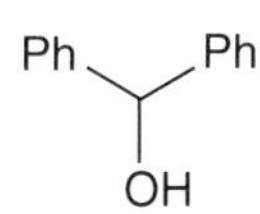	77
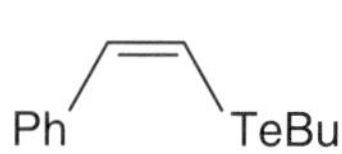	n-C_8H_{17}Br	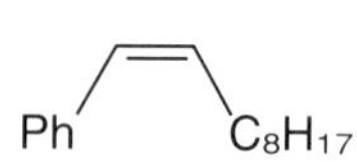	70
	PhCHO	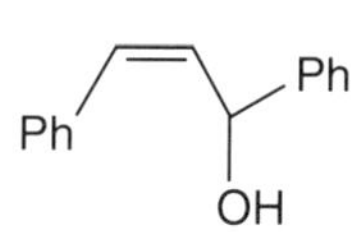	89
	cyclohexanone	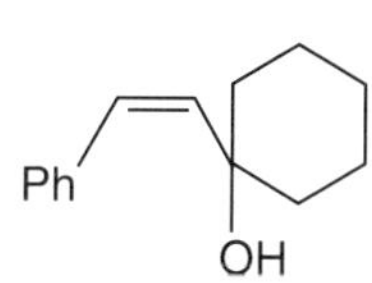	79

Telluride	E	Product	Percentage Yield%
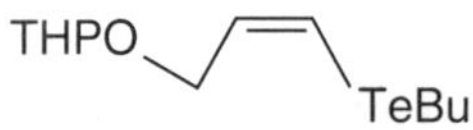	PhCHO	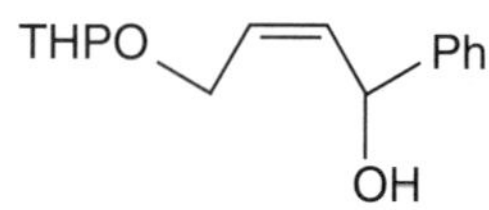	72
Ph—≡—TeBu	PhCHO	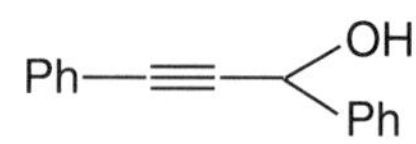	89
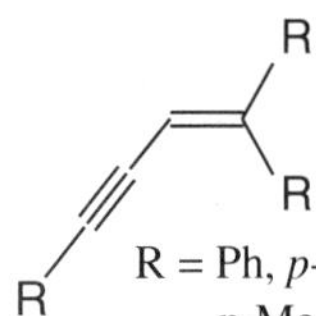 R = Ph, *p*-Me, *p*-$MeOC_6H_4$	H^+	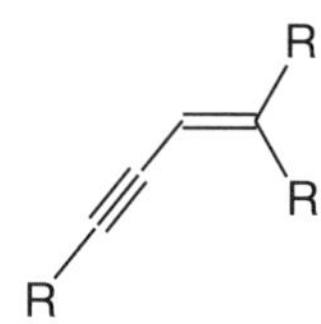	70–90
R = Ph	Me_2SO_4	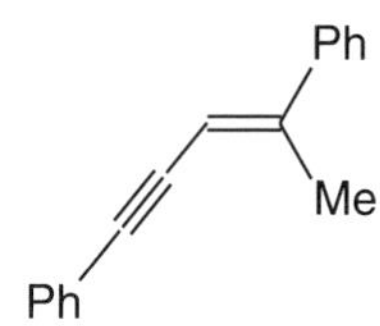	93
R = Me	PhCHO	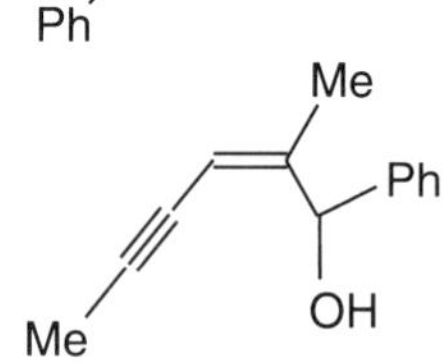	63
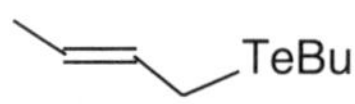	PhCHO	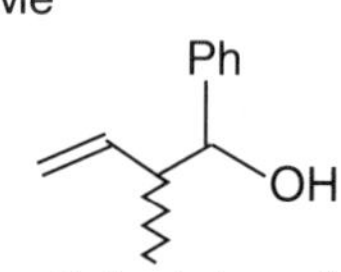 (1:1 mixture pf diastereoisomer)	73
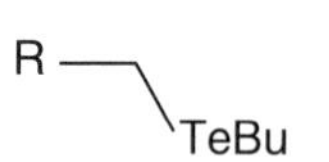 R = *p*-Me, p-BrC_6H_4[a]	n-C_3H_7CHO	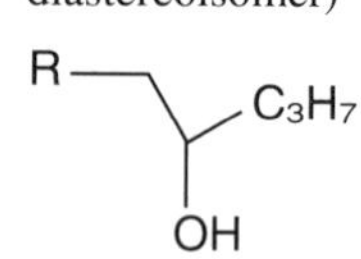	88, 68
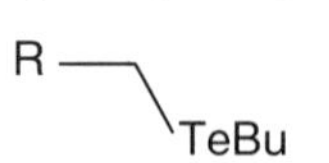 R = *o*-I-[a, b], m-NCC_6H_4[a, b]	pinacolone	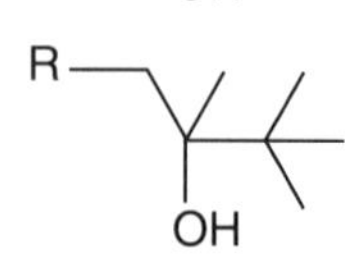	65, 96

a Solvent: ether.

b n-BuLi added to the telluride in the presence of pinacolone at −105°C.

$$\mathrm{RTeR + RR^1Li \xrightarrow{THF,\ -78^\circ C} 2RLi \xrightarrow{E} 2RE} \quad \text{(method B)}$$

Telluride	R[1]	E	Product	Percentage yield%
Ph_2Te	n-Bu, t-Bu	PhCHO	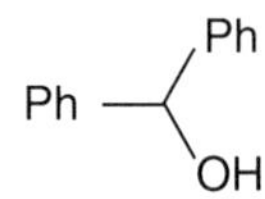	80, 72
Ph_2Te	s-Bu	PhCHO	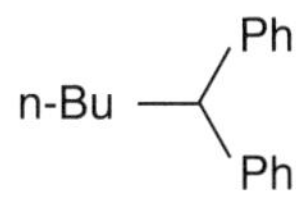	86
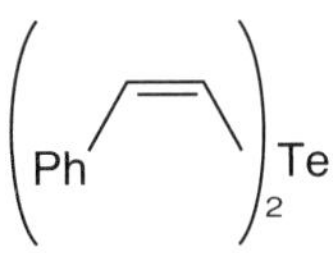	n-Bu	$n\text{-}C_8H_{17}Br$	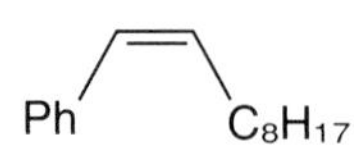	70
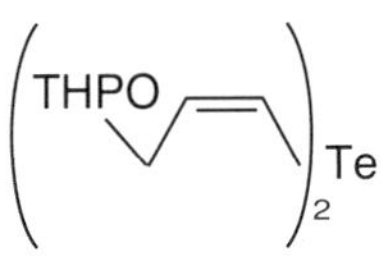	n-Bu	PhCHO	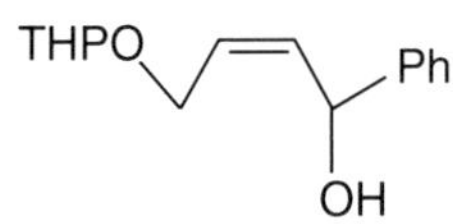	70
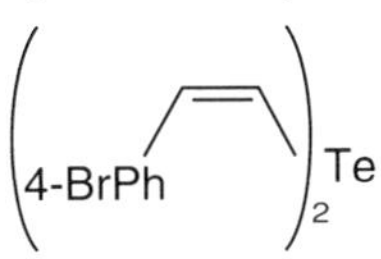	n-Bu	PhCHO	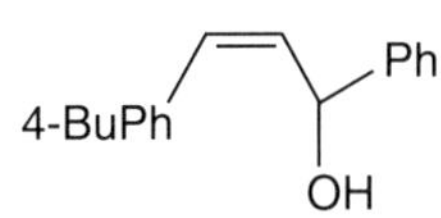	66
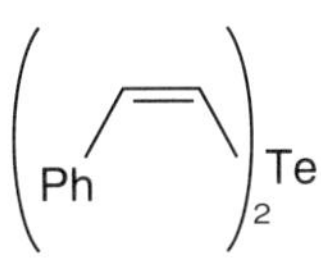	n-Bu	cyclohexanone	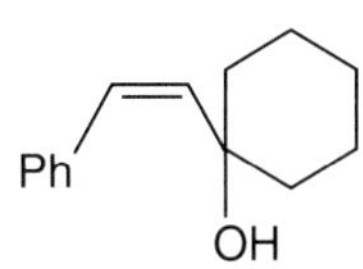	82

$$\mathrm{RX + R^1TeLi \xrightarrow{THF,\ -78^\circ C} RTeR^1 \xrightarrow[-R^1TeR^1]{R^1Li} RLi \xrightarrow{E} RE} \quad \text{(method C)}$$

RX	R[1]	E	Product	Percentage yield%
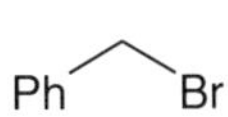	Ph	PhCHO	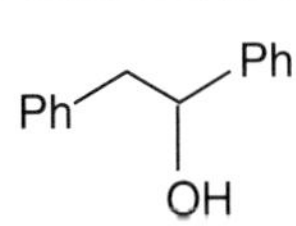	72

RX	R^1	E	Product	Percentage yield%
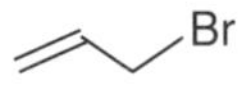	Ph	PhCHO	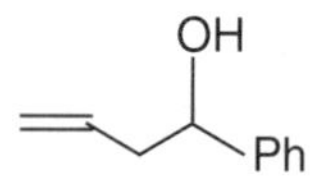	96
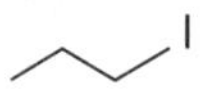	s-Bu	PhCHO	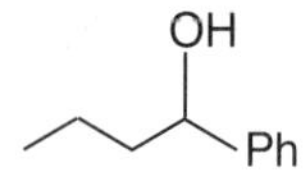	84

4.9.1.2 Acyl- and aroyllithium compounds

Acyl- and aroyllithiums are generated by submitting telluroesters, easily accessible from the corresponding acid chlorides and tellurolate anions [6], to the above-described lithium–tellurium exchange reactions [7].

The exchange step and the capture of the formed lithium intermediate with the electrophile take place in a high yield with aromatic telluroesters (employing n-BuLi in THF/ether at −105°C) and with aliphatic telluroesters bearing no α-hydrogen atoms (employing t-BuLi in THF at −78°C).

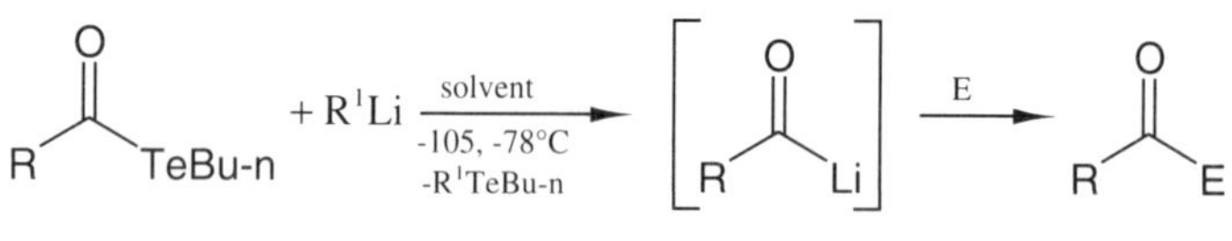

R	R^1Li (eq)	solvent (temperature)	E	Product	Percentage yield%
Ph, p-$F_3CC_6H_4$ o,o-$F_2C_6H_3$	n-BuLi (1.2)	THF/ether (-105°C)	pinacolone	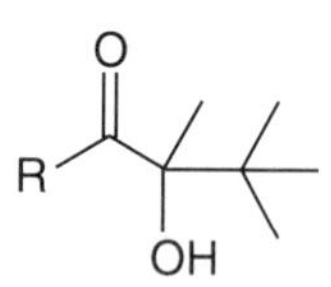	60–85
t-Bu	t-BuLi (1.6)	THF (-78°C)	pinacolone	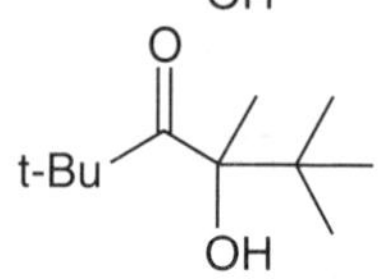	66
			t-BuCHO	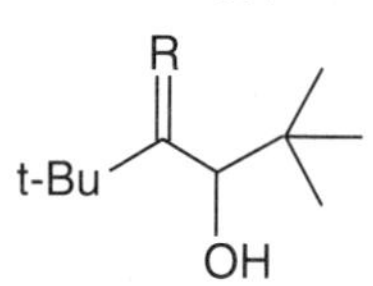	25

R	R^1Li (eq)	solvent (temperature)	E	Product	Percentage yield%
				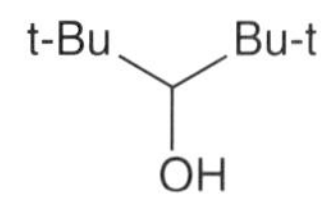	40
1-adamantyl	n-BuLi (1.6)	THF (-78°C)	pinacolone	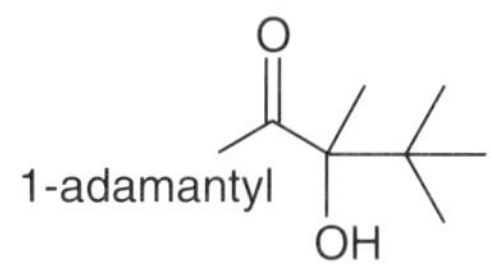	57
t-Bu	n-Bu (1.2)	THF/ether/ HMPA	Me_3SiCl	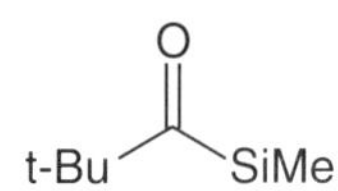	70

It must be pointed out that the corresponding selenoesters and acylstannanes are inadequate for similar reactions.

4.9.1.3 Heterosubstituted methyllithium compounds

Lithium–tellurium exchanges can be put to further use, allowing the generation of a variety of heteroatom-substituted methyllithiums suitable for trapping with electrophiles [8].

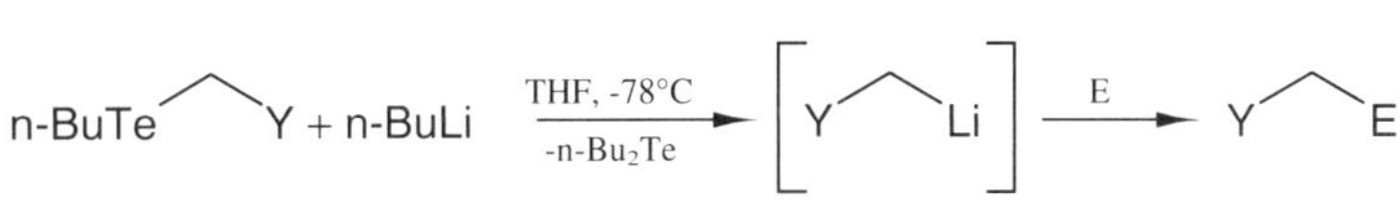

Run	Y	E	Product	Percentage yield%
1	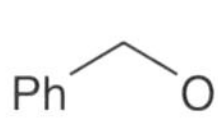	PhCHO	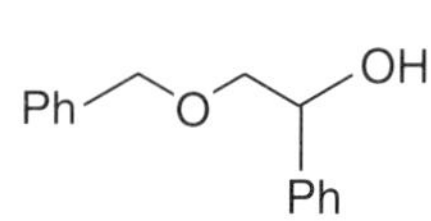	66
2	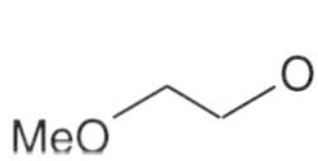	PhCHO	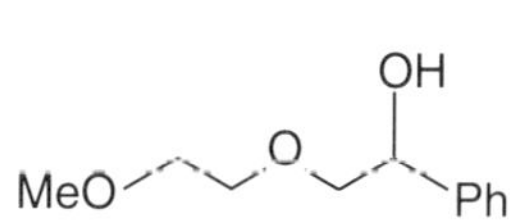	70

Run	Y	E	Product	Percentage yield%
3		Ph_2CHO	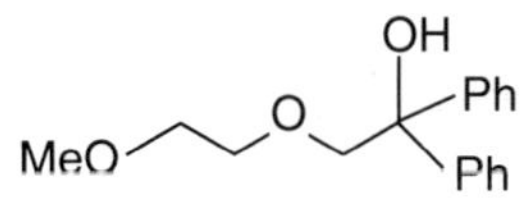	60
4		PhCOCl	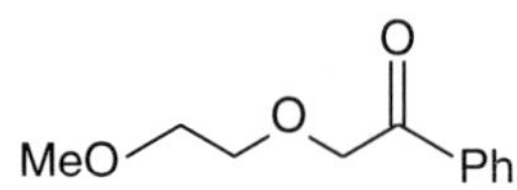	40
5		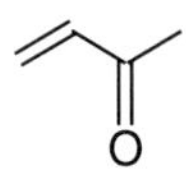	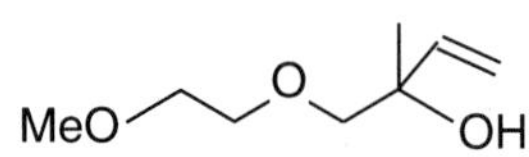	51
6	Me_2N	PhCHO	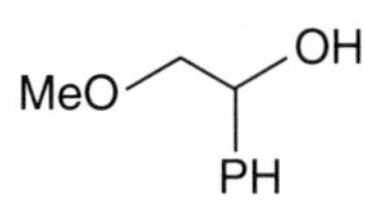	73
7	Me_3Si	PhCHO	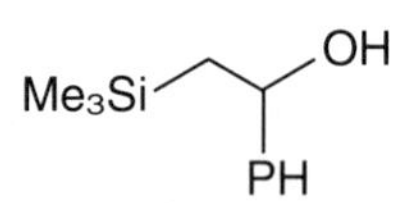	89
8	BuTe	PhCHO	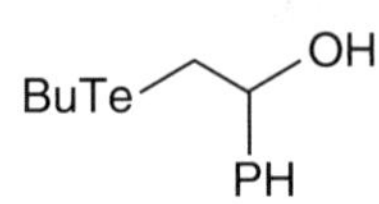	81[a-b]
9		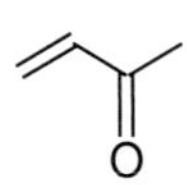	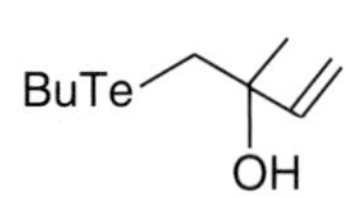	65[a-b]
10		Me_3SiCl		98[a](NMR yield)
11	Bu_3Sn	Me_3SiCl	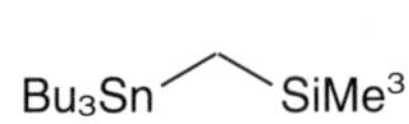	85[b,c](NMR yield)
12	PhSe	H^+	PhSeMe	90

a: 1 eq LiCl is added.

b: Exchange reaction at –95°C.

c: Solvent: ether.

The reaction can also be carried out satisfactorily in different solvents such as ether, toluene and hexane (in contrast to the previous reported lithium–tin exchange, which is sluggish in solvents of low polarity [9]).

The following important remarks should be noted:

- Only 1,2 adducts are obtained with methyl vinyl ketone (runs 5 and 9).
- Tin and selenium derivatives (runs 11 and 12) undergo lithium–tellurium exchange preferentially to lithium–tin and lithium–seleniuim [10] exchange.
- The formation of $BuTeCH_2Li$ from $BuTeCH_2TeBu$ is improved markedly by the addition of LiCl (runs 8–10). Therefore, it can be generated *in situ* from CH_2Cl_2 and BuTeLi due to the concomitant formation of LiCl.
- Two different electrophiles can be introduced successively by a consecutive one-pot lithium–tellurium exchange/trapping sequence.

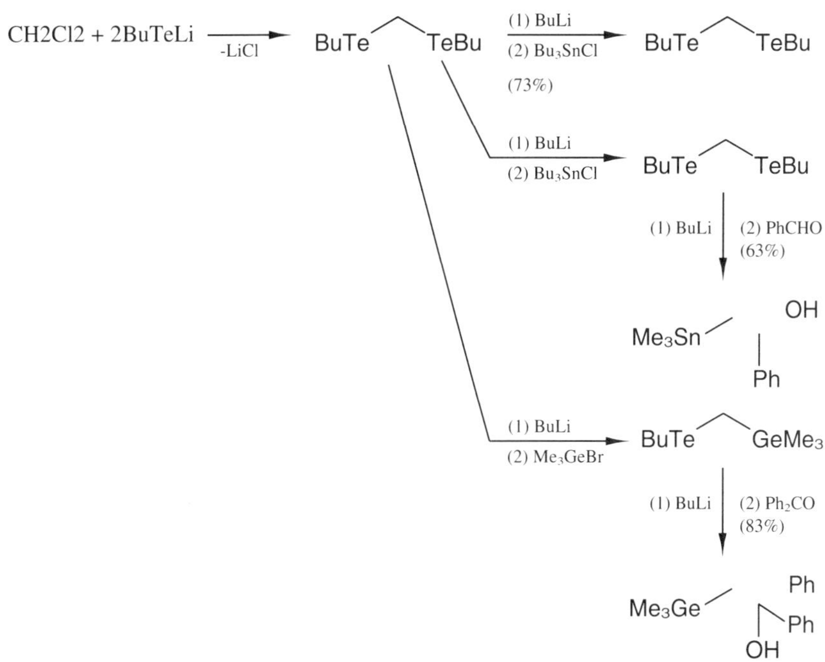

Heteroatom-substituted methyllithiums (typical procedure) [8] To a solution of $MeO(CH_2)_2OCH_2TeBu$-n (0.274 g, 1.0 mmol) in THF (5 ml) is added n-BuLi (1.6 M in hexane, 0.63 ml, 1.0 mmol) at −78°C. After 15 min, benzaldehyde (0.117 g, 1.0 mmol) is added in one portion. The reaction mixture is warmed at 25°C over a period of 1 h, and then poured into saturated aqueous NH_4Cl (30 ml). The product is extracted with ether (3 × 30 ml) and dried ($MgSO_4$). Radial layer chromatography on SiO_2 gives $MeO(CH_2)_2OCH_2CH(OH)Ph$ (0.137 g (70%)) together with dibutyl telluride (0.183g (76%)).

4.9.2 Vinyl cuprates by copper–tellurium exchange

Vinyl cuprates are valuable reagents for the conjugated addition of vinylic groups to α,β-unsaturated carbonyl compounds, a very important reaction in organic synthesis. Vinyl cuprates can be generated by copper–tellurium exchange at vinylic tellurides by means of higher-order organic cuprates, and effect the above-mentioned addition with retention of the olefin geometry [11].

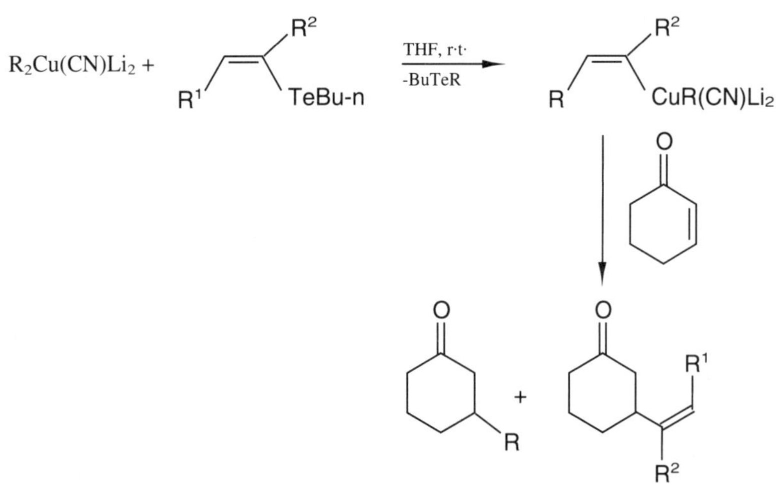

R = Me, n-Bu	R^1 = Ph; R^2 = H	trace	88–90%
R = Me	R^1 = H; R^2 = Ph	23%	75%
R = s-Bu	R^1 = H; R^2 = Ph	trace	89%

As can be seen from the results, no appreciable difference is observed on changing the methyl group for the n-butyl or s-butyl group in the original organocuprate, since little transference of these groups to the β-carbon occurs in competition with the vinylic transference.

If, however, a phenyl group replaces the butyl group in the telluride, the transference of the phenyl group becomes important. Therefore, phenyl vinyl tellurides are inadequate in these additions.

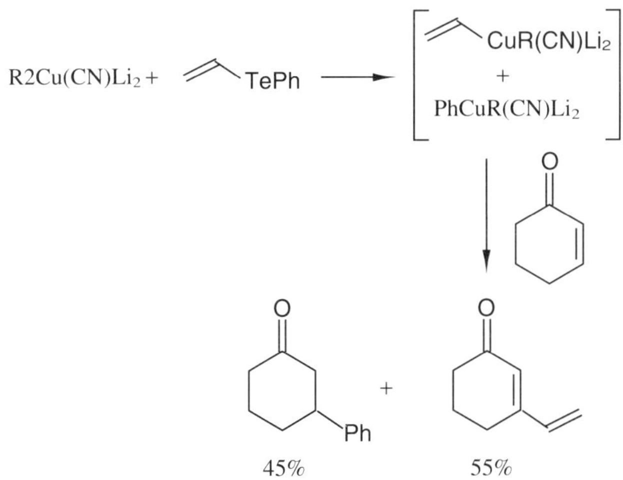

Bisvinylic tellurides are very useful reagents for transmetallation reactions since only 1 eq is required to generate 2 eq of the mixed vinylic organocuprate [12].

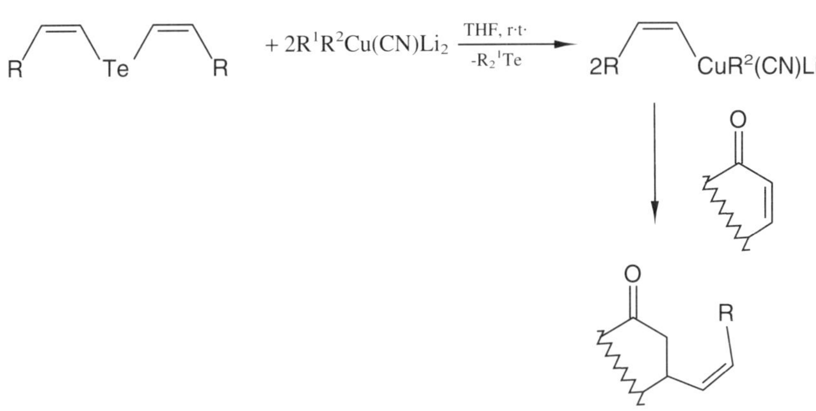

This reaction is sensitive to steric factors, as shown by the low yields of 1,4-adducts obtained with β,β-disubstituted enones (mesityl oxide).

An especially efficient cyanocuprate reagent is the thienyl derivative n-Bu(2-thienyl)CuLi$_2$, since the presence of the non-transferable 2-thienyl group avoids the formation of by-products, resulting from the transference of the butyl group to thc cnonc.

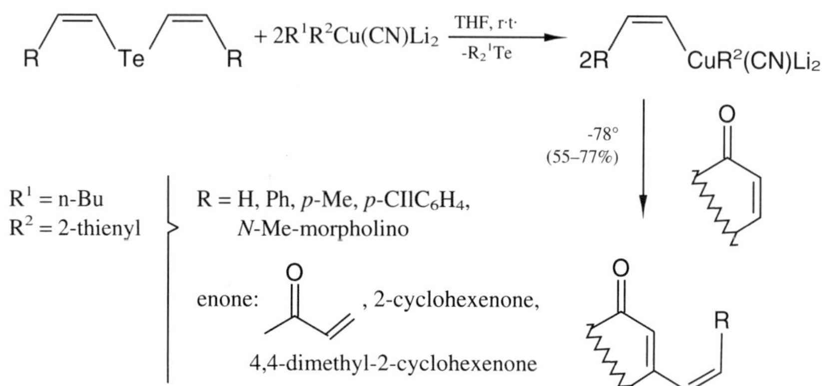

The reagent $M_2Cu(CN)Li_2$ is also successful in several cases.

The non-transferability of the 2-thienyl group makes vinyl-2-thienyl tellurides (see Section 3.11) advantageous over the corresponding phenyl derivatives, which are susceptible to the previously discussed phenyl transference.

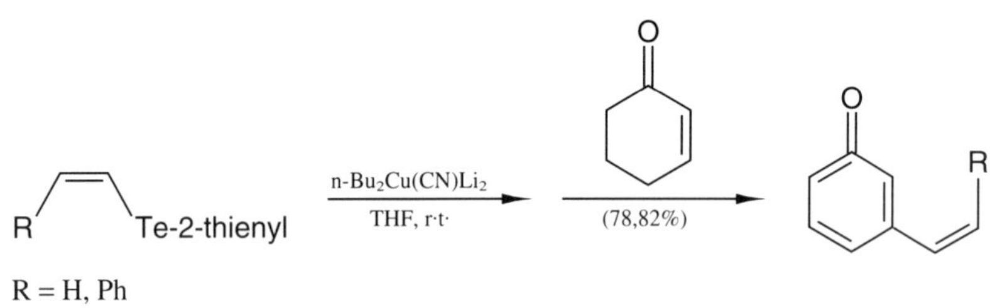

Tellurodienes and telluroenines (see Section 3.11.2.4) are also suitable for transmetallation reactions.

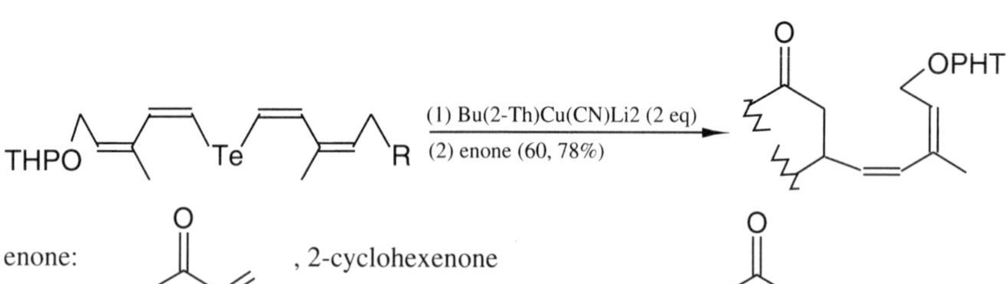

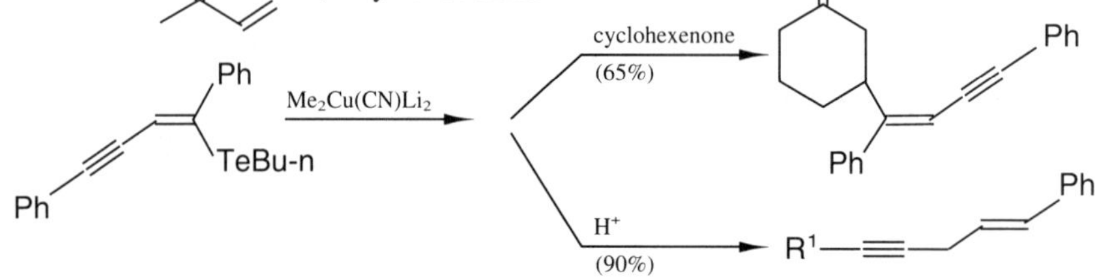

Cu–Te exchange with divinyl telluride and addition to enones (general procedure) [12] n-BuLi (1.35 ml; 2.5 mmol of a 1.84 M solution in hexane) is added to thiophene (0.21 g; 2.5 mmol) in THF (2 ml) in a 10 ml two-necked flask under N_2 at −70°C. The temperature is allowed to reach 0°C, while stirring is maintained for 30 min. The yellowish solution is then transferred via a cannula into a 25 ml, two-necked flask containing CuCN (0.18 g; 2.0 mmol of a commercial (Aldrich) sample, dried (P_2O_5) under vacuum in an Abderhalden apparatus) and THF (3 ml), previously purged with N_2 and cooled to −70°C. Warming to 0°C produces a homogeneous solution which is recooled to −70°C. n-BuLi (1.1 ml; 2.0 mmol of a 1.84 M solution in hexane) is added. Stirring is continued for 15 min and then the mixture is allowed to reach room temperature. The bisvinyl telluride (1.1 mmol) in THF (2 ml) is added, the solution stirred for 1 h, and then cooled to −70°C, and the enone (2.0 mmol) added rapidly via a syringe. The cooling bath is removed and the solution is stirred for 20 min at room temperature. The reaction is quenched with a 4:1 mixture of saturated aqueous NH_4Cl and NH_4OH, and extracted with EtOAc (30 ml). The organic layer is separated, washed with brine (2 × 50 ml) and dried ($MgSO_4$). The solvent is evaporated and the residue chromatographed on SiO_2, eluting first with hexane and then with hexane/EtOAc (5:1), giving the desired 1,4 addition product.

Cu–Te exchange with vinyl 2-thienyl telluride and addition to enones (general procedure) [12] To a stirred suspension of CuCN (0.089 g, 1.0 mmol of a commercial (Aldrich) sample dried (P_2O_5) under vacuum in an Abderhalden apparatus) in THF (3 ml), under N_2 cooled to −70°C, is added n-BuLi (1.1 ml, 2.0 mmol of a 1.84 M solution in hexane). The resulting homogeneous solution is stirred for 30 min at −70°C and then allowed to reach room temperature. A solution of vinyl thienyl telluride (1.0 mmol) in THF (2 ml) is added. After 1 h the reaction mixture is cooled to −70°C, the neat enone (1.0 mmol) is added rapidly and the mixture stirred for 20 min at room temperature. The reaction is quenched with a 4:1 mixture of saturated aqueous NH_4Cl and NH_4OH and extracted with EtOAc (15 ml). The organic layer is separated, washed with brine (2 × 25 ml) and dried ($MgSO_4$). The solvent is evaporated and the residue purified by SiO_2 chromatography, eluting first with hexane and then with hexane/EtOAC (5:1), giving the 1,4 addition product.

REFERENCES

1. T. Hiiro, N. Kambe, A. Ogawa, N. Miyoshi, S. Murai and N. Sonoda, *Angew. Chem.* **26**, 1187 (1987) (in English).
2. S. M. Barros, J. V. Comasseto and J. N. Berriel, *Tetrahedron Lett.* **30**, 7353 (1989).
3. M. J. Dabdoub, V. B. Dabdoub and J. V. Comasseto, *Tetrahedron Lett.* **33**, 2261 (1992).

4. See M. Pereyre, J. P. Quintard and A. Rahm, *Tin in Organic Synthesis*, pp. 153, 23. Butterworth, London, 1987.
5. T. Kanda, T. Sugino and N. Sonoda, *Phosphorus Sulfur Silicon* **67**, 103 (1992).
6. J. Bergman and L. Engman, *Z. Naturforsch.* **35B**, 217 (1980); S. L. Piette and M. Renson, *Bull. Soc. Chem. Belg.* **79**, 383 (1970); S. L. Piette, D. Debergh, M. Baiwir and G. Llabses, *Spectrochim. Acta* **36A**, 769 (1980); G. A. Gardner and H. J. Gysling, *J. Organomet. Chem.* **197**, 111 (1980); T. Kanda, S. Nakaiida, T. Murai and S. Kato, *Tetrahedron Lett.* **30**, 1829 (1989).
7. T. Hiiro, Y. Morita, T. Inoue, N. Kambe, A. Ogawa, I. Ryn and N. Sonoda, *J. Am. Chem. Soc.* **112**, 455 (1990).
8. T. Hiiro, Y. Atarashi, N. Kambe, S. I. Fujiwara, A. Ogawa, I. Ryn and N. Sonoda, *Organometallics* **9**, 1355 (1990).
9. D. Seyferth and M. A. Weiner, *J. Am. Chem. Soc.* **84**, 361 (1962).
10. C. A. Brandt, J. V. Comasseto, W. Nakamura and N. Petragnani, *J. Chem. Res. (S)*, 156 (1983).
11. J. V. Comasseto and J. N. Berriel, *Synth. Commun.* **20**, 1681 (1990).
12. F. C. Tucci, A. Chieffi and J. V. Comasseto, *Tetrahedron Lett.* **33**, 5721 (1992).

4.10 ORGANYL TELLURIDES AS EXCHANGERS OF CARBON RADICALS

In the last few years organyl tellurides have attained an important role in free radical reactions, a subject in rapid development in modern organic synthetic chemistry.

It was discovered that the acetyl derivative of *N*-hydroxy-2-thiopyridone is especially suitable as a source of methyl radicals, being photolysed by irradiation with a simple tungsten lamp. On this basis, and in view of the high radicophilicity of tellurides (especially anisyl tellurides), a radical exchange occurs when the reagent is irradiated in the presence of an alkyl anisyl telluride, to afford methyl anisyl telluride and a new alkyl radical. In the presence of an electrophilic olefin, the alkyl radical is trapped, giving a relatively electrophilic radical, which in turn reacts with the thiocarbonyl function of the starting reagent to afford a tandem adduct, and reforming the methyl radical, thus beginning a new cycle [1, 2].

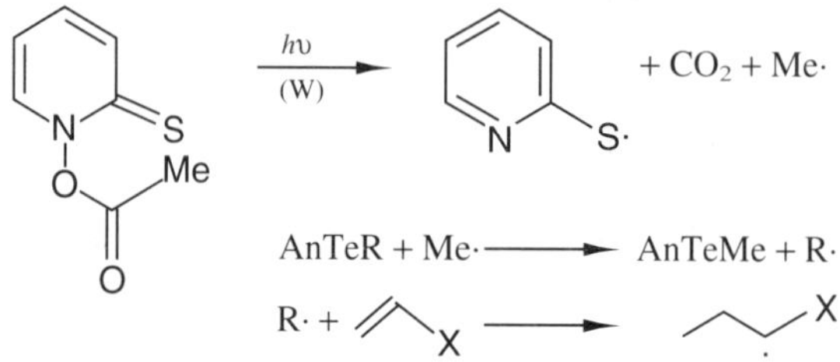

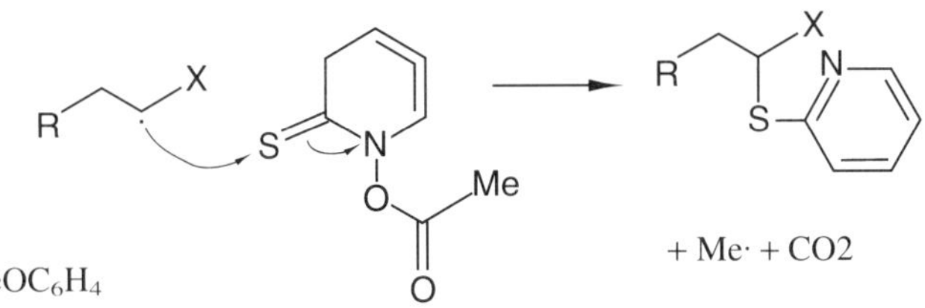

An = p-$MeOC_6H_4$

The above-discussed radical chemistry is very useful for the manipulation or synthesis of complex products.

4.10.1 Tellurium-mediated addition of carbohydrates to olefins

Carbohydrate anisyl tellurides are easily prepared by treatment of the corresponding mesylates or tosylates with the anisyl tellurolate anion. By irradiation of these tellurocarbohydrates in the presence of *N*-acetoxythiopyridone and the electrophilic olefin, the tandem adduct is formed. The oxidative elimination of the thiopyridine moiety leads to the *trans*-olefins [2].

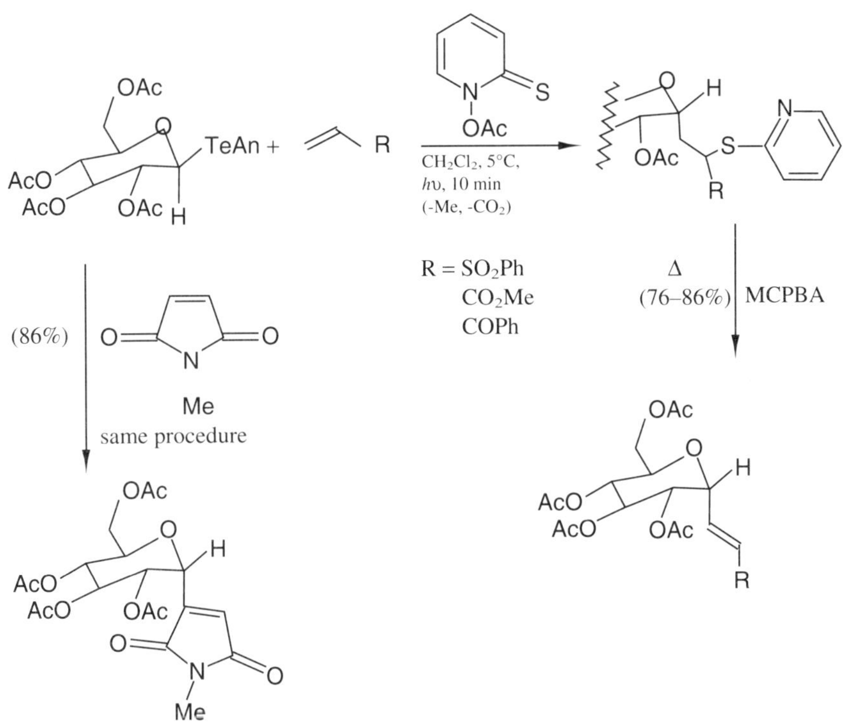

Radical addition (general procedure) [2] To the tellurocarbohydrate (1 mmol) and the appropriate olefin (5 mmol) in dry CH_2Cl_2 (4 ml) under argon at 5°C is added acetoxy-2-thiopyridone (0.5 mmol). Photolysis with a 150 W tungsten lamp for 10 min is followed by further addition of the reagent (0.25 mmol). This is repeated until all the carbohydrate disappears (TLC; about 1.5 mmol of the reagent is usually required).

The synthesis of the antibiotic showdomycin, starting from a ribose derivative, is illustrative [2].

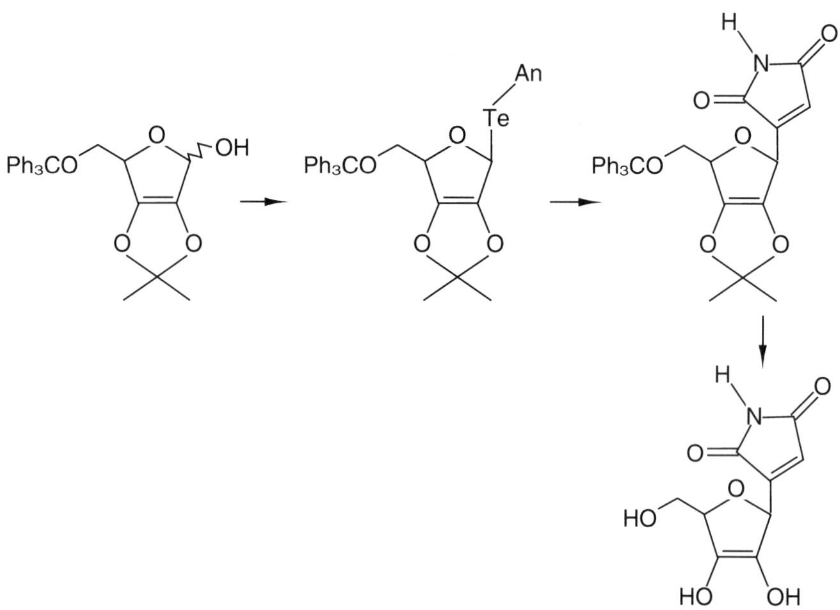

4.10.2 Intramolecular radical cyclization

4.10.2.1 Synthesis of six-membered carbocycles

The intramolecular version of the described radical reaction provides an easy approach to the synthesis of six-membered carbocycles [3].

Aryltellurium-mediated free radical cyclization (general procedure) [3] A mixture of *N*-acetoxy-2-thiopyridone (0.60 mmol, prepared by the reaction of equimolar amounts of *N*-hydroxythiopyridone sodium salt with acetyl chloride, with exclusion of light) and the telluride (0.2 mmol) in CH_2Cl_2 (2 ml) is stirred

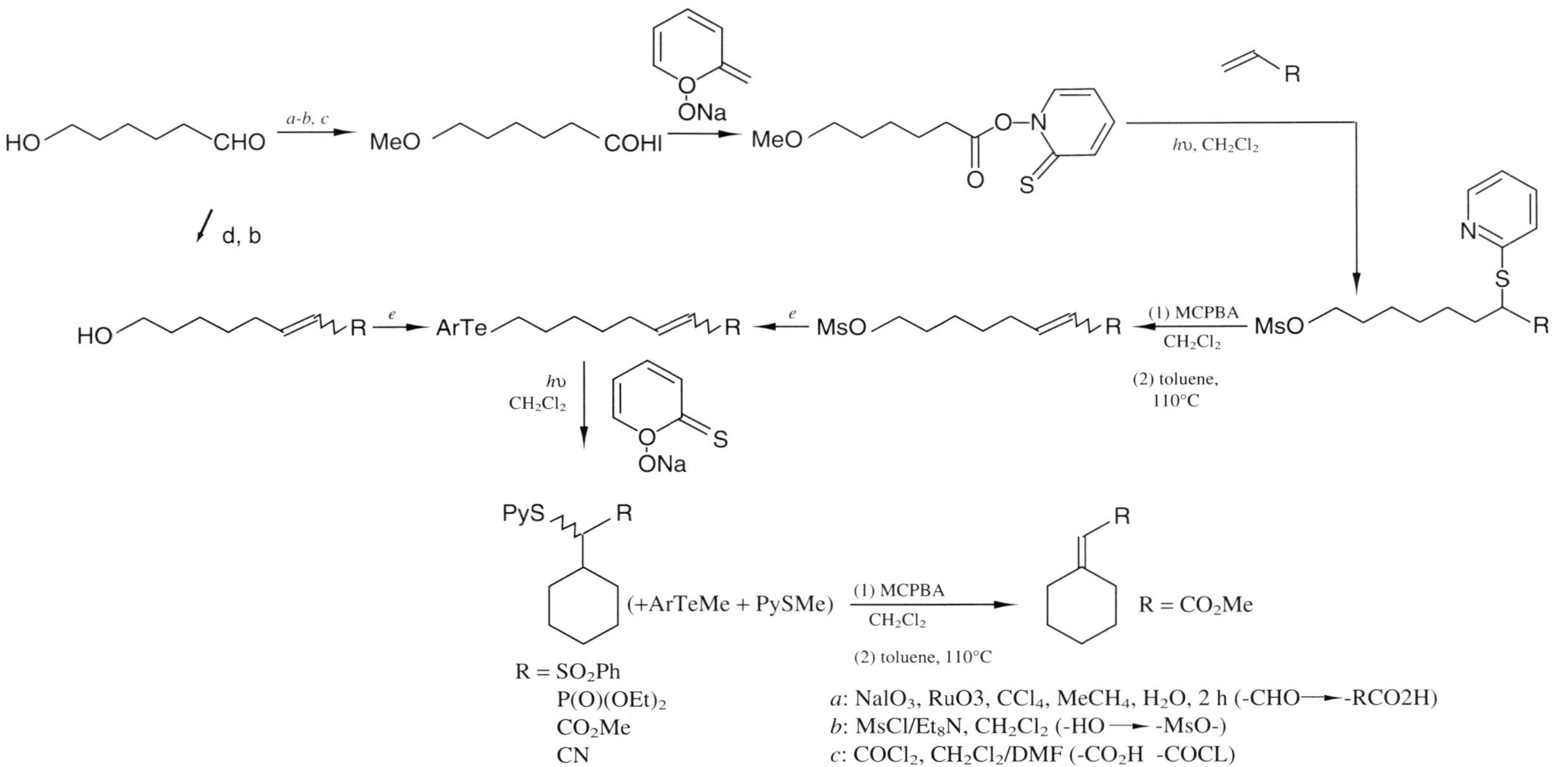

a: $NalO_3$, RuO3, CCl_4, $MeCH_4$, H_2O, 2 h (-CHO ⟶ -RCO2H)
b: $MsCl/Et_8N$, CH_2Cl_2 (-HO ⟶ -MsO-)
c: $COCl_2$, CH_2Cl_2/DMF (-CO_2H -COCL)
d: Ph_3P-CHR/toluana, 110°C, overnight (Wittig reaction)
e: $(Ar_2Te_2)NaBH_4$/EtOH Ar = Ph, *p*-$MeOC_6H_4$

vigorously under argon. The mixture is irradiated using a 250 W halogen lamp for 15 min at room temperature. The course of the reaction is monitored by TLC. If the reaction is not complete after the consumption of the 2-thiopyridone ester, a further quantity of this reagent is added, and the irradiation is continued. This procedure is repeated until all the starting material is converted. The solution is evaporated, and by chromatography of the residue on a short SiO_2 column the cyclized product is separated from methyl aryl telluride and the thiopyridyl derivative.

4.10.2.2 Synthesis of cyclonucleosides

The following synthesis of cyclo-5,6-dihydrouridine is self-explanatory [4].

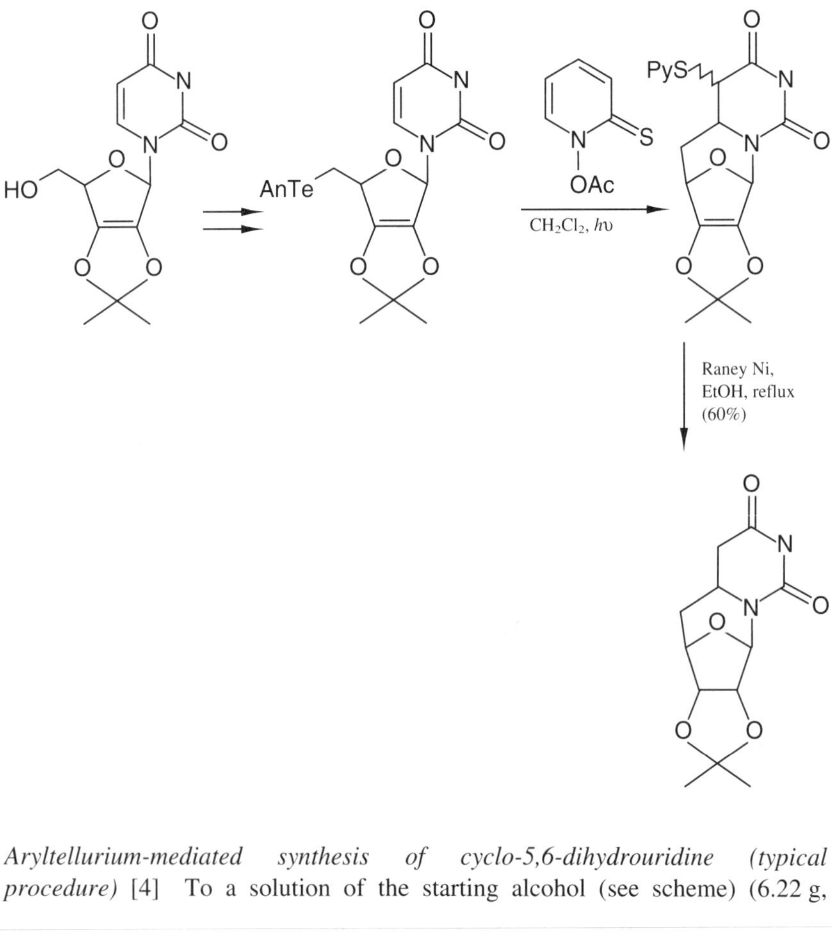

Aryltellurium-mediated synthesis of cyclo-5,6-dihydrouridine (typical procedure) [4] To a solution of the starting alcohol (see scheme) (6.22 g,

22 mmol) in pyridene (100 ml) is added dropwise mesityl chloride (2.55 ml, 33 mmol). The mixture is stirred overnight at room temperature, then H_2O is added and the solution evaporated to dryness. The residue is purified on an SiO_2 column (eluting with EtOAc/heptane 1:1) to yield the mesylate as crystals (6.34 g (80%)). The mesylate (1.81 g, 5 mmol) in dry and degassed THF (50 ml) is added slowly to sodium anisyltellurolate (from $(AnTe)_2$ (1.29 g, 0.55 eq) in degassed THF (2.5 ml) and $NaBH_4$ (0.380 g, 2 eq) in EtOH (8 ml)) under argon. The reaction mixture is stirred at room temperature for 6 h, and then H_2O (40 ml) is added. The solvent is evaporated to dryness. After extraction with EtOAc, the organic layer is dried ($MgSO_4$), filtered and evaporated. The residue is purified on an SiO_2 column (eluting with EtOAc/heptane 6:4) to yield the telluride as crystals (2 g (80%)). To a solution of the telluride (0.252 g, 0.5 mmol) in dry and degassed CH_2Cl_2 (5 ml) is added *N*-acetoxythiopyridone (0.338 g, 2 mmol) under argon. The reaction mixture is irradiated with a tungsten lamp (150 W) for 3 h and the solution is allowed to warm at 40°C. The reaction is purified on an SiO_2 column (eluting with ether/pentane 8:2) to yield the thiopyridyl cyclonucleotide as a white foam (0.14 g (75%)). This product (0.2 g, 0.53 mmol) is heated overnight under reflux in EtOH in the presence of Raney nickel. The mixture is filtered through Celite and evaporated. The residue is purified on an SiO_2 column (eluting with EtOAc/heptane 6:4) to yield the final product, cyclo-5,6-dihydrouridine (0.085 g (60%)).

REFERENCES

1. D. H. R. Barton, N. Ozbalik and J. C. Sarma, *Tetrahedron Lett.* **29**, 6581 (1988).
2. D. H. R. Barton and M. Ramesh, *J. Am. Chem. Soc.* **112**, 891 (1990).
3. D. H. R. Barton, P. J. Dalko and S. D. Gero, *Tetrahedron Lett.* **32**, 4713 (1991).
4. D. H. R. Barton, S. D. Gero, B. Q. Sire, M. Samadi and C. Vincent, *Tetrahedron* **47**, 9383 (1991).

4.11 MISCELLANEA

4.11.1 α-Hydroxyalkylation of α,β-unsaturated carbonyl compounds

Diisobutylaluminium phenyl tellurolate, a highly air- and moisture-sensitive reagent, prepared by reaction of diisobutylaluminium hydride with diphenyl ditelluride, undergoes an *in situ* addition to α,β-unsaturated carbonyl compounds, leading to the corresponding β-phenyltelluroaluminium enolate. This intermediate is hydrolysed by aqueous HCl into β-phenyltelluro compounds, or smoothly affords an aldol reaction with aldehydes to give α-hydroxyalkyl-β-phenyltellurocarbonyl compounds.

These adducts, submitted to MCPBA oxidation, undergo telluroxide elimination, regenerating the original C=C bond. The overall transformation therefore provides α-hydroxyalkylation of α,β-unsaturated carbonyl compounds [1].

$$PhTeTePh + 2i\text{-}Bu_2AlH \xrightarrow{THF} 2PhTeAlBu_2\text{-}i + H_2$$

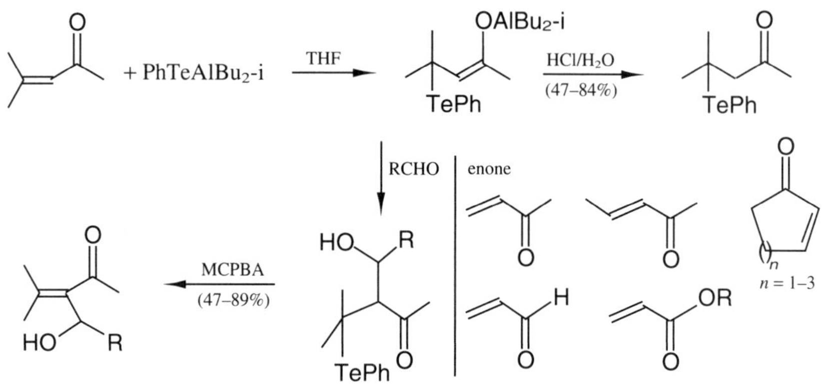

enone 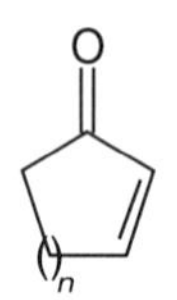 n = 1,2; R = n-C_3H_7, Ph

4.11.2 Dehydrotelluration of β-trichlorotelluroketones: preparation of α-methylene ketones

β-Trichlorotelluroketones, prepared *in situ* from siloxycyclopropanes and tellurium tetrachloride, undergo dehydrotelluration by DMSO at 0°C, giving the corresponding α-methylene ketones in good yields [2]. The reaction also proceeds with bases such as TMEDA, triethylamine or pyridine.

Bis(β-oxoalkyl)tellurium dichlorides, obtained by a similar 2:1 ratio reaction, are inert to DMSO but undergo a double dehydrotelluration when TMEDA is used.

Since the siloxycyclopropanes are easily prepared from ketones via cyclopropanation of the corresponding silyl enol ethers [3], the overall procedure provides the α-methylenation of ketones under mild conditions.

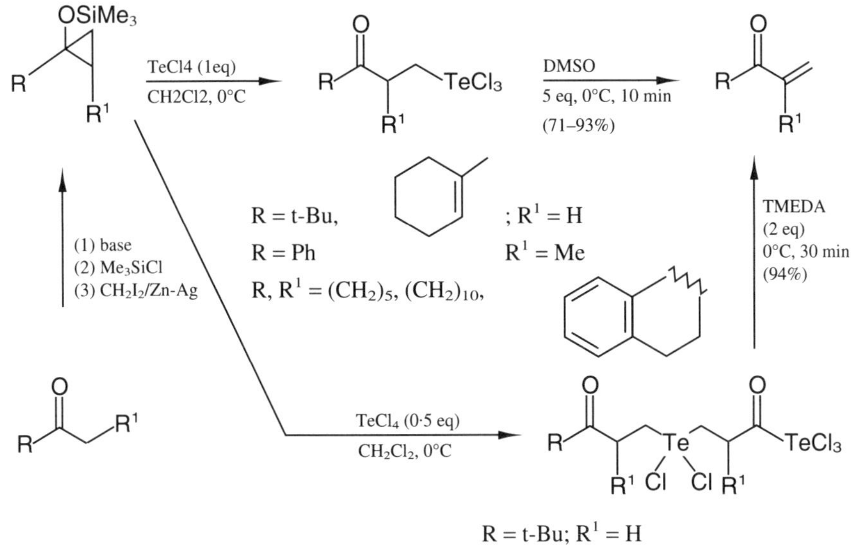

Similar dehydrotrichlorostannation of β-trichlorostannyl ketones by DMSO requires prolonged heating at 60°C.

Preparation of α-methylene ketones (typical procedure) [2] Siloxycyclopropane (from cyclododecene) (1.34 g, 5 mmol) is added to a suspension of $TeCl_4$ (1.34 g, 5 mmol) in CH_2Cl_2 (10 ml) at 0°C. After stirring for 10 min DMSO (1.8 ml, 25 mmol) is added, and stirring continued for a further 10 min. The black precipitate is filtered off and aqueous treatment (pentane/H_2O) is carried out. The organic phase is dried ($MgSO_4$) and evaporated under vacuum. The residue is chromatographed on SiO_2 to give α-methylene cyclododecanone (0.903 g (93%)).

4.11.3 Preparation of telluroesters and their conversion into ketones

Telluroesters are easily accessible by reacting phenyl trimethylsilyl telluride with aroyl chlorides [4].

$$PhTeSiMe_3 + ArCOCl \xrightarrow[(62\text{–}91\%)]{THF,\ r.t.} PhTeCOAr + Me_3SiCl$$

Ar = Ph, *p*-Cl, *o*-Cl, *p*-Br, *p*-NC, *p*-NO_2, *p*-MeO,
p-MeC_6H_4, *p* $ClCOC_6H_4$ (giving $(PhTe)_2CO$)

The starting phenyl trimethylsilyl telluride is prepared from lithium or bromomagnesium phenyl tellurolate and chlorotrimethylsilane [5].

Phenyl trimethylsilyl telluride (typical procedure) Lithium phenyl tellurolate (18.4 mmol) is treated with chlorotrimethylsilane (2.7 g, 25 mmol) at 25°C. A vigorous exothermic reaction occurs during which the mixture turns from brown to off-white. After 1 h the volatile materials are pumped out and the bulk fraction fractionally distilled under vacuum (b.p. 70°C/0.1 torr).

Telluroesters (typical procedure) [5] Phenyl trimethylsilyl telluride (2.10 ml, 10.9 mmol) is added to a stirred solution of benzoyl chloride (1.42 g, 10.1 mmol) in dry THF (30 ml) under argon. The mixture is stirred for 3 h at room temperature and then evaporated. The residue is purified by recrystallization from pentane, giving tellurophenyl benzoate as yellow needles (2.71 g (87%)).

By treatment with dialkyl cuprates, telluroesters are easily converted into ketones in high yields [4].

$$\text{ArCOTePh} \xrightarrow[\substack{\text{R=Me (1,1 eq, 15 min) (87\%)} \\ \text{R=n-Bu (1,2 eq, 30 min) (97\%)}}]{\text{R}_2\text{CuLi, THF/ether, }-78^\circ\text{C}} \text{ArCOR}$$

4.11.4 Conversion of allylsilanes into allylamines via phenyltellurinylation

Allylamines are obtained by sequential treatment of allylsilanes [6] with benzenetellurinyl trifluoroacetate in the presence of $BF_3 \cdot Et_2O$ and then with an alkylamine [7].

The reaction intermediate is an unstable and not isolated phenyl allyl telluroxide. Arylamines behave similarly, but require longer reaction times.

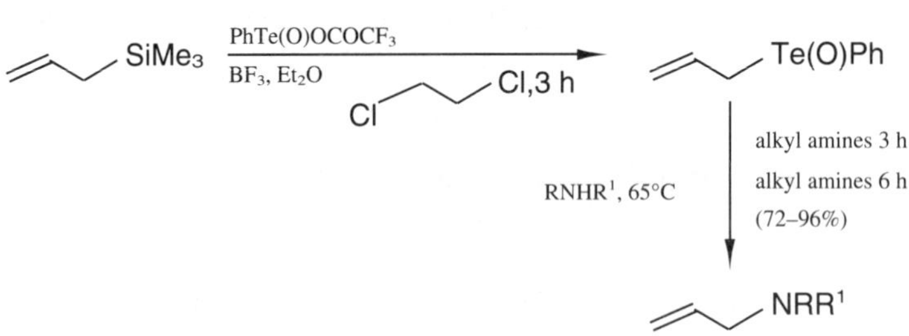

$R = H$; $R^1 = Ph$, *p*-Cl-, *p*-$NO_2C_6H_4$, n-C_8H_{17}

$R = Me$; $R^1 = Ph$

R, $R^1 = (CH_2)_5$, ($CH_2)_2$

Substituted allylsilanes react with aryl- and alkylamines, giving γ-*trans*- and α-*trans*-substituted allylamines, respectively.

This selectivity is independent of the *cis/trans* geometry and the α or γ isomerism of the starting allylsilanes.

This result is consistent with an equilibrium between an initially formed α-substituted allyl telluroxide (resulting from the well-known regio-controlled electrophilic attack at the γ carbon) and the γ isomer (which is really the main product, as detected by NMR). In the subsequent reaction with amines, the more nucleophilic alkylamines attack the central carbon atom, giving rearranged products, whereas the less nucleophilic arylamines attack the terminal carbon, giving the thermodynamic γ product.

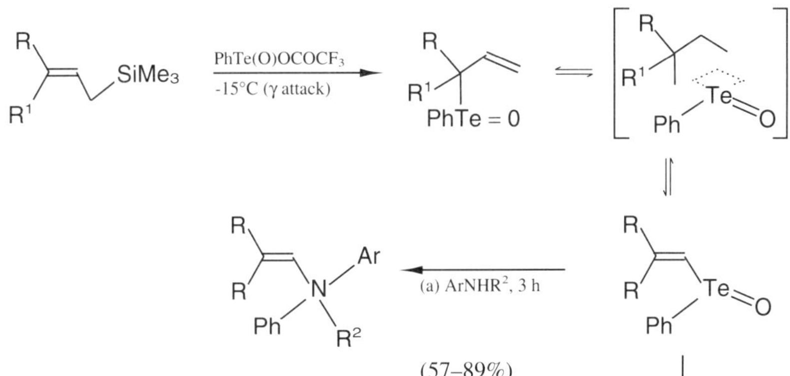

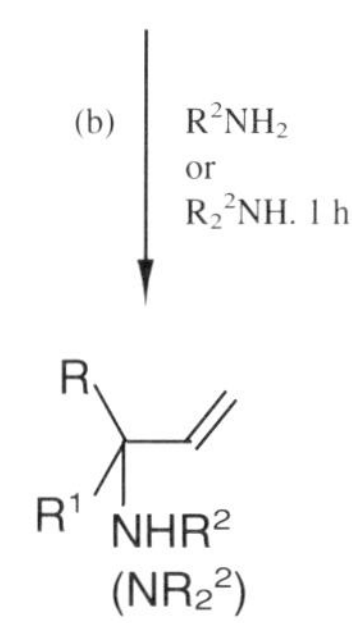

Representative examples

$R = H$; $R^1 = n\text{-}C_6H_{13}$ (*cis/trans* = 84/16) — amine:
(a) *p*-ClC$_6$H$_4$, PhNHMe
(b) n-C$_4$H$_9$NH$_2$, cyclo-C$_6$H$_{11}$NH$_2$, piperidine

R, $R^1 = (CH_2)_5$:
(a) *p*-ClC$_6$H$_4$, NH2
(b) n-C$_4$H$_9$NH$_2$

Allylamines from allylsilanes (typical procedure) [7] To a solution of PhTe(O)OCOCF$_3$ (1 mmol) (generated from $(PhTeO)_2O$ (0.229 g, 0.5 mmol) and $(F_3CCO)_2O$ (0.120 g, 0.55 mmol)) in $ClCH_2CH_2Cl$ (4 ml) are added $BF_3 \cdot Et_2O$ (0.156 g, 1.1 mmol) and then allyltrimethylsilane (0.125 g, 1.1 mmol) in $ClCH_2CH_2Cl$ (2 ml). The solution is stirred at room temperature for 1 h. *p*-Chloroaniline (0.32 g, 2.5 mmol) is then added and the mixture heated at 65°C with stirring for 6 h. The reddish-black mixture is poured into saturated aqueous $NaHCO_3$ and extracted with CH_2Cl_2. The extract is dried ($MgSO_4$) and evaporated under vacuum. The residue is chromatographed on SiO_2 (eluting with hexane/EtOAc 95:5), giving n-allyl-*p*-chloroaniline as a pale yellow oil (0.151 g (90%)).

4.11.5 Olefin inversion by *syn*-chlorotelluration/*anti*-dechlorotelluration

The inversion of the olefin configuration can be attained by employing a one-pot procedure involving a sequential *syn* chlorotelluration/*anti* dechlorotelluration [8].

Reaction of the appropriate alkene with 1 eq of tellurium tetrachloride in chloroform or acetonitrile gives a 2-chloroalkyltellurium trichloride adduct in a high yield (see Section 3.5.1.2). Reduction of this adduct with aqueous sodium sulphide results in the instantaneous separation of elemental tellurium and formation of the starting alkenes which, in most cases, exhibit an inverted configuration (with relevant variations depending on the solvent and olefin used). The proposed mechanism involves an intermediate epitelluride.

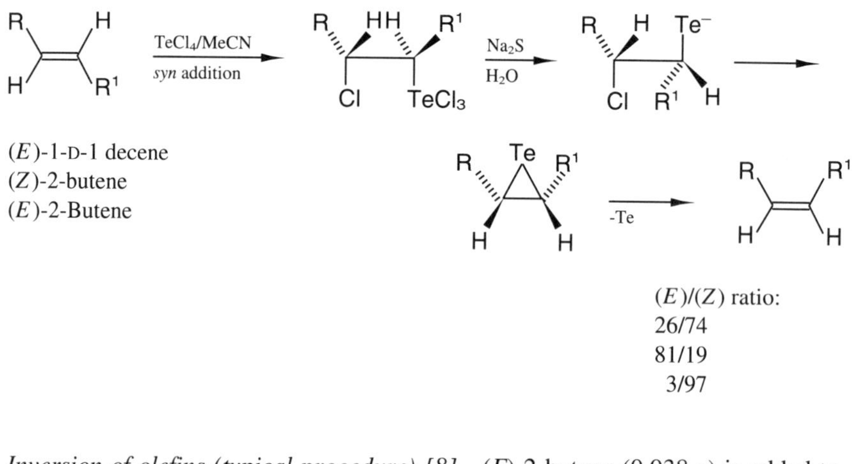

(*E*)-1-D-1 decene
(*Z*)-2-butene
(*E*)-2-Butene

(*E*)/(*Z*) ratio:
26/74
81/19
3/97

Inversion of olefins (typical procedure) [8] (*E*)-2-butene (0.038 g) is added to a solution of $TeCl_4$ (0.220 g, freshly sublimed) in MeCN (5 ml). The mixture is stirred at 0°C for 3 h and then evaporated at room temperature under vacuum. $Na_2S \cdot 9H_2O$ (14% aqueous solution, 10 ml) is added dropwise (through a septum), producing evolution of gas and precipitation of elemental Te. Analysis of the gaseous 2-butene shows an (*E*/*Z*) ratio at 3/97.

4.11.6 Tellurium tetrachloride as a catalyst for dithioacetalization and ketalization

Tellurium tetrachloride is an efficient Lewis acid catalyst for the thioacetalization of aldehydes and thioketalization of aliphatic ketones [9].

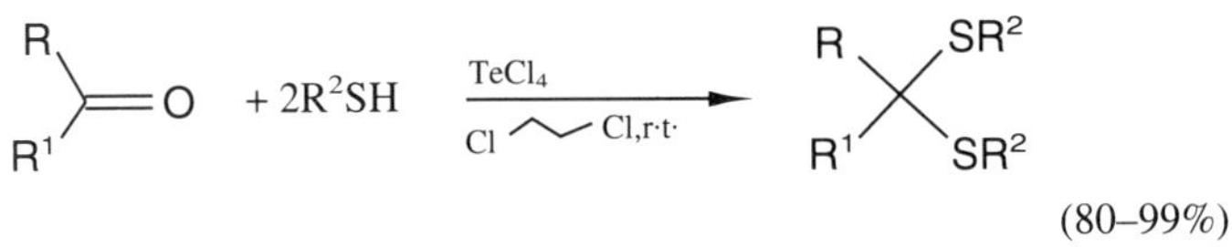

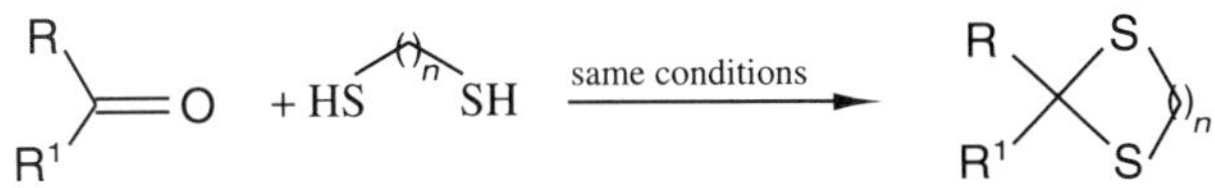

R = H; R^1 = Ph, *p*-HO, *p*-Me, *p*-ClC_6H_4, PhCH = CH_2;
R^2 = Et and *n* = 2, 3

R, R^1 = $(CH_2)_5$; *n* = 2

R, R^1 = 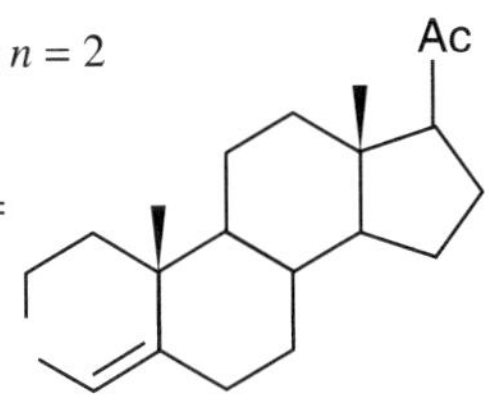 ; *n* = 2 R, R^1 = 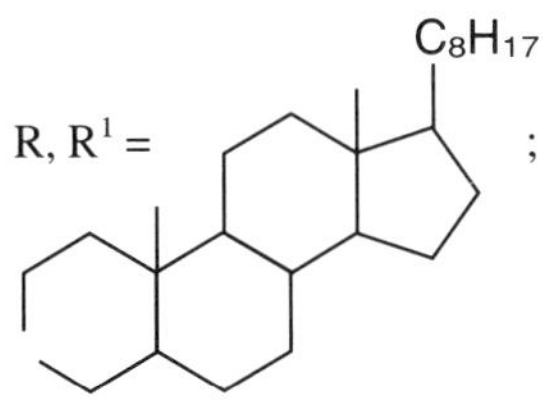 ; *n* = 2, 3

Selective ketalizations are achieved as illustrated.

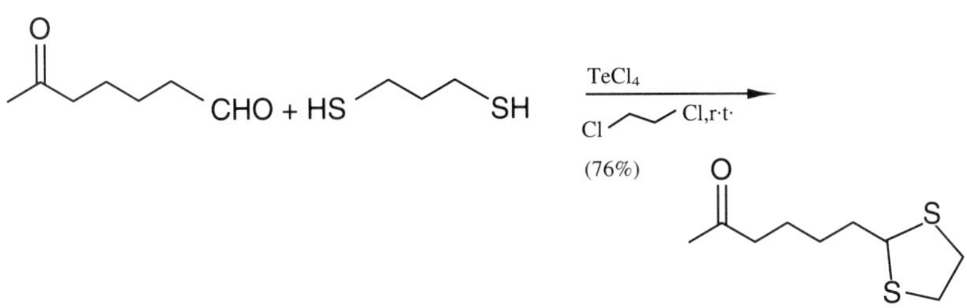

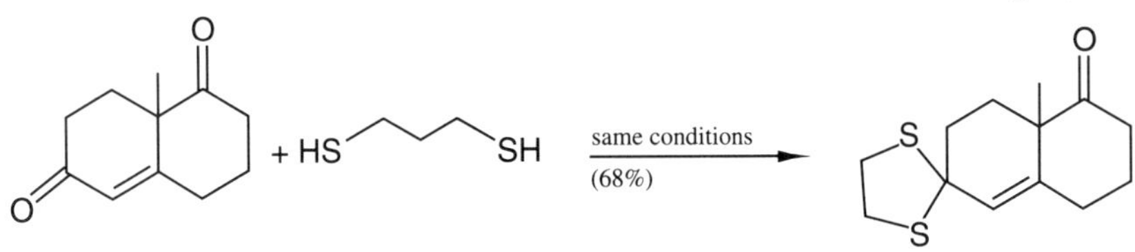

$TeCl_4$-catalysed dithioacetalization (or dithioketalization) (general procedure) [9] To a solution of the carbonyl compound (5 mmol) and the thiol (10 mmol) or dithiol (5 mmol) in $ClCH_2CH_2Cl$ (10 ml) is added powdered $TeCl_4$ (0.25 mmol), and the resulting mixture is stirred for 2–3 h at room temperature, during which period black tellurium gradually separates. The reaction is quenched by adding $NaHCO_3$ (about 0.2 g), and the insoluble material is filtered off. The filtrate is dried (Na_2SO_4), and the solvent evaporated under vacuum,

giving the crude dithioacetal or dithioketal, which is purified by SiO_2 chromatography or by Kugelrohr distillation.

4.11.7 Tellurium tetrachloride as a Lewis acid catalyst for aromatic alkylation

In the presence of tellurium tetrachloride, aromatic hydrocarbons are alkylated with reactive alkylating agents such as benzylic or t-butyl alcohols and chlorides [10]. The yields are high with toluene but only moderate for benzene, *p*-xylene and anisole. Equivalent and catalytic amounts of tellurium tetrachloride, respectively, are required for the alcohols and chlorides. Tellurium tetrachloride therefore works as a reagent for the conversion of the alcohols into chlorides, and then as a Lewis catalyst for the aromatic substitution (*para* isomers are formed preferentially to *ortho* isomers).

$$\mathrm{ROH} + \mathrm{TeCl_4} \xrightarrow[\text{-HCl}]{} [\mathrm{R{-}OTeCl_3}] \xrightarrow[\text{-TeOCl}_3^-]{} [\mathrm{R^+}] \xrightarrow[\text{-TeOCl}_2]{\text{TeOCl}_3^-}$$

$$\mathrm{RCl} \xrightarrow[\text{TeCl}_4\text{ cat.}]{\text{ArH, r·t·}} \mathrm{Ar{-}R}$$

$R = PhCHCH_3, PhCH_2$, t-Bu; ArH = toluene (83, 87, 34%)

$R = PhCHCH_3$; ArH = benzene, *p*-xylene, anisole (3, 40, 59%)

$$\mathrm{RCl} + \mathrm{ArH} \xrightarrow{\text{TeCl}_4\text{ cat.}} \mathrm{Ar{-}R}$$

$R = PhCHCH_3$, t-Bu; Ar = toluene (93, 71%)

$TeCl_4$-catalysed Friedel–Crafts aromatic alkylation (typical procedure) [10] To a solution of 1-phenylethanol (3.7 g, 30 mmol) in toluene (30 ml) is added slowly $TeCl_4$ (9.7 g, 36 mmol), keeping the temperature at 25°C (exothermic reaction). Small amounts of white precipitate appear immediately, and after a few minutes the colour of the mixture becomes dark brown. The mixture is stirred for 3 h and then quenched with H_2O (20 ml). The organic layer is separated, washed with brine (2 × 20 ml) and dried ($MgSO_4$). Evaporation of the solvent leaves an oily residue which is distilled under vacuum, giving a mixture of 1-phenyl-1-tolylethanes (4.9 g (83.3%); b.p. 117–128°C/1 torr). GLC analysis (silicone OV-101, 0.24 mm × 30 m capillary column at 100–260°C, 4°C min^{-1}) reveals an *ortho*/*para* ratio of 12/88.

4.11.8 Additional reactions

4.11.8.1 Tellurium tetrachloride as a catalyst for cationic oligo- and polymerization reactions

Tellurium tetrachloride, a source of $TeCl_3^+$ ions, behaves as a cationic catalyst for oligo- and polymerization reactions [11]. Two types of monomers, phenyl-substituted ethylenes* and benzyl chlorides have been submitted to these reactions.

4.11.8.2 Tellurium tetrachloride-promoted rearrangement of cycloheptatrienes to benzylic chlorides

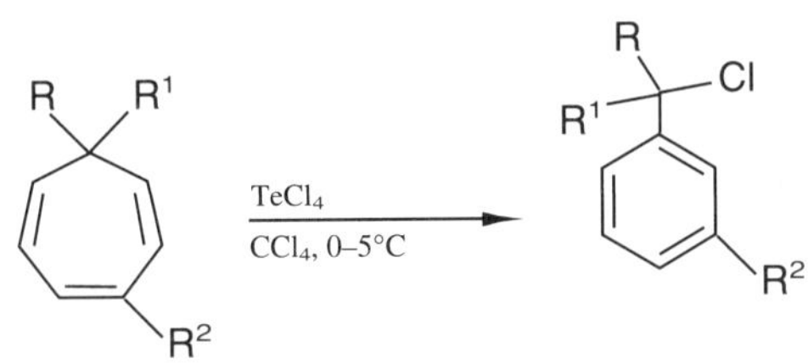

[12]

4.11.8.3 Allylic alcohols by oxidation of allylic phenyl tellurides

On treatment with an oxidizing agent (H_2O_2, $NaIO_4$, t-BuOOH), allylic tellurides give the rearranged allylic alcohols as the main product (via a [2,3]-sigmatropic

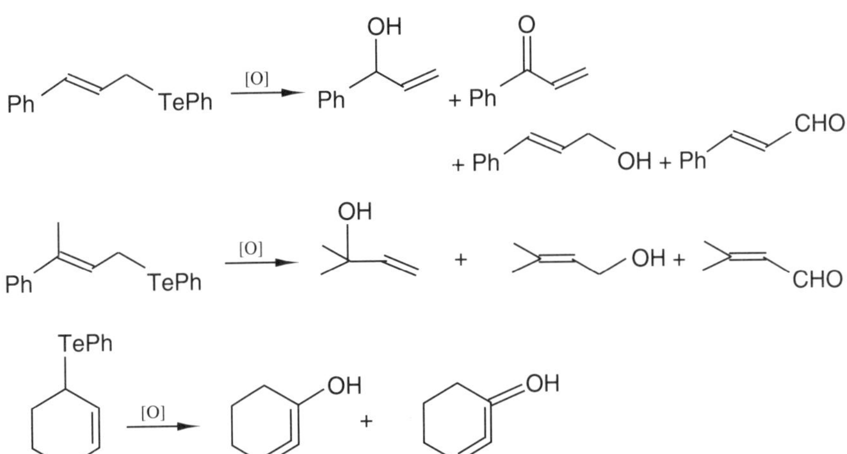

*The capacity of $TeCl_4$ to oligomerize *trans*-stilbene is peculiar, since this olefin is traditionally inactive towards polymerization.

shift of the telluroxide, in analogy to the well-known rearrangement of allylic selenides) together with variable amounts of non-rearranged allylic alcohols and α,β-unsaturated carbonyl compounds [13].

REFERENCES

1. K. Sasaki, Y. Aso, T. Otsuba and F. Ogura, *Chem. Lett.*, 607 (1989).
2. H. Nakahira, I. Ryu, L. Han, N. Kambe and N. Sonoda, *Tetrahedron Lett.* **32**, 229 (1991).
3. E. S. Colvin, *Silicon Reagents in Organic Synthesis*, p. 99. Academic Press, London, 1988; W. P. Weber, *Silicon Reagents for Organic Synthesis*, p. 235. Springer-Verlag, Berlin, 1983.
4. K. Sasaki, Y. Aso, T. Otsubo and F. Ogura, *Chem. Lett.*, 977 (1986).
5. J. E. Drake and R. T. Hemmings, *Inorg. Chem.* **19**, 1879 (1980). K. Sasaki, Y. Aso, T. Otsubo and F. Ogura, *Tetrahedron Lett.* **26**, 453 (1985).
6. E. W. Colvin, *Silicon Reagents in Organic Synthesis*, p. 25. Academic Press, London, 1988.
7. N. X. Hu, Y. Aso, T. Okubo and F. Ogura, *Tetrahedron Lett.* **29**, 4949 (1988).
8. J. E. Backwall and L. Engman, *Tetrahedron Lett.* **22**, 1919 (1981).
9. H. Tani, K. Masumoto and T. Inamasu, *Tetrahedron Lett.* **32**, 2039 (1991).
10. T. Yamaguchi, K. Hattori, S. Mizutaki, K. Tamaki and S. Uemura, *Bull. Chem. Soc. Jpn* **59**, 3617 (1986).
11. M. Albeck and T. Tamari, *J. Organomet. Chem.* **238**, 357 (1982).
12. M. Albeck, T. Tamari, M. Sprecher and K. Ohe, *J. Org. Chem.* **48**, 2276 (1983).
13. S. Uemura, S. I. Fukuzawa and K. Ohe, *Tetrahedron Lett.* **26**, 921 (1985).

Index

C

D

E